AF553678

Medicinal Plants

Aspects and Prospects

Medicinal Plants

Aspects and Prospects

— Editors —

Dr. MUKESH KUMAR
Associate Professor, Department of Botany
Sahu Jain P.G. College, Najibabad – 246 763, U.P.
(Affiliated to MJP Rohilkhand University, Bareilly)

Dr. ANJALI KHARE
Assistant Professor, Department of Botany
Advance Institute of Science and Technology, Dehradun – 248 001, U.K.
(Affiliated to HNB Garhwal Central University, Srinagar)

Dr. C.P. SHUKLA
Chief Executive Officer
LVP Shodh Sansthan, Koraon, Allahabad – 212 306, U.P.

2014

BIOTECH BOOKS®

© 2014 EDITORS
ISBN 978-81-7622-309-6

All rights reserved. Including the right to translate or to reproduce this book or parts thereof except for brief quotations in critical reviews.

The views expressed in various chapters are those of the authors and not of publisher of the book.

Published by: **BIOTECH BOOKS®**
4762-63/23, Ansari Road, Darya Ganj,
New Delhi - 110 002
Phone: +91-011-23262132
E-mail: biotechbooks@yahoo.co.in

Printed at: **Chawla Offset Printers**
Delhi - 110 052

PRINTED IN INDIA

Preface

India represents one of the richest diversity expanses of medicinal plants. They have provided the means of curing disease perhaps since the evolution of mankind on the earth. In the traditional health care systems, the medicinal plants form an important constituent of modern medicines. The forest wealth caters to the needs of about 80 per cent of Ayurvedic, 46 per cent Unani and 33 per cent Allopathic systems of medicine, thus contributing a pre-eminent share to the economy of the rural people.

During recent times, the medicinal plants have attracted global interests and attention since these constitute a rich treasure trove of cultural information and are source of natural products which provides health and security to millions in rural communities. They help to generate additional employment and income, offer opportunities for processing enterprises, contribute to foreign exchange earnings, and are basic to biodiversity and conservation objectives. The demand for Ayurvedic formulations is increasing by the day, both in the domestic market as well as internationally. Fast multiplying population of human has created a great pressure on the existing medicinal flora. Destruction of biotic communities is going on so fast that valuable biological information is lost, and already many species have become extinct.

Worldwide the number of species used for medicinal purposes is estimated to be more than 50,000 which covers about 13 per cent of all flowering plants. The tremendous growth of pharmaceutical industry considerably affect the popularity of herbal medicines, but the traditional herbal system of medicines still remains intact as cultural heritage in different regions of the world and continues to cater the needs of 80 per cent of the population of the world mainly in poor countries. The herbal drugs are low priced, harmless and easily available to common man. These are available even to those people where modern health care systems are yet to be developed. Medicinal plants are an important part of our natural wealth. They serve as important therapeutic agents as well as valuable raw materials for manufacturing numerous traditional and modern medicines. Since ancient times, plants with therapeutic properties have secured an important place in the healing practices and treatment of diseases.

Ethnobotany, the study of relationship between people of a primitive society and their plant environment deals with the total natural relationship of man with plants. Usually it is defined as an anthropological approach to botany. Ethnobotanists aim to document, describe and explain complex relationships between cultures and (uses of) plants, focusing primarily on how plants are used, managed and perceived across human societies. This includes use for food, clothing, currency, ritual, medicine, dyes etc. In the beginning, ethonobotanical specimens and studies were not very reliable because the botanists and the anthropologists did not always collaborate in their work. The botanists focused on identifying species and how the plants were used instead of concentrating upon how plants fit into people's lives. On the other hand, anthropologists were interested in the cultural role of plants and not the scientific aspect. In the early 20th century, botanists and anthropologists finally collaborated and the collection of detailed data began. Beginning in the 20th century, the field of ethnobotany experienced a shift from the raw compilation of data to a greater methodological and conceptual reorientation. This is also the beginning of academic ethnobotany. The so-called "father" of this discipline is Richard Evans Schultes even though he did not actually coin the term "Ethnobotany".

Today the field of ethnobotany requires a variety of skills: botanical training for the identification and preservation of plant specimens; anthropological training to understand the cultural concepts around the perception of plants; linguistic training, at least enough to transcribe local terms and understand native morphology, syntax, and semantics. There are several methods of ethnobotanical research and those relevant to medicinal plants are archaeological search in literature, herbaria and the field studies. "Man, ever desirous of knowledge, has already explored many things, but more and greater still remains concealed; perhaps reserved for far distant generations, who shall prosecute the examination of their creator's work in remote countries and make many discoveries for the pleasure and convenience of life..." The above quotation of Linneaus deals with the relationship between medicinal plants and the total filed of ethnobotany. A great deal of information about the traditional uses of plants is still intact with tribal peoples. But the native healers are often reluctant to accurately share their knowledge to outsiders.

Medicinal plants abundant in supply are not infinite and with the wide spread use and extraction, they are on the merge of depletion. There are no visible and concerted efforts made towards conservation and wide use of medicinal plants. At the current rate of consumption and use, the status of medicinal plants is threatened risking our own future benefits and knowledge. India has a rich wealth of medicinal plants, a large number of species and types being found in different parts of the country. Although a fairly good knowledge is now available on the use of drugs obtained from plant, not much attention has been paid to their cultivation in this country. It is now well-known fact that the medicinal plant sector possesses great potential to uplift the economy of India. The research activities in medicinal plants have expanded in several directions and there has been a natural explosion of information on various aspects of the medico-magico-religious plants. The rapid advances and development in the research of medicinal plants prompted us to assemble the up-to-date information in the form of present book targeted to both post-graduate students and research scholars.

The present volume consists of widely ranging 25 chapters which encompass topics on general aspects, diversity, biotechnology and evaluation of herbals with reference to medicinal plants. The medicinal values of lower plants *i.e.* marine algae and Cyanobacteria have also been described in a separate chapter. The authors have tried to extract all spheres of useful information and hope that the present book would be informative, useful and stimulating to the students, teachers, scientists, research scholars, farmers, foresters and all those interested in the pursuit of medicinal and ethnobotanical aspects of vegetation wealth, especially working in the developing world.

The editors feel privileged to thank all the authors whose contributions have enriched the present volume. We also express our deep sense of gratitude to our parents whose blessings have always prompted us to pursue academic activities deeply. It is possible that in a work of this nature, some mistakes might have crept in text inadvertently and for these we owe undiluted responsibility. We are also thankful to Professor G. S. Paliwal and Professor R. D. Gaur (former Heads), Professor S. C. Tiwari (Dean, Faculty of Science) and Professor A. B. Bhatt, Head, Department of Botany, HNB Garhwal Central University, Srinagar (Uttarakhand) for their cooperation and valuable suggestions for the preparation of this volume more informative and useful. We thank them from the core of our heart. The editors are also thankful to the authorities of their respective colleges where they could work peacefully and bring this academic endeavor to the light of the day.

The editors profoundly thank to our publisher, M/S Biotech Books, Ansari Road, Darya Ganj, New Delhi- 110 002 for kindly accepting our proposal to publish the present volume so nicely. Their prompt action and enthusiastic cooperation during the compilation of this volume is highly appreciated and acknowledged.

The author (MK) is thankful to the UGC, New Delhi for providing financial assistance in the form of a major research project entitled, "Inventorization, Documentation and Survey of the Cyanobacterial Forms of Three Himalayan States- UK, H.P. and J & K". The experience gained during the research work of this project has been incorporated in the body of the volume of this book. His wife Dr (Smt) Kiran Sharma, PGT Hindi, Kendriya Vidyalaya, BHEL, Ranipur, Haridwar (U.K.) and two sons Ruchir Sharma and Nishant Sharma deserve special thanks for their affection and multifarious cooperation extended during the preparation of this academic and scientific pursuit.

Mukesh Kumar

Anjali Khare

C.P. Shukla

Contents

Medicinal Plants: Aspects and Prospects (2014) **Pages 1–22**
***Editors:* Mukesh Kumar, Anjali Khare and C.P. Shukla**
ISBN: 978-81-7622-309-6
***Published by:* BIOTECH BOOKS, NEW DELHI**

Chapter 1
Medicinal Plants: General Aspects and Prospects

***Mukesh Kumar*[1] *and C.P. Shukla*[2]**

[1]*Department of Botany, Sahu Jain P.G. College, Najibabad – 246 763, U.P.*
[2]*L.V.P. Shodh Sansthan, Koraon, Allahabad – 212 306, U.P.*

ABSTRACT

Nature has provided us with bountiful gifts in the form of biodiversity, with which humans have experimented. Indigenous peoples lives are based on the resources found in their local environment. Traditional knowledge has been passed on from one generation to another only by the words. It is very important to understand the local biodiversity of a particular region to use it effectively for drug designing programmes. The traditional aspects in this regard play a vital role. Effective ethnobotanical survey in some of the rural and socio-economically backward areas of Uttar Pradesh has revealed an interesting dependency of the local on the medicinal biodiversity of the region. This approach has resulted in a new aspect of drug plants from natural resources. The dependency on the medicinal flora obtained from the ethnic/local groups form the starting material for the search of the potential bioactive leads. In the present chapter, whole gamut of issues related to Traditional and modern knowledge regarding medicinal, cosmetic and other uses of Indian plants have been discussed and future strategies for strengthening alternative therapy have been suggested with enough emphasis on the public sensitization on the utility of medicinal plants, IPR of plant products, *in situ* and *ex situ* conservation especially of endangered plants.

Keywords: Ethnobotany, Plant biodiversity, Herbal therapy.

Introduction

Glimpses of Indigenous Knowledge

Many traditional societies, often referred to as indigenous or tribal people, have accumulated a lot of empirical knowledge on the basis of their experience whilst dealing with nature and natural resources. This traditional wisdom is based on the intrinsic realisation that man and nature form important part of an indivisible whole, and therefore should live in partnership with each other. This eco-centric view of traditional societies is widely reflected in their attitudes towards plants, animals, rivers and the earth. This whole body of knowledge centered around economic value of plant and animal species is a part of ethnobiology and has potential value for the society at large.

India is one of the greatest store-house of traditional knowledge and wisdom. It has the potential of becoming a major player in the global trade in herbal formulations, ethnomedicine and products, but the tribal wisdom and their indigenous knowledge, unfortunately lacks due recognition and adequate modern documentation and is diminishing fast.

Therefore, there is an urgent need to promote and product these fragile knowledge systems and the tribals wisdom. Further, these is need to revive, update and link this wisdom with modern science in order to get value added product like medicines, natraceuticals, cosmaceuticals etc. for human welfare. Harshberger (1985) coined the term 'Ethnobotany' for the indigenous knowledge and wisdom that existed for ages in India. Ethnobotany is defined as the study of relationship which exists between people of a primitive society and their plants environment. Jain (1987) defined it as the total nature and traditional relationship and the interaction between man and his surrounding plant wealth.

Plant Genetic Resources in Relation to Livelihood

Ethnobiology must have been the first knowledge on plants and animals, which the early man had acquired by sheer necessity, intuition, observation and experimentation in the forests. It is now almost universally taken as the total direct relationship between man and plants. It includes the use of plants by both tribals and non-tribal communities without any implication of primitive or developed societies. Further, Ethnobotany by nature is an interdisciplinary science and has relevance to sociology, anthropology, taxonomy, phytochemistry, archaeology, ecology, agriculture, medicine, linguistics etc. Today, Ethnobotany has become an important and crucial area of research and development in resource management, sustainable utilization and conservation of biodiversity and for socio-economic development. All over the world, there has been an increasing interest towards the scientific study of man-plant interaction in the natural environment among the botanists, social scientists, anthropologists, and practitioners of indigenous systems of medicine. The Indian tribes and their future have been the subject of vigorous discussion during the past several decades and great concern has been expressed for their protection, welfare, development and upliftment by the Government as well as voluntary organisations, Tribal Sub Plan (TSP) was implemented by the Government and the strategy has been in operation since the fifth five year plan for improving the quality of life of the tribal communities. There are 500 tribal communities of different ethnic groups inhabiting India, which constitute about 8 per cent of the total population of the country. The tribal population in India has gone up from about 20 million (Census, 1951) to 66.5 million (Census, 1981). Among these the most important tribal groups are Gonds, Bhils and Santhals inhabiting central and western zones of the country.

The state of Uttar Pradesh presents unique physical and ethnic diversity and is inhabited by large number of tribals, *viz.* Bhotiya, Raji, Jaunsari in the hill regions, Tharu and Bhoxa in the terai region of Uttarakhand and Gond, Kol, Baiga, Chero, Kharwar, Korwa, Ghasias, Agaria, Panika, Oraon, Mushar, Sahariya etc. in the plains of Uttar Pradesh. Among them there are food gatherers and hunters (Raji, Korwa, Kol), nomads (Bhotiyas) and cultivators (Tharus and Bhoxas). The tribes of Uttar Pradesh belong to two great stocks the Mongoloid and Dravidian. The ethnobotanical studies were conducted among the tribes Agaria, Baiga, Bhoxa, Bhotia Kharwar, Kol, Korwa, Jaunsari, Tharu, Raji, Sahariya, Panika, Oraon, Gond and other hill communities, who live in the tribal tracts and forest areas and villages of Lakhimpur Kheri, Sonbhadra, Allahabad, Varanasi, Banda, Garhwal, Jhansi, Lalitpur, Nainital, DehraDun, Gonda, Bahraich, Pithoragarh and Gorakhpur districts of Uttar Pradesh and Uttarakhand. The tribal areas are botanically rich on account of its diverse forests, river, valleys and hillocks, which provide them their basic needs and livelihood. They practice primitive agriculture, observing traditional practices like fishing, hunting, gathering foods, religious ceremonies, taboos, and ethnoherbal therapy. They collect and utilize various herbs, leaves, roots, rhizomes, tubers, flowers, fruits and seeds of many wild as well as cultivated plant species grown in ambient vegetation for their day-to-day needs and healthcare.

Ethnobotany and Traditional Knowledge

The knowledge on the use of bioresources hidden among the traditional culture is vital not only for ethnic societies but also for the urban cultures. India is fortunate in having more than 550 ethnic tribes with a vast indigenous knowledge coupled with an amzing diversity of flora and fauna.

Traditional societies who are the custodians of such knowledge systems are to be amply rewarded by sharing benefits arising from out of sale of drug products based on traditional knowledge. Confidence and cooperation of tribal communities at all stage is vital. National and International level conventions for exchange of information among all tribal resource persons are essential. Such cross-cultural ethnic knowledge among different developing nations helps in establishing the species of plants and animals that are used for a specific ailment among different ethnic communities. The number of species common to treat a particular ailment within and between communities, or where diverse species are used which ones are better? Collective and cooperative efforts of all developing nations in the study of cross-cultural ethnobotany can provide solutions for many such issues and result in establishing safer, cheaper and locally acceptable drugs for many of the ailments rampart among the ethnic communities in the entire region. The vital role of taxonomist in such studies and the need for strengthening the taxonomic base in all developing nations is emphasized.

With the modernity, primitive tribal societies are fast dwindling, hence, traditional knowledge possessed by them must come out for further conservation and development of food and herbal medicines so that it is not forgotten and lost in antiquity.

Vedic Era of Ethnobotany

Ethnobotany is holistic living and alternate system(s) of medicine emphasize both physical and mental well-being of an individual while, Vriksayurveda and Homeopathy believe in the harmony with the nature. India has one of the oldest, richest and diverse folk traditional societies associated with the use of medicinal plants in its healthcare system to promote safeguard and relieve pain, heal wounds and refresh mind, muscles and nerves. One of the oldest records in traditional medicine is Vriksayurveda compiled by Parashara and Manohar (1994). Glorious traditions in this context date back to Rigveda (4500-2500 BC), Atharveda (2500-2000 BC), Samhita era *i.e.* Brihad Samhita, Charak Samhita (2000 BC) and Sushruta Samhita (1000-800 BC). Later came the Ayurveda, a practice in Sanskrit. The vedic and post-vedic periods (pre-Budhist) had celebrated Indian herbalists and physicians like Bhikshu, Atrey, Mahabharata, Nagarjuna, Vaghbhatta Jr. (800 AD), Chakradatta, Sharangadhar, Bangasen and Dhanvantari. In Mughal period, Hindu, Raja Nighantu and Naradhan Pundia appeared. In British period, Madana Pala Nighantu by Madan Pala and Bhava Prakash by Bhava Mishra came into being. Since then, many authentic and excellent publications like Pharmacopoea of India and books of Homeopathy and Materia Medica have come out on the subject.

Ethnobotany at the Service of Mankind

During the past few years, a substantial wealth of knowledge has accumulated using knowledge collected from Sadhus, Maharajas, Rishies, Giri, Puri and Nath etc. in groves, forests and hills as well as from those involved in witchcrafts, magic folklores, herbal practitioners as doctors and villagers of ethnic communities, foresters, shepherd, migratory people and classical literature, travelogues, notebooks, gazetteers, herbarium etc. and pertinent previous literature which existed in various languages in the form of poems, exaggerated eulogy, religious taboos and beliefs, archaeological data, sacred groves, questionnaire, field work, grandma's prescriptions, voucher specimens etc. It has advanced from an agrobased tribal society to an industry and technology based society and involved conservation of biodiversity, discoveries of new herbal drugs, characterization of drugs, updating and documentation of existing knowledge and marketing of herbs and their products, etc. Some basic issues of ethnobotany in general human welfare have been described in Table 1.1. Major Indian contributions on Medicinal plants have been reviewed in Table 1.2. Herbal cure of population growth and major diseases have been depicted in Table 1.3. An annoted list of some important plants used to treat a plethora of diseases has been given in Table 1.4. The systematic exploration of plant wealth in conjunction with work on ethnobotanical data is needed so as to access the plant genetic resources of diffrent area.

Table 1.1: Contribution of ethnobotany in human welfare

Sl.No.	*Economic Value*	*Author(s) and Year (s)*
1.	Hut [*Butea monosperma* (Lam.) Kuntze]	Singh (2002)
2.	Timber and Agricultural implements (*Albizia lebbeck* Benth.)	Tosh (2004)
3.	Edible plants (*Brassica compestris* Linn. emend Metzger)	Sharma (2004)
4.	Food from forests (*Cordia dichotoma* Forst.)	Soni (1995)
5.	As Fodder (*Indigofera astragalina*)	Bhatt and Joshi (2004)
6.	For making Brooms (*Aristida adscensionis* Linn.)	Bhatt *et al.* (2002)
7.	Gums and Resins (*Gardenia latifolia* Ait., *Acacia chundra* Willd.)	Saini (2004)
8.	Tannins and Dyes (*Acacia chundra* Willd., *Cassia fistula* Linn)	Sen *et al.* (1983)
9.	As Detergents [*Madhuca longifolia* (Koeniz) Macbr.]	Prasad *et al.* (2002)
10.	Beverages and Narcotics (*Coffea arabica* Linn., *Cannabis sativa* Linn.)	Tosh (2004)
11.	Plants as fertilizers (*Arachis hypozea* Linn.)	Kavitha (2004)
12.	Ethnomagico religious beliefs (*Mimosa pudica* Linn.)	Jain (1963)
13.	For making ropes, mats, hats and baskets (*Corchorus capsularis* L., *Crotalaria juncea* L.)	Thakur and Thakur (1994)
14.	Fibre yielding (*Helictere sisora* Linn.)	Tosh (2004)
15.	For musical instruments (*Ailanthus excelsa* Roxb.)	Parmar (1978)
16.	For yield of Oils [*Madhuca longifolia* (Koeniz) Macbr.]	Mahapatra (1983)
17.	Spices and Condiments (*Piper nigrum* Linn.)	Gopinatham (2004)
18.	Bio-diesel formation (*Jatropha curcas* Linn.)	Shankaran *et al.* (2003)
19.	As antifertility drugs (*Abroma angusta* Linn., *Laccardia lacca*)	Jain *et al.* (2005)
20.	For stuffing pillows and mattresses (*Bombax ceiba* Linn.)	Singh (2002)
21.	Bidi wrappers (*Diospyros melanoxylon* Roxb.)	Shukla (2006)
22.	Bows and Arrows (*Bambusa arundinacea* Willd.)	Shukla (2008)
23.	Tea preparation (*Anogeissus latifolia* Wall. en Bedd.)	Shukla (2006)
24.	Used for Tooth Brushes (*Azadirachta indica* A. Juss.)	Bhatt *et al.* (2000)
25.	As modulators of Radiation (*Ocimum americanum* L., *Withania somnifera* Dunal.)	Goyal (2004)

Role of National Biodiversity Strategy and Action Plan (NBSAP)

To evolve a proper and precise methodology for conserving the ecology, a national biodiversity strategy and action plan (NBSAP) is required to cover areas like national ecosystems, wild species and their relatives, agriculture ecosystem and domestic species. Issues such as conservation, sustainable use, ethical practices and cultural, scientific and economic aspects of biodiversity assume importance herein involving villagers, NGOs, ethnic groups, floristics, and government officials to make joint forest programmes (JFM) aimed at people participation. Role of biodiversity need not be overstressed for the life on Earth still remains largely unexplored at the species and intraspecies level, diversity varies in different ecosystems, geographical areas and also among different taxonomic groups, human beings have already obtained enormous economic benefits from diversity and Germplasm centres of biodiversity which are so very much required for the upliftment of country's economy and betterment. Indeed with a mosaic of geoclimatic conditions India possesses a rich diversity of plants, animals and microbes, as well as of the ecosystems, related as one amongst the world's megadiversity centres.

Threats to Plant Biodiversity

- Over-exploitation of plants
- Unscientific exploration
- Environmental degradation- floods, cyclones, typhoons, tsunamis, shifting sand dunes, earth quakes, soil erosion, volcanoes etc.
- Ever increasing human population
- Intensive cultivation
- Genetically modified organisms

In view of the threats to plant biodiversity, board has engaged itself into the recording of endangered, vulnerable, rare, threatened and one of danger species in various habitats using remote sensing and other similar techniques. Such data have been presented in Red Data Books. Hence, there is priority based needs for exploration, collection, maintenance, evaluation, conservation, plant domestication and introduction of Plant genetic Resources (PGR) or exchange of PGR through National Bureau of Plant Genetic Resources (NBPGR) per policy and seed development of plants, fruits and seeds requirement of import into India as per World Health Organization (WHO, 1995) provisions.

Conservation Strategies of Plant Genetic Resources (PGR)

Since, 95 per cent of the medicinal plants are harvested from the wild, to revive these renewable natural resources both *in-situ* and *ex-situ* conservation strategies are required. The former aims at conserving plants in their natural habitats, for that we have 140 Botanical Gardens, Biosphere Reserves, Sanctuaries and National Parks while the latter does so through Gene bank, Seed bank, Pollen bank, DNA libraries, Plant tissue culture repositories and Cryopreservation (Malik, 2004).

Following eight Government departments and 40 NAGS (National Active Germplasm Structures) deal with strategies to conserve PGR: DAC (Department of Agriculture and co-operation); DARE (Department of Agriculture Research and Education); DOEF (Department of Environment and Forestry); DOC (Department of Commerce); MHFW (Ministry of Health and Family Welfare); MTX (Ministry of Textiles); CSIR (Council of Scientific and Industrial Research); DBT (Department of Biotechnology) and NGOs.

Table 1.2: Applications of ethnobotany in human welfare

Sl.No.	*Place*	*Author(s) and Year*
1.	Sacred Plants of Gujarat	Bhatt *et al.* (2001)
2.	Ethnobotany of Bastar	Sahu and Sahu (2002)
3.	Antivenomous plants of East Khandesh	Bagul and Yadav (2003)
4.	Ethnobotanical studies in East Sikkim	Janmeda (2004)
5.	Herbal treatment of children among tribals of Shadol district	Khan *et al.* (2005)
6.	Harnessing the Dewghat Forest Biodiversity for medicinal uses	Shukla (2006)
7.	Ethnobotanical Flora of Dewghat Forest	Shukla (2008)

Aims of Sustainability

With the aim of sustainability, Indian Council of Agricultural Research, New Delhi (ICAR) society in its 74th general meeting claimed to have attained following measures towards making India a developed nation through sustainability efforts, over 2.5 lakh germplasms have been stored in the National Gene Bank, for posterity, Nutrient deficiencies in the feed and fodder mapped for whole of the country. Soil erosion maps in different studies printed, Soil map of India on 1:1 million scale have been prepared and published. Perfect resource conserving technologies adopted and watershed management technology have developed, Integrated plant nutrient system evolved, Shifting cultivation banned, Large capacity, evaporatively cooled storage structures for enhancing shelf-life of fruits, vegetables and efforts are on for herbal products too. Tractor mounted harvester for turmeric have been developed and about 6 lakh farmers, rural youths and extension functionaries trained by Krishi Vigyan Kendras (KVK).

Distributional Pattern of Medicinal Plants

From a perusal of Tables 1.1–1.5, it is clear that life forms ranging from herbs, shrubs, climbers, lianas to trees have medicinal properties which are distributed to almost all the parts of the plant *viz.* Herbs (28 per cent), herbaceous climbers (7 per cent), lianas (3 per cent), shrubs (21 per cent), woody climbers (5 per cent) and trees (36 per cent) (Jakhar *et al.*, 2004)

Macroanalysis of the distribution of medicinal plants shows that they are distributed diverse ecosystems and landscapes elements. Western and Eastern ghats, the Vindhyas, Chota Nagpur plateau, Aravalis and Himalayas (with dry and moist deciduous vegetation of north-eastern) are ideal hotspots for such plants in India. Of the 386 families and 2200 genera of medicinal plants of India, the family-wise larger proportion of plant species are given in Table 1.3 (Rawat and Uniyal, 2003).

Since India has centres of origin and diversity of more than 20 major Agro-botanical crops and over 8000 species are of ethnobotanical interest (Mukerjee, 2004). Bioprospecting, therefore is of immense value of the total angiospermic plants, utilitywise distribution of plants is as Medicinal (7500), edible (3900), culturally important (700), used for fibre (525), fodder plants (400), insecticidal and pesticidal (300), gums, resins and dye plants (300) and incense and perfume plants (100).

Table 1.2a: Plant parts used for medicinal purposes

S.No.	Plant Parts	Percentage Used
1.	Whole plant	16 per cent
2.	Roots	30 per cent
3.	Rhizome	4 per cent
4.	Stem	6 per cent
5.	Bark	14 per cent
6.	Wood	3 per cent
7.	Leaves	6 per cent
8.	Flowers	10 per cent
9.	Fruits	7 per cent
10.	Seeds	4 per cent

Table 1.3: Family-wise distribution of medicinal plants

S.No.	Families	No. of Medicinal plants
1.	Asteraceae	419
2.	Euphorbiaceae	214
3.	Laminaceae	214
4.	Fabaceae	214
5.	Rubiaceae	208
6.	Poaceae	168
7.	Acanthaceae	141
8.	Rosaceae	129
9.	Apiaceae	118

Table 1.3: Indian work on medicinal plants

Sl.No.	Work	Author(s) and Year
1.	A Dictionary of economic products of India	Watt (1889-1896)
2.	A Textbook of Pharmacognosy	Shah and Qadry (1989)
4.	Adivasi Aushadh (Homeopathy) Parts 1-6	Hembrom (1994)
5.	Agro's Dictionary of medicinal plants	Prajapati and Kumar (2003)
6.	Anti-carcinogenic effect of some plant products	Aruna and Shivaramakrishnan (1992)
7.	Astanga Hridaya Samhita	Aruhdutt (1925)
8.	Astanga Sangrah	Chhangani (1945)
9.	Bhavprakash	Vaish (1997)
10.	Bhavprakash Nighantu	Chunekar (1988)
11.	Bio-energy	Elamathi, Thomas and Singh (2005)
12.	Black Pepper	Gopinathan (2004)
13.	Brihad Samhita	Vaharmihir (1984)
14.	Commercially important medicinal plants of India	Chadha, Gupta and Nagarajan (1980)
15.	Compendium of Indian medicinal plants	Rastogi and Mehrotra (1993)
16.	Dictionary of Indian folk medicine and Ethnobotany	Jain (1991)
17.	Drugs from plant resources: An overview	Sivarajan and Balachandran (1998)
18.	Ethnomedicinal plants	Trivedi and Sharma (2004)
19.	Ethno-veterinary medicinal plants in Udaipur district	Deora *et al.* (2004)
20.	Glimpses of Indian Ethnobotany	Jain (1991)
21.	Glossary of Indian medicinal plants	Chopra *et al.* (1956)

Contd...

Table 1.3–*Contd...*

Sl.No.	*Work*	*Author(s) and Year*
22.	Homeopathic medicinal plants	Suresh Baburaj (2003)
23.	Indian Materia Medica	Nadkarni and Nadkarni (1954)
24.	Indian Medicinal Plants	Kirtikar and Basu (1945)
25.	Kalpvriksha	Panara (1996)
26.	Medicinal Pteridophytes: Ann overview	Dixit and Singh (2004)
27.	Modulators of Radiosensitivity	Goyal (2004)
28.	Neem for disease management	Venkatesan and Shannlingam (2004)
29.	New vistas in ethnobotany in South Asia	Maheshwari (1996)

Table 1.4: Herbal drugs for major diseases

Sl.No.	*Attributes*	*Author(s) and Years*
1.	Contraceptives	Sharma *et al.* (2003)
2.	Ethnogynaecological Plants	Kartikayani (2003)
3.	Healthcare during Pregnancy and Post pregnancy	Nargas and Trivedi (2004)
4.	Jaundice	Singh *et al.* (2004)
5.	Preventive herbals against Malaria	Bhatt *et al.* (2001)
6.	Herbal treatment of Diabetes	Shukla (2012)
7.	Dental Protectors	Gupta and Mishra (2000)
8.	Phytotherapy of Ophthalmic diseases	Jadeja *et al.* (2005)
9.	Insecticidal Plants	Roy and Saraf (2005)
10.	Anticancer Plants	Chowdhury and Sarkar (2004)
11.	Pedaledima (Leg swelling, bone fracture)	Karthikayani (2003)
12.	Abodminal Ulcer, Mouth and Stomach Ulcers	Karthikayani (2003)
13.	Leucorrhoea (*Gymnema sylrestrus*)	Karthikayani (2003)
14.	Menstrual Troubles (*Polygonum glabrus*)	Gopal (2004)
15.	Leucoderma	Misra (2004)
16.	Rheumatism	Misra (2004)
17.	Influenza	Misra (2004)

Table 1.5: List of some plants used against various diseases

Sl.No.	*Diseases*	*Plants*
1.	Abscesses	*Sida rhombifolia* Linn.
2.	Amaebiosis	*Syzygium cuminii* (Linn.) Skeels
3.	Anorexia	*Crataeva nurvala* Buch.-Ham., *Piper nigrum* Linn.

Contd...

Table 1.5–*Contd...*

Sl.No.	*Diseases*	*Plants*
4.	Antibacterial against *Pseudomonas aeruginosa*	*Musa paradisiaca* L., *Emblica officinalis* Gaertn.
5.	Anticancer	*Taxus* sp.
6.	Antifertility in woman	*Hyptis suaveolens* Poit.
7.	Antifungal activity	*Argyrea speciosa, Argyreia speciosa* Sweet.
8.	Antiguinea worm	*Crotalaria juncea* Linn., *Commiphora weightii*
9.	Antiseptic to Cattle worm	*Phyllanthus emblica* Linn., *Ipomoea carnea* Jacq.
10.	Aphrodisiacs (enhance sexual potential)	*Asparagus racemosus* Willd., *Abrus precatorius* Linn., *Aloe barbedensis* Mill.
11.	Arthritis	*Withania somnifera* Dunal.
12.	As Antidotes of Scorpion and Snake bite	*Allium cepa* Linn., *Berberis asiatica* D.C., *Cyperus rotundus* Linn.
13.	Asthma	*Solanum surattense* Burm., *Calotropis procera* (Ait.) R.Br., *Adhatoda vasica* Nees.
14.	Blood in urine	*Euphorbia thymifolia* Linn.
15.	Blood purifier	*Imperata cylindrica* Beauv.
16.	Body Pain and Rheumatism	*Achyranthes aspera* Linn.
17.	Body swelling	*Plumbago zeylanicum* Linn.
18.	Bone Fracture	*Terminalia arjuna* (Roxb.) Wight & Arn.
19.	Bronchitis	*Acacia nilotica* (Linn.) Delile
20.	Cardiac stimulant	*Terminalia arjuna* (Roxb.) Wight & Arn.
21.	Catarrh	*Hyptis suaveolens* Poit.
22.	Cervical spondylgysis	*Actea racimosa* L,
23.	Chicken pox	*Ficus religiosa* Linn.
24.	Cholera	*Calotropis procera* (Ait.) R.Br.
25.	Constipation	*Murraya koenigii* (Linn.) Spreng., *Piper betle* Linn., *Senna alata* L.
26.	Cuts and Burns	*Beta vulgaris* Linn.
27.	Dental ailments	*Madhuca indica* J.F. Gmel., *Salvadora oleoides* Decne.
28.	Diabetes	*Acacia nilotica* J.F. Gmel., *Adiantum caudatum* Linn., *Dillenia indica* Linn.
29.	Diarrhoea	*Anthocephalus cadamba* Miq., *Ficus bengalensis* Linn.
30.	Dropsy	*Mucuna prusiens* (L.) DC
31.	Eczema	*Solanum surattense* Burm.
32.	Epilepsy	*Benincasa hispida* (Thunb.) Cogn.
33.	Eye Problems	*Albizzia lebbeck, Desmodium gangeticum* DC.
34.	Filaria	*Caesalpinia decapetata*
35.	For Nervous system	*Cannabis sativa* Linn., *Datura metel* Linn., *Celastrus paniculatus* Willd.
36.	Gastric troubles and Stomachache	*Chrysanthemum americanum* (L.) Vatke

Contd...

Table 1.5–*Contd...*

Sl.No.	Diseases	Plants
37.	Goitre	*Butea monosperma* (Lam. Kuntze)
38.	Gout	*Piper nigrum* Linn.
39.	Hair growth	*Evolvulus alsinoides* Linn.
40.	Headache	*Artocarpus heterophyllus* Lam.
41.	Hydrocele	*Mimosa pudica* Linn.
42.	Impotency	*Anellema scapiflorum*
43.	In Livestock diseases	*Cassia tora* Linn., *Cassia fistula* Linn.
44.	Insect repellant	*Pongamia pinnata* Pierre.
45.	Jaundice	*Boerhaevia reflexa*
46.	Leprosy	*Albizzia odoratissima* Benth., *Anagalis arvensis* Linn.
47.	Leucoderma	*Argemone mexicana* Linn.
48.	Malaria	*Cleome gynandra, Ricinus communis* Linn., *Cinchona* sp., *Mellia azadirachta* Linn.
49.	Mastitis	*Zingiber officinale* Rosc.
50.	Migraine	*Ocimum cannum* Sims.
51.	Mouth and Teeth Diseases	*Dalbergia sissoo* Roxb.
52.	Mumps	*Tragia involucrata* Linn.
53.	Muscular pain	*Clausena excavate* Burm. f.
54.	Nausea	*Glochidion lanceolarium* Voigt, non Dalz.
55.	Nose bleeding	*Phyllanthus niruri* Hook. f., non Linn.
56.	Paralysis	*Terminalia chebula* Retz.
57.	Peptic ulcer	*Achyranthes aspera* Linn.
58.	Piles	*Abutilon indicum* (Linn.) Sweet.
59.	Pimples	*Bombax ceiba* Linn.
60.	Pinworm	*Benincasa hispida* (Thunb.) Cong.
61.	Pregnancy	*Butea superba* Roxb.
62.	Prevention of conception	*Ficus religiosa* L.
63.	Pyorrhoea	*Mimusops elengi* Linn.
64.	Rheumatism	*Grewia tiliaefolia* Vahl.
65.	Ringworm	*Ocimum cannum* Sims.
66.	Scabies	*Argemone mexicana* Linn.
67.	Spleen	*Gloriosa superba* Linn.
68.	Tapeworm	*Carica papaya* Linn.
69.	Thread worm, Filaria, Colic pain	*Andrographis paniculata* Wall. ex Nees.
70.	Tuberculosis	*Actiniopteris radiate, Barleria prionitis* Linn.
71.	Typhoid	*Portulaca castrim* L.
72.	Urinogenital Disease	*Bixa orellana* Linn.

Contd...

Table 1.5–*Contd...*

Sl.No.	*Diseases*	*Plants*
73.	Uterine infection Menorrhagia	*Saraca asoca* (Roxb.) De Wilde
74.	Veneral diseases	*Acacia catechu* Willd. (*gonorrhoea*), *Abelmoscus esculentus* (Linn.) Moench (*syphilis*), *Acacia nilotica* (Linn.) *Detile.*
75.	Vermicidal	*Aegel marmelos* Correa ex Roxb., *Ficus bengalensis* Linn.
76.	Whooping cough	*Achyranthes aspera* (whole plant) Linn.

Important Plant Products with Radioprotective Properties

Rubia cordifolia

The roots of this plant are valued in the Ayurveda for the maintenance of general health and normal blood circulation. The other important plants of high altitudes are: *Cinchona officinalis* (against Malaria), *Acorus calamus* (brain tonic), *Andrographis paniculata* (hepatoprotectant, anthelmentic, blood purifier), *Oroxylum indicum* (fever), *Rauwolfia serpentina* (treatment of blood pressure), *Tinospora cordifolia* (vitaliser), *Coptis trifolia* (Antipyretic, Jaundice), *Koelpinia linearis* (for lymphatic diseases, gastrointestinal disorders, bonefracture, inflammation), *Juniperus macropoda* (kidney diseases, joint pains), *Codonopsis clematidea* (Arthritis, anti-infectious), *Swertia chirayata* (bite disorder) and *Sida cordifolia* (antibacterial and antifungal).

Liv.-52

It is a non-toxic herbal preparation composed of *Capparis spinosa, Cinchosium intybus, Solanum nigrum, Cassia occidentalis, Terminalia arjuna, Achillea millefolium* and *Tamarix gallica* and is clinically active in *hepatotoxicity*, other hepatic disorders and modulates radiosensitising (Saini *et al.*, 1984).

Miscellaneous

Herbal gulal, herbal fragrance, digestive churna etc. The list is indeed endless includes Mederna Diabacure and Yoghurt etc.

Cosmetic Properties of Some Important Plants

Aloe Vera: A Vital Ingredient in Beauty Product

Aloe barbadensis is used as a dietary supplement. The gel of this plant is very potent with 200 components and 74 known nutrients including B-12, 21 amino acids, iron, manganese, calcium, zinc, minerals and enzymes. *Aloe vera* products are used as health drinks and antacids. The medicinal properties of this gel are useful in arthritis, diabetes and have antibacterial, antifungal and antioxidant and other properties. The calcium isocitrate salts present in this plant prevent hardening of arteries and improves blood circulation. It decreases the bad cholesterol and increases the good cholesterol. Potassium maintains the heart rhythm and stabilizes the blood pressure. Antioxidants like Vit. A, C and E present in this plant neutralize the free radicels and reduce thereby heart diseases. The prescribed dose is 60 ml. every morning on an empty stomach. The beauty industry swears by the cosmetic properties of *Aloe vera.* It increases (1) Cell regeneration (2) Healing and boosting the natural immune system (3) It is a very powerful moisturiser and cleanser. Now-a-days, wound healing creams containing *Aloe vera* have flooded the market.

Anti-diabetic Plants

Stevia is a shrub whose leaves are 20 times more candied than sugarcane. It is free of carbohydrates and fat. It doesn't affect blood sugar level and doesn't taste bitter, ideal for patients of diabetes, who will be 10 crore by 2010. Its market with the next 5 year is going to be 1000 crore in rupees. *Stevia* is available in the market as leaf paste, dry powder and herbal tea.

Bio-energy Product: *Jatropha*

Jatropha is a pollution free bio-diesel producing petrocrop with many medicinal and industrial uses. In medicine extracts of this plant are used in folk remedies for cancer, alopecia, burns, cough, dermatitis, purgative, diuretic, vermifuge, dysentery, dyspepsia, eczema, fever, gonorrhoea, inflammation, jaundice, paralysis, scabies, rheumatis, sores, stomachache, syphilis, ulcers and uteroses. Besides, latex of the plant can be applied topically to bee and wasp stings. Decoction of root is used to cure dysentery. Leaves are antiparasitic, applied to scabies. Juice of the leaves is used externally for piles. Fruits are anthelmintic and useful in anaemia, fistula and heart diseases. Oil of the plant is used for cleaning wounds and sores. Curcin is the active principle. Oil contains oleic and linoleic acid, bark contains tannins, wax, resins, saponins and reducing sugars.

Other uses of *Jatropha* include as a substitute for oil in engines, preparation of soaps and candles. Latex of plant inhibits water melon mosaic virus. Bark is a fish poison. Seeds are used as contraceptive and insecticide. Plant prevents soil erosion and can be grown on waste lands. Uttarakhand and Chattisgarh Governments have taken first initiative to grow the crop commercially. The advantage of the biodiesel this plant can be used as a substitute for oil engines, it neither pollute the atmosphere nor does it add to the cause of global warming (Somasundaram and Manjunath, 2004).

Guggul: Mahaushadi of Ayurveda

The gum oleoresin extracted from *Commiphora wightii* is well known for its anti-rheumatic, anticholesterolaemic and anti-inflammatory action. It is widely distributed in Gujarat, Rajasthan and parts of Madhya Pradesh, Maharashtra and Karnataka. Due to over exploitation, it is going to be endangered plant, calling for it's propagation which is now being done, in Guggulu Herbal farm at Mangliawas (Ajmer district) and Central Council for Research in Ayurveda and Siddha (Department of AYUSH, Government of India).

Versatile Neem

Almost every part of neem is bitter and useful as timber, fuel wood and fodder. Neem oil (Margosa) is used as lubricant and is also used in soaps and cosmetics. De-oiled neem cake has multiple uses in agriculture, poultry and animal husbandry.

Neem can reclaim wasteland and can utilize industrial effluent water. Limnoids like Azadirachtin A, Selanin, Nimbin and Melinatriol are effective against insects as biopesticides. Azadirachtin is potent locust anti feedant; Nimbin is effective against fungi like *Tinea rubrurn* and *Coccidiodes immunities* while Selanin has been reported to use for feeding of striped cucumber, beetle, horse flies and Japanese beetle (Das, 2004). Neem products are effective against following diseases: Tungro virus; Sheath blight (*Acrocylindrium oryzae*); Blast (*Pyricularia oryzae*); Sheath blight (*Rhizoctonia solani* of Paddy; Rust (*Puccinia arachidis*) and Foot rot (*Sclerotium rolfsii*) of groundnut; Ganoderma wilt (*G. lucidum*) of coconut; Powdery mildew (*Erysiphae polyconi*); Root rot (*Macrophomina phaseolind*) of Black gram and wilt of Betelvine (*Phytophthora parasitis*). Neem wood is resistant to termites. Information chemistry and structure activity relationship of chemicals from Neem and allied species like China Berry tree *Melia azadirach'* work as potential pesticide. Thus, Neem is good enough for biodynamic farming.

Safed Musli (*Chlorophytum borivilianum*)

Safed Musli is distributed in almost all parts of India. The plant has high aphrodisiac values and acts on Central Nervous System. Its dry roots are known to remove impotency. Chemical composition of the plant induces steroids, resin phonetics, tannins, carbohydrates, calcium, magnesium and potassium. Tubers contain steroid sapoginin and are fat free. It is useful in Pitta and Vata. It is effective on fatigueness and dahaa and purifies blood. It is useful in certain diseases like renal calculus, dhupani, sangrahani, leucorrhoea and diabetes.

Black Pepper (*Piper nigrum L.*): Perennial Woody Climber

Pepper is known to be an effective cure for dyspepsia, malaria and hemorrhoids. It has also been used as aphrodisiac and as insect repellent. Dry pepper is effective against asthma, cough, heart diseases and pains and is also advised in the treatment of diabetes and piles. The use of pepper in homeopathy is for treating malaria, nausea, vomiting and inflammation of penis with burning pain while coconut oil with pepper soothes itching of the body. Pepper with milk is good for dandruff. Fruit of pepper is used in vertigo and coma. Volatile oil containing alkaloids and piperin are active principles.

Aswagandha (*Withania somnifera*)

The plant is found in almost all parts of India. Its stem bark is taken in migraine to increase masculine and seminal strength. It relieves asthma and cough. Fine powder of the stem with mustard oil is applied locally on genitals to increase potency and masculine strength. It is anti-cancerous and is also used in arthritis and constipation. As a rasayana, the plant promotes health and longevity by increasing disease resistance, arresting ageing and revitalizing body. It improves memory and has significant anxiolytic and antidepressant effects. Roots are anti-inflammatory, anti-cancerous and antioxidant.

Tulsi (*Ocimum sanctum*): Herbal Queen

It is an important religious and medicinal plant belonging to Labiatae. It's leaves when boiled with tea and consumed orally helps in cold and cough. It has anti-stress properties. Flavanoids, orientin and vicenin have been reported to have radio-protective effects. Extract of it is antioxidant when taken with tea.

Bhuiawla (*Phyllanthus niruri*)

It is effective against hepatitis. Its grinded plant parts are used in diffrent liver disorders.

Sarpagandha (*Rauwolfia serpentina*)

It grows in waste places and in shady forests from Punjab east wards to Skkim, Bhutan, Nepal and Western Ghats. The root of the plant is very useful. Root decoction is used for uterine contraction and for expulsion of foetus, to treat intestinal disorders, anti-helminthic and induces hypotension. Also used in psychosis and schizophrenia (Trivedi and Nehra, 2004).

Anti Oxidative Plants

Free radicals are highly reactive chemical species which possess an unpaired electron in the outer shell of the molecule and can react with proteins, lipids, carbohydrates and DNA.They attack the nearest stable molecules, stealing its electron (Dubey *et al.*, 2005) leading to many destructive diseases. Free radicals are produced in the body as by-products of normal cellular metabolic activities such as prostaglandin synthesis, mitochonodrial electron transport, endoplasmic reticulum enzyme activity, oxy-haemoglobin, auto-oxidationn and phagocytosis as well as exposure to pollutants and ionizing radiation. A balalnce is maintained between oxidative attack of free radicals and the anti-oxidants.

The anti-oxidative defence system present in human body may be of two types, *viz.*, enzymatic, *e.g.* catalse, glutathione peroxidase and reductase, s-transferase, sequential removal of ROS and non-enzymatic, *e.g.* metal binding proteins and low molecular weight non-enzymatic antioxidants like cytochrome-C, caretinoids, ascorbic acid, riboflavin, lipoic acid, selanin and zinc, which modulate radiation damage (Jain, 2002). Many plants with Vitamin A, E and C are rich in antioxidants *e.g. Withania, Ocimum, Mentha, Phyllanthus, Cassia* etc.

Selected Plants for Herbal Gardens (Arogyavan)

Andrographis paniculata (Kalmegh), *Plantago ovata* (Isabgol), *Aloe barbadensis* (Ghrit kumari), *Gloriosa superba* (Kali hari), *Acorus calamus* (Bach), *Bacopa monari* (Brahmi), *Asparagus racemosus* (Satawar), *Glycyrrhiza glabra* (Mullati), *Curcuma amada* (Amahaldi), *Puereria tuberosa* (Vidarikand), *Catharanthus roseus* (Sadabahar), *Mentha arvensis* (Mentha), *Piper nigrum* (Pepper), *Tagetus minuta* (Gainda), *Ocimum sanctum* (Tulsi), *Cymbopogon pendulus* (Lemongrass), *Phyllanthus officinalis* (Amla), *Jatropha curcas* (Ratan jyot), *Chlorophytum arundinaceum* (Safed Musli), *Withania somnifera* (Aswagandha), *Rauwolfia serpentina* (Sarpgandha) prioritized are some plants suitable to grow in our country or establishment of Arogyavan. Other plants as *Saraca asoca* (Ashok), *Aconitum heterophyllum* (Atees), *Aegle marmelos* (Bael), *Santalum album* (Chandan), *Swertia chirata, Berberis aristata, Tinospora cordifolia* (Giloy), *Commiphora wightii* (Guggul), *Nardostachys jatamansi* (*Jatamansi*), *Crocus sativus* (*Kesar*), *Garcinia indica* (*Kokun*), *Solanum nigrum* (*Makoy*), *Coleus barbeters* (*Pahterchur*), *Cassia angustifolia*

(*Senna*), *Embelia ribes* (Vaividang) and *Aconitum ferox* (Vatsnabha) are also good establish of Herbal Garden.

Commercial Scale of Herbal Trade

It is daunting, vast, secretive and largely unregulated trade of medicinal plants, mainly from the wild. India has 7000 plant species for traditional medicines. The annual production of medicinal and aromatic plant raw materials is worth about 200 crores. Besides, about 300 natural products are used as raw materials in flavour and fragrance industry in the form of essential oils, extracts, oleoresin, concrete, absolutes, resinoids and tinctures. As per WHO estimate, medicinal plant related trade in India is estimated to be 550 crore per year. According to Mukherjee (2004), India's total turnover of Rs. 2,300 crore of Ayurvedic and herbal products, major products contribute around Rs. 1,200 crores. Other formulations fetch around 650 crore and classical Ayurvedic formulations contribute remaining 450 crores. Indian systems of Medicinal market in India is being reproduced below:

Sl.No.	*Systems*	*Per cent*	*Market Size (Rs. Crores)*
1.	Ayurveda	84	3,500
2.	Unani	02	100
3.	Siddha	14	5
4.	Homeopathy	0	600
5.	Tibetan Medicines	-	-
	Total		4,205

Global Market

According to WHO, EXIM bank of India, in its report has reported the value of medicinal plant related trade in India of the order of 5.5 billion dollars and is growing rapidly. **Export:** India's share in International market is about 0.5 per cent. The quantity (kg) of medicinal plants export in 2001 was 47,477,464 kg whose value in rupees was 3,15,77,40,878. The countries to which India exports crude drugs are USA, Germany, France, Switzerland, UK and Japan who share between them 80 per cent of the total export of crude drugs from India. Principal drugs exported are *Aloe, Belladona, Acorus, Cinchona, Cassia tora, Dioscorea, Digitalis, Ephedra, Plantago, Senna* etc.

Constraints and Future Strategies

Following constraints are associated with the use, development and marketing strategies: Poor agricultural practices, Poor harvesting and post harvest practices, Poor propagation of herbs, Poor quality control, Lack of cross cultural comparisons of herbs, Similarity and dissimilarity of uses, Lack of research on high yielding varieties, Paucity of pharmacognosity evaluation, Inefficient processing, High energy losses during processing, Standardization of raw and final products, Toxicological studies, Chemistry of raw materials, Lack of analytical procedures to detect residues of underground pollutants and heavy metals, Validation of plants by ethnic tribes,

Testing the efficacy of active ingredients, Storage and shelf life of herbs and their products, Difficulties in marketing, Lack of local market, Lack of trained personnel and equipment, Lack of access to latest technology for market information, Safety, stability of the product, Quality control to check losses in impotency, Lack of knowledge of optimal stage of harvest, drying, grinding, processing, packing and transport of herbs and their products.

Intellectual Property Rights (IPR)

IPR are protected in several ways: Trade secrets, patents, plant breeder's right, farmers right and copyright. Indian Patents Act was passed in 1970 (Sahai, 2003). In 1983, Paris convention established equal protection of Industrial IPR. India jointed it in 1998. TRIPS is administered by WTO. Benefits of IPR are many like transfer, exchange, encourage and safe guard new ideas and to encourage investment in Research and development (R&D). As to the Indian scenario, there are (1) Fast alterations in IPR (2) Indians believe in heritage, tradition and benefits to society at the cost of individual interest (3) Continued and systematic efforts to survival and (3) Checking of biopiracy. TRR (Traditional Resource Rights) and TPR (Traditional Plant Resources Rights) and their database are equally important. (Mashelkar, 2001).

Documentation of Traditional Knowledge (TK)

Karnataka Peoples Biodiversity Registers (PBRs) and NGO initiatives like FRLHT (Foundation for the Revitalization of Local Health Traditions, Banglore) SRISTI (Society for Research and Initiatives for Sustainable Technologies, Ahmedabad), BBA (Beej Bachao Andolan, Tehri Garhwal) and BCPP (Biodiversity Conservation and Prioritization Programmes- 56 sites in 7 states) and TKDL (Digital Library of Traditional Knowledge) are some of the initiatives in this direction.

Healthcare awareness and research developments about preventive and curative herbs which are safe and affordable have come out in a big way from agro based ethnobotany to industry-based Pharma industry both in concept and compass. Herbal medicines and products laying emphasis on mind-body relationship have been successfully and increasingly used to cure stress-related, anxiety-related, psychosomatic, post-traumatic stress related, digestion-related, radiation-related, toxin-related, disability-related and infection-related disorders. Herbal gardens and Spas, the centres of alternative healing on picturesque locale using herbal oils (Abhyanga), fragrant essential oils (Aromatherapy), Seaweed wrap, Shiastu (Acupressure) and Swedish massages and circular bath soothe the nerves and muscles rejuvenating mind, body and soul relationships. 7,500 medicinal plants, 3900 edible plants and 100 perfume plants have a domestic market of Indian systems of medicine of the tune of 4,205 crores and International market of Rs. 315,77,40,878. Seabuckthorn, *Aloe*, Mushrooms, Liv.-52, Brahm-rasayana etc. are excellent cosmetic and dietary supplements.

Over and unscientific exploitation of medicinal and other utility plants, denudation of the forest cover and degradation of the ecosystems are the harbingers of a disaster that is already upon us in the form of habitat losses and, endangered and vulnerable biodiversity. Therefore, we should treat forests with respect, care, concern and consideration for sustainable progress. Since to undo the damage caused to nature is difficult, to conserve natural fauna and flora with responsibility and sensitivity is easy for sustainable development and has been discussed. Tissue culture techniques, cryopreservation, forest management etc. should be tried, along with the development and strengthening of herbal gardens and medicinal, aromatic and spices boards based on existing constraints. Standardization and quality control of herbal drugs, management of herbal trade and documentation of Traditional Knowledge (TK) by establishing Digital Library. Future strategies in this context have been suggested. But for exports of our herbal medicines/preparations like Hamdard's Safi, Dabur's Shilajit, Himalaya Drug's Karek capsules, Zandu Pharma's Maha Sudarshan Churna, etc. component labelling has been made compulsory from Jan. 2006 by Health Ministry, Government of India.

Reference

Aruhdutt 1925. *Astanga Hridaya Samhita*. Nirnay Sagar Press, Mumbai.

Aruna K and Sivaramkrishnan V 1992. Anticarcinogenic effect of some plant products *Food Chem Toxicol* **30:** 935.

Bagul R M and Yadav S S 2003. Antivenomous traditional medicines from Satpuda, East Khandesa. *Plant Archives* **3:** 319-320.

Bhatt D C, Mehta D K and Parmar R P 2002. Studies on some ethnomedicinal plants from Talaja Taluka of Bhavnagar distt., Gujrat. In: *Ethnobotany* (ed. Trivedi P C) Aavishkar Publ Jaipur 295-310.

Bhatt D C, Mitaliya K D, Mehta D P and Prajapati M M 2001. Preventive herbal medicine against Malaria. *Ad Plant Sc* **14:** 7-10.

Bhatt D C, Mitaliya K D, Parmar R P, Lashkari P I and Dodia S K 2003. Observations on foliage employed as vegetable by tribals and rurals in Gujarat *Ad Plant Sci* **16:** 413-416.

Bhatt D C and Joshi P N 2004. Ethnobotanical Plants of Pachham hills of Kutch district, Gujrat. *Ethnomedicinal Pl* 117-135.

Bhatt D C, Mitaliya K D and Mehta S K 2000. Plant twigs as tooth brushes. *Ad Plant Sci* **13:** 19-22.

Chadha Y R, Gupta R and Nagarjan S 1980. Scientific appraisal of some commercially important medicinal plants of *India. Indian Drugs* and *Pharma Industry* **15:** 7

Chaudhuri A B and Sarkar D D 2004. *Biodiversity-Endangered:* India's Threatened and Medicinal Plants *Scientific Publ* Jodhpur.

Chhangani G S 1945. *Astanga Sangrah.* Varanasi: Chowkhamba Sanskrit Series.

Chopra R N, Nayar S L and Chopra T C 1956. *Glossary of Indian Medicinal Plants.* New Delhi: CSIR.

Chunekar K C 1988. *Bhavprakash Nighantu.* Varanasi: Chowkhamba Bharti Academy.

Das A 2004. Versatile uses of Neem. *Agrobios* **3**: 19-20.

Deora G S, Jhala S G P and Rathore M S 2004. Ethnoveterinary medicinal plants in north-west part of Udaipur distt. In: *Ethnomedicinal Plants* (Eds Trivedi P C and Sharma N K) Pointer Publ Jaipur 74-81.

Dixit R D and Singh S 2004. Medicinal Pteridophytes- An Overview. *In: Medicinal Plants* (Ed Trivedi P C) Aavishkar Publ Jaipur 269-298.

Dubey P K, Edwin E and Sheeja E 2005. Oxidative Stress. *Pl Archives* **5:** 1-8.

Elamathi S, Thomas A and Singh S S 2005. Bioenergy - A new hope for the millennium. *Agrobios News letter* **3:** 15-16.

Gopal G V 2004. Medicinal plants in India and their conservation: An ethnobotanical approach. In: *Ethnomedicinal Plants* (Eds. Trivedi P C and Sharma N K) 1-17.

Gopinathan K M 2004. Black Pepper: The King of spices. *Agriobios* **3:** 55-58.

Goyal P K 2004. Modulation of Radiosensitivity by certain plants and plant products. *In: Ethnomedicinal Plants* (Eds. Trivedi P C and Sharma N K) *Pointer Publ* Jaipur 226-239.

Gupta A K and Mishra S K 2000. Folk-lore dental protector plants of Chhattisgarh, M.P. India *Ad Plant Sci* **13:** 501-503.

Gupta R 1964. Survey record of medicinal and aromatic plants of Chamba Forest Division Himachal Pradesh. *Indian Forester* **90:** 454-458.

Harshberger J W 1895. The purpose of Ethnobotany. *Bot. Gazette* **31:** 146.

Hembrom P P 1994. *Adivasi Aushadh* (Homeopathy) Parts 1-6. Satiya Distt. Pakur, Bihar, Paharia Seva Samity.

Jadeja B A, Bhatt D C and Odedra N K 2005. Plants used in abscess in the Saurashtra region of Gujrat. *Plant Archives* **5:** 113-117.

Jain M 2002. Evaluation of antioxidative efficacy of certain plant extracts- A study on mice liver. Ph.D. Thesis, Univ. of Rajasthan, Jaipur.

Jain S K 1963. Observations on ethnobotany of the tribals of M. P. *Vanyajati,* **11:** 177-183.

Jain S K 1987. Endangered species of medicinal herbs in India. *Medicinal Herbs Indian Life* **16:** 44-53.

Jain S K 1991a. *Dictionary of Indian Folk Medicine and Ethnobotany.* Deep Publications, New Delhi.

Jain S K 1991b. *Contributions of Indian Ethnobotany.* Jodhpur *Scientific Publ.,* Jodhpur.

Jain V, Sharma T, Dashone K, Saraf S and Saraf S 2005. Phytosome: A novel approach for herbal drug delivery system. *Plant Archives* **5:** 325-330.

Jakhar M L, Kakralya B L, Singh S J and Singh K 2004. Enhancing the Export Potential of Medicinal Plants through Biodiversity. Conservation and Development under Multi-adversity-environment. *In: Medicinal Plants. Aavishkar Publ Jaipur* 36-94.

Janmeda B S 2004. Ethnobotanical Studies in East-Sikkim. Ph.D. Thesis, C.C.S. Univ. Meerut.

Karthikayani T P 2003. Studies on Ethnogynaecological plants used by the Irulars of Siruvani hills, Western Ghats. *Plant Archives* **3:** 159-166.

Kavitha K 2004. *Azolla - A Nature Friend. Agrobios News Letter,* **3:** 30-31.

Khan A A, Singh P and Pandey R 2005. Herbal treatment curing children disease among tribals of Shakdol Distt M.P. **5:** 159-163.

Kirtikar K R and Basu B D 1933. *Indian Medicinal Plants.* Allahabad : Lalit Mohan Basu. Vols **1-4.**

Mahapatra A K 1983. Essential oil yielding grasses in Thar desert. *J Econ Taxon Bot* **4:** 173-175.

Maheshwari J K 1996. New vistas in Ethnobotany (editorial). In: Ethnobotany in South Asia. *J Econ Taxon Bot* 1-11.

Malhotra S K and Moorthy S 1973. Some useful medicinal plants of Chandrapur district, Maharashtra state. *Bull Bot Surv India* **15:** 13-21.

Malik C P 2004. Intellectual Property Rights, Growth and Competitiveness of Indian Pharmaceutical Industries. *In: Medicinal Plants. Aavishkar Publishers,* Jaipur 27-35

Mashelkar R A 2001. Intellectual Property Rights and the Third World. *Curr Sci,* **81:** 955-965.

Misra M K 2004. Ethnomedicobotany of Orissa-A Review. In: *Ethnomedicinal Plants* 82-115.

Mukherjee T 2004. Medicinal Plants: Need for protection. In : *Medicinal Plants.* Aavishkar Publ Jaipur 391-404.

Nadkami K M and Nadkami A K 1954. *Indian Materia Medica.* Bombay: Dhootpapeshwar Prakashan Ltd. Vols **2.**

Nargas J and Trivedi P S 2004. Traditional plant based medicinal system in health care during pregnancy and post-pregnancy. In: *Ethnomedicinal Plants* 164-172.

Panara B P 1996. *Kalpvriksha.* Limdo Gola Publ, Ahmedabad.

Parmar S 1978. A note on folk musical instruments of Madhya Pradesh. *Folk lore,* 26-27.

Prajapati N D and Kumar V 2003. Agros Dictionary of medicinal plants. *Agrobios,* Jodhpur, India.

Prasad V K, Rajgopal T and Badrinath K V S 2002. Ethnobotany of Kondareddis of Rampur Agency, Eastern Ghats, Andhra Pradesh. In: *Ethnobotany* (Ed. P. C. Trivedi). Aavishkar Publ Jaipur 205-220.

Rastogi R P and Mehrotra B N 1993. *Compendium of Indian Medicinal Plants.* Vols. 1-4. CDRI, Lucknow and PID.New Delhi.

Roy A and Saraf S 2005. Insecticidal Plants and Screening Methods: A Review. *Plant Archives* **5:** 9-21.

Sahai S 2003. Indias Plant Variety Protection and Farmer's Right Act, 2001. *Curr Sci* **84:** 407.

Sahu T R and Sahu P 2002. Ethnobotanical Scenario of Madhya Pradesh: Knowledge and Culture in Tribal belt of Bastar. **In:** Ethnobotany. Aavishkar Publ Jaipur 73-97.

Saini D C 2004. Ethno-phyto-toxicological studies in Sidhi district of Madhya Pradesh. **In:** *Ethnomedicinal Plants,* Pointer Publ Jaipur 192.

Saini M R, Kumar S, Jagetia G C and Saini N 1984. Effect of Liv.52 against radiation sickness and mortality. *Ind Pract* **37:** 1133-1138.

Sankaran M, Dutta M, Singh N P and Chanda S 2003. *Jatropha* - A boon to waste lands. *Agrobios News Letter* Jodhpur. **2:** 53.

Sen R, Pal D C and Guha A 1983. *Kuchila: A Tribal Dye. Vanyajati* **31:** 3-4.

Shah C S and Qadry J S 1989. *A Textbook of Pharmacognosy.* Shah Prakashan, Ahmedabad.

Sharma N K 2002. Ethno-medico-religious plants of Hadoti plateau (S.E. Rajasthan) In: *Ethnobotany.* Aavishkar Publ Jaipur 394-411.

Sharma R K, Chopra G, Bhushan G G and Munjal K 2003. Use of Indian Plants in fertility regulation in Mammals. *J Exp Zool* **6:** 57-74.

Sharma S K 2004. *Medicinal Plants:* A Probe in the forests of Rajasthan. In: *Medicinal Plants* (Ed. Trivedi P C). Aavishkar Publ Jaipur 181-216.

Shukla C P 2006 Harnessing the Dewghat forest biodiversity for medicinal uses. *Proc Nat Acd Sci* **76:** 182-193.

Shukla C P 2008. Floristic Investigation on Ethnomedicianal flora of Dewghat Forest Koraon Range (U.P.) Proceedings in National Seminar on Natural Resources held at 30-31 August 2008, Sahu Jain P.G. College Najibabad (U.P.).

Singh G S 2002. Minor forest products of Sariska National Park: An Ethnobotanical Profile. In: *Ethnobotany* (Ed. Trivedi P C). Aavishkar Publ Jaipur 221-238.

Singh L, Vats P, Garg S and Ranjana 2004. Medicinal incense and edible Gymnosperms, Pteridophytes and Mushrooms of East Sikkim. *Acta Botanica Indica* **32:** 75-80.

Sivarajan V V and Balachandran I 1994. *Ayurvedic Drugs and Their Plant Sources.* New Delhi: Oxford and IBH Publishing Co. Pvt. Ltd.

Somasundaram E and Manjunath G 2004. *Jatropha curcas-* A pollution free power tree. *Agrobios* **3:** 43-44.

Soni P L 1995. Food from forest. *Ind For* **121:** 837-884.

Thakur D and Thakur D S 1994. *Tribal Life and Forests.* Deep Publ., New Delhi.

Tosh J 2004. Ethnobotany - Green-Gold branch in botanical sciences. In: *Ethnomedicinal Plants* (Eds: Trivedi P C and Sharma N K). Pointer Publ Jaipur 1-77.

Trivedi P C and Nehra S 2004. Potential Medicinal Plants: Botany, Medicinal uses and chemical constituents. In: *Medicinal Plants,* Aavishkar Publ Jaipur 405-424.

Vaharmihir 1984. *Brihad Samhita* (Ed: Thakur) Pravin Pustak Bhandar, Rajkot.

Vaish S G 1997. *Bhavprakash.* Khem Raj Shree Krishan Das, Mumbai.

Vanathi D and Rathika S 2004. Neem-A Boon to Biodynamic Farming. *Agrobios* **3:** 40-41.

Vartak V D and Ghate V S 1990. Ethnobotany of Neem. *Biol Ind* **1:** 55-59.

Verma V 2005. Ethnobotanical studies in East Sikkim. Masters Thesis, Sikkim Manipal University of Health, Medical and Technological Science.

Watt G 1889-1893. *The Dictionary of Economic Products of India.* Govt. Press, 1-4.

Medicinal Plants: Aspects and Prospects (2014) ***Pages* 23–36**
***Editors:* Mukesh Kumar, Anjali Khare and C.P. Shukla**
ISBN: 978-81-7622-309-6
***Published by:* BIOTECH BOOKS, NEW DELHI**

Chapter 2

Biotechnology for Medicinal and Aromatic Plants

B.N. Pandey

Government PG College, Chunar,
Mirzapur – 231 304, Uttar Pradesh

ABSTRACT

Biotechnology is being used either as central tool or as an adjunct to other methods in plant propagation, genetic modification and improvement, product formation and germplasm storage. Plant propagation through biotechnology involves plant tissue culture technology for rapid clonal multiplication and is useful for those plants that are in high demand, or those that are slow to cultivate naturally. Plant modification and improvement through biotechnology involves somaclonal breeding, haploid production,protoplast culture, genetic transformation, etc. for the qualitative and quantitative improvements in medicinal and aromatic plants. The role of biotechnology in product formation involves development of systems for *in vitro* production of secondary metabolites and precursor biotransformation, and also up-scaling of process for their extraction. Germplasm conservation through biotechnology involves storing plant genetic material of *in vitro* for future revival and propagation.

Keywords: ***Biotechnology, Biotransformation, Genetic transformation, In vitro conservation, Micropropagation, Secondary metabolites.***

Introduction

Medicinal and aromatic plants (MAPs) have been used as sources of important therapeutic aid for alleviating human ailments. Approximately 80 per cent of the people in the world's developing countries rely on traditional medicines for their primary health care needs, and about 85 per cent of traditional medicines involves the use of plants. With the increasing realization of the health hazards and toxicity associated with the indiscriminate use of synthetic drugs and antibiotics, interest in the use of plants and plant-based drugs has revived throughout the world (Nalawade *et al.*, 2003).

There is growing interest of natural products as a source of new chemical entities not only for the development of modern drugs but also as dietary supplements (nutraceuticals), ingredients to food and beverages, phyto-cosmetics and other herbal products.Global market for herbal products which includes medicine, health supplement, herbal cosmetics is around US$65 billion. According to World Bank report, the international herbal medicine market is expanded to reach US$ 5 trillion in 2050 with an annual growth rate of between 10 to 20 per cent.

However, the alarming and distressing fact about the present use of medicinal and aromatic plants is that in lack of proper cultivation practices, most of the pharmaceutical industry is highly dependent on wild populations for the supply of raw materials for extraction of medicinally important compounds. This leads to destruction of plant habitats, and the illegal and indiscriminate collection of plants from these habitats, many medicinal plants are severely threatened.

The above observation is based on the fact that about 90 per cent of medicinal plants used by industries are collected from the wild. While over 800 species are used in production by industry, less than 20 species of plants are under commercial cultivation. Over 70 per cent of the plant collections involve destructive harvesting because of the use of parts like roots, bark, wood, stem and the whole plant in case of herbs. This poses a definite threat to the genetic stocksand to the diversity of medicinal plants. Recently some rapid assessment of the threat status of medicinal plants using IUCN designed CAMP methodology revealed that about 112 species in southern India, 74 species in Northern and Central India and 42 species in the high altitude of Himalayas are threatened in the wild (Sharma *et al.*, 2010).

To exploit the full potential and sustainable use of medicinal and aromatic plants multi-pronged approach should be adopted. On one hand, there is required development of improved cultivars by conventional methods and improvement in agro-technological methods so that profitability is increased and more and more farmers can be attracted towards cultivation of medicinal and aromatic plants. On the other hand, biotechnological interventions for conservation, micropropagation, production of secondary metabolites, biotransformation of intermediates into pharmaceutically important products and genetic improvement are also required. This chapter describes the application of biotechnology in plant propagation, genetic modification and improvement, product formation and germplasm storage.

Micropropagation

With resurgence of public interest in plant-based medicine and the rapid expansion of pharmaceutical industries, these *in vitro* techniques and innovative approaches will be useful. Using *in vitro* propagation techniques it is now possible to produce a large number of pathogen-free uniform clones of elite, rare, and important native medicinal plants for reintroduction in their natural habitat and safe exchange of germplasm across international borders (Pandey 2010).

There are many advantages of 'micropropagation' (the *in vitro* method of clonal propagation) over 'macropropagation' (the conventional vegetative method of clonal propagation). These are: (i) Small initial size of plant material, (ii) Short time requirement, (iii) Small space requirement, (iv) High multiplication rate, (v) Production of pathogen-free plants, (vi) Year-round production, (vii) Maintenance and movement of germplasm, (viii) Distribution of commercial plant material in international trade.

There are also problems associated with micropropagation. Some of these are: (i) hyperhydration (thick, brittle, and water-soaked shoots with short internodes), (ii) somaclonal variation (appearance of genetic variation in plants regenerated from *in vitro* cultures), (iii) contamination, (iv) oxidative browning of culture medium, (v) recalcitrance of adult trees, and (vi) high cost.

However, despite these problems, from the commercial viewpoint, micropropagation is the most important and pragmatic aspect of plant tissue culture. Micropropagation (plantlet regeneration through plant tissue culture) generally involves four stages (Murashige 1974). Later, Debergh and Maene (1981) introduced one more stage, the 'preparative stage', making micropropagation a five-stage process.These stages are: Stage 0 (Preparative stage), Stage I (Establishment stage), Stage II (Multiplication stage), Stage III (Rooting stage), and Stage IV (Transplantation)

- **Stage 0** (Preparative stage) includes exposing the mother plant (from which explants are obtained), in glasshouse, to suitable light, temperature and growth regulator treatments to improve the quality of 'explants' (small organs or pieces from which the cultures are started).
- **Stage I** (Establishment stage)i ncludes procurement of explants from the mother plants, surface sterilization and transfer of explants to culture medium. The nature of the explant to be used for *in vitro* propagation depends on the kind of culture to be initiated, the purpose of the proposed culture, and the plant species to be used. The common explants used are nodal cutting (for enhanced axillary bud proliferation), meristem tips (for virus eradication), immature zygotic embryos (for multiplication through somatic embryogenesis), etc.
- **Stage II** (Multiplication stage) is the most crucial stage where most of the failures in micropropagation occur. Micropropagation is achieved by three approaches of multiplication. These are: (i) axillary bud proliferation (growth and proliferation of existing buds), (ii) organogenesis (adventitious shoot formation) directly or indirectly via callus, and (iii) somatic embryogenesis (somatic embryo formation) directly or indirectly via callus. Although axillary bud proliferation produces the smallest number of plantlets, it is the most popular approach to clonal propagation because it produces 'true to type' plants as the cells of the shoot apex are least susceptible to genetic changes under culture conditions (Thorpe 1990). Adventitious shoot formation has a greater potential for multiplication, as shoots may arise

from any part of inoculation (Thorpe 1990). Indirect adventitious shoot formation is, however, not desirable because it leads to somaclonal variation (George 1993). Somatic embryo formation has the potential for producing the greatest number of plantlets (Thorpe 1990). It is appealing from commercial angle because it reduces the cost by eliminating the rooting stage and it is amenable (a) to automation at multiplication stage, (b) for field planting as synthetic seed and (c) for germplasm storage and transportation (Bhojwani and Rajdan 1996). Plants that normally form storage organs can also be induced to form such organs under *in vitro* conditions for direct planting in soil, cutting down the rooting stage as they can be formed on unrooted shoots (George, 1993). They do not require acclimatization and reduce the transplantation losses. They are easy to store and transport. *In vitro* storage organ formation has been reported in several species, such as corm formation in *Crocus* and *Gladiolus*, tuber formation in *Dioscorea* and potato, and bulblet formation in lilies.

- ☆ **Stage III** (Rooting stage) involves rooting in the axillary or adventitious shoots. Somatic embryos have a pre-formed radicle and it does not require this stage. Similarly, storage organs, produced *in vitro*, also do not require rooting stage. In some species, the Stage II shoots (adventitious or axillary) produce roots spontaneously and defined rooting stage is not required but in most of cases a separate rooting stage is required (Debergh and Maene 1981, George, 1993/1996).
- ☆ **Stage IV** (Transplantation stage) involves the transfer of shoots/plantlets/ somatic embryos *ex vitro* from aseptic cultures to soil. The plants multiplied *in vitro* are exposed to a unique set of growth conditions (high humidity, low light, high levels of inorganic and organic nutrients and growth regulators, sucrose as carbon source and poor gaseous exchange), which may support rapid growth and multiplication but also induce structural and physiological abnormalities in the plants, rendering them unfit for survival under *in vivo* conditions (low humidity, high light intensity, and autotrophic mode of nutrition) (Bhojwani and Rajdan 1996). Therefore, before these plants are transferred to the greenhouse or field conditions, they are put under the process of 'acclimatization' or 'hardening' to gradually adapt or adjust physiologically to the new environment.

Rout *et al*. (2000) have given an overview of regeneration of medicinal plants by direct and indirect organogenesis and by somatic embryogenesis from various types of explants. A review by Nalawade *et al*. (2003) summarizes the information about tissue culture studies on Chinese medicinal plants and related species reported earlier. Also, it describes mass propagation of valuable medicinal herbs through (i) shoot morphogenesis in *Limonium wrightii*, *Adeno phoratriphylla*, *Anoectochilus formosanus*, *Scrophularia yoshimurae*, *Pinellia ternata*, *Bupleurum falcatum*, *Zingiber zerumbet*, *Dendrobium linawianum*, and *Fritillaria hupehensis* and (ii) somatic embryogenesis in *Angelica sinensis* and *Corydalis yanhusuo* done at Taiwan Agricultural Research Institute and Chaoyang University of Technology, Taiwan.

A review by Tripathi and Tripathi (2003) reports plant regeneration through (i) axillary buds and stem meristems in medicinal plants like *Catharanthus roseus, Cinchona ledgeriana, Digitalis* spp., *Rehmannia glutinosa, Rauwolfia serpentina, Isoplexis canariensis, Atropa belladonna, Picrorhiza kurroa, Nothapodytes foetida, Plantago ovata, Zingiber spectabile*, and *Clerodendrum colebrookianum*; (ii) callus mediated organogenesis in *Plumbago rosea, Dioscorea alata, Cephaelis ipecacuanha, Asparagus cooper, Solanum melongena, Hyoscyamus muticus, Psorale acorylifolia, Zingiber officinale, Mentha arvensis, Centella asiatica, Plumbago zeylanica, Solanum laciniatum, Echinacea pallida*, and *Lepidium sativum*; and (iii) somatic embryogenesis in *Podophyllum hexandrum, Asparagus cooperi, Medicago sativa, Cayratia japonica, Acacia catechu, Aesculus hippocastanum, Acanthopanax koreanum, Typhonium trilobatum, Catharanthus roseus*, and *Psoralea corylifolia*.

Chaturvedi *et al.* (2007) have mentioned that although there are a number of reviews published on micropropagation of medicinal plants but most of the pharmaceutically important medicinal plants have not been micropropagated on large scale of commercial significance. They provided the success story of large-scale rapid cloning of *Dioscorea floribunda*, an important drug plant intractable-to-multiply and also discussed the clonal multiplication of some other important medicinal plants like *Rauwolfia serpentina, Withania somnifera, Aloe vera, Curcuma domestica, Allium sativum, Zingiber officinalis, Dioscorea deltoidea, Costus speciosus, Solanum khasianum, Podophyllum peltatum*, and *Picrorhiza kurroa*.

Sharma *et al.* (2010), in their review, have reported tissue culture protocols that have been developed for a wide range of medicinal plants of India like *Saussaurea lappa, Picorrhiza kurroa, Ginkgo biloba, Swertia chirata, Gymnema sylvestre, Tinospora cordifolia, Salaca oblonga, Holostemma* spp., *Celastrus paniculata, Oroxylum indicum, Glycyrrhiza glabra, Tylophora indica, Bacopamo oniera*, and *Rauwolfia serpentina*.

Aromatic plants have been used commercially as spices, natural flavours, raw material for essential-oil industry, and medicinal purposes. Gantait *et al.* (2011) have given a review of the regeneration by *in vitro* organogenesis from various types of explants of aromatic plants – *Ammomum subultum, Curcuma aromatica, C. domestica, C. zedoaria, Cinnamommum vernum, Crocus sativus, Ellatoria cardamomum, Fragaria vesca, Kaempferia galanga, Lavendula angustifolia, Mentha piperita, Murraya koengii, Myristica fragarance, Piper nigrum, Vanilla planifolia, Zingiber officinale*, etc.

Genetic Modification and Improvement

Plant breeders have always used the best available technology for developing new varieties with desirable traits. The major objectives include: increased yield, improved quality, resistance to pests and diseases, tolerance for abiotic stresses, change in agronomic traits for better cultural practices, etc. The innovative methods of plant biotechnology have provided opportunity (a) for creation of a wide range of useful genetic variability and (b) in improving efficiency of selecting desired traits.

For increasing genetic variability following biotechnological tools are employed: (i) useful alien gene transfer by inter-specific crossing and then overcoming hybrid incompatibility using embryo rescue technique, a plant tissue culture tool; (ii) somatic hybridization by protoplast fusion for production of cybrids and organelle recombinants for transfer of cytoplasmic genes for male sterility and disease resistance; (iii) induction of somaclonal variation (appearance of genetic variation in plants regenerated from *in vitro* cultures) and selection for useful traits like disease resistance and abiotic stress tolerance; and (iv) production of transgenic plants by genetic engineering for transferring genes into economic plants from diverse sources like bacteria, animals, unrelated plants, etc. using molecular tools.

For improving selection efficiency following biotechnological tools are employed: (i) reducing breeding cycle by shortening the time in the production of homozygous lines using the anther culture technique; (ii)application of biochemical markers like isozyme markers and molecular markers like restriction fragment length polymorphism (RFLP), random amplified polymorphic DNA (RAPD), DNA amplification fingerprinting (DAF), amplified fragment length polymorphism (AFLP), etc. instead of conventional morphological markers for selection of complex traits; and (iii) *in vitro* screening of cultured cells for selection of desirable cell lines even occurring in low frequency.

Rout *et al.* (2000) have reviewed the use of the various biotechnological approaches to improve medicinal plants through somaclonal variations and genetic transformation and reported that the genetic manipulation of plants together with the establishment of *in vitro* plant regeneration system facilitates efforts to engineer secondary product pathways. Advances in the cloning of genes involved in relevant pathway, the development of high throughput screening system for chemical and biological activity, genomic tools and resources, and the recognition of a higher order of regulation of secondary plant metabolism operating at the whole plant level facilitate strategies for the effective manipulation of secondary products of plants (Gomez-Galera *et al.*, 2007).

For transformation to be successful, DNA must first be introduced into the target cell. After passage through the plant cell wall and membrane, the introduced DNA must then proceed to the nucleus, pass through the nuclear membrane and become integrated into the genome. It is believed that the introduced DNA can function for a short time in the nucleus as an extra-chromosomal entity, but integration into the genetic material of the target cell is necessary for long-term functionality and expression (Finer 2010).

Transformation has been reported in more than 120 species of at least 35 families, including the major economic crops, vegetables, ornamental, medicinal, fruit, tree and pasture plants, using *Agrobacterium* mediated or direct transformation methods (Birch 1997). However, *Agrobacterium*-mediated transformation offers several advantages over direct gene transfer methodologies (particle bombardment, electroporation, etc), such as the possibility to transfer only one or few copies of DNA fragments carrying the genes of interest at higher efficiencies with lower cost and the transfer of very large DNA fragments with minimal rearrangement (Shibata and Liu 2000).

The two gram-negative rod-shaped soil bacteria *Agrobacterium tumefaciens* and *A. rhizogenes* (family Rhizobiaceae), causing plant diseases crown gall and hairy root, respectively, are the natural genetic engineers. They are able to transform or modify mainly dicotyledonous plants. The mobile segment – T-DNA – of their plasmids, Ti (tumor inducing) and Ri (root inducing), respectively, plays the key role in transformation. Genetic transformation, using *Agrobacterium tumefaciens*, has been reported for various medicinal plants like *Azadirachta indica* (Naina *et al.*, 1989), *Atropa belladonna* (Yun *et al.*, 1992), etc.

Plant infection with *Agrobacterium rhizogenes* induces the formation of proliferative multibranched adventitious roots at the site of infection, 'hairy roots'. This infection is followed by the transfer of a portion of DNA *i.e.* T-DNA, known as the root inducing plasmid (Ri-plasmid), to the plant cell chromosomal DNA. Transformed hairy roots mimic the biochemical machinery present and active in the normal roots, and in many instances transformed hairy roots display higher product yields. Hairy roots, transformed with *Agrobacterium rhizogenes*, have been found to be suitable for the production of secondary metabolites because of their stable and high productivity in hormone-free culture conditions. Tripathi and Tripathi (2003) have reported a number of medicinal plants that have been successfully transformed with *Agrobacterium rhizogenes*: *Artemisia annua, Aconitum heterophyllum, Digitalis lanata, Papaver somniferum, Eschscholzia californica, Atropa belladonna, Solanum aviculare, Pueraria phaseoloides.*

Product Formation

The value of medicinal and aromatic plants is due to the phyto-chemicals compounds collectively called secondary metabolites. These secondary metabolites are not required for the vital functions of the plant; rather they play secondary roles such as attractants of pollinators and chemical defense against pests and diseases. However, these chemicals are of great economic value to mankind as pharmaceuticals, agricultural compounds and food additives. As it is already mentioned that most of the pharmaceutical industries depend on wild populations for the supply of raw materials and this leads to destruction of plant habitats, and the illegal and indiscriminate collection of plants from these habitats, many medicinal plants are severely threatened. Biotechnology again can play a definite role by (a) direct production of plant secondary metabolites *in-vitro*, and (b) biotransformation.

Production of secondary metabolites *in vitro* takes place through plant cell suspension culture (a culture consisting of cells or cell aggregates in an agitated liquid culture). There are two types of cell suspension cultures: (I) batch culture, and (II) continuous culture.

In batch culture, cells are grown in a fixed volume of nutrient culture medium and they follow sigmoid pattern of growth. In contrast to batch culture, the cultures are maintained at a constant sub-maximal growth rate indefinitely in continuous culture. In continuous culture, cells are continuously supplied with fresh nutrient culture medium but the volume of culture medium remains the same because the addition of fresh medium is balanced by removal of equal volume of the spent medium (closed continuous culture) or the culture *i.e.* old medium plus cells (open continuous culture).

The commercial application of *in vitro* secondary metabolite depends on the successful cultivation of plant cells on a large scale, which is possible with the use of 'bioreactors'. Bioreactor is a glass or steel vessel in which plant cells are cultured in large scale. Ideally, bioreactors are fitted with probes to monitor the pH, temperature and dissolved oxygen in the culture and have provisions to sample the cultures, add fresh medium, adjust pH, air supply, mixing of cultures and controlling the temperature, without endangering the aseptic nature of the culture. It, thus, allows closer control and monitoring of culture conditions (Bhojwani and Rajdan 1996).

There is increasing number of reports of use the of 'immobilized cell system, as an alternative to 'free cell culture systems' for the production of secondary metabolites (Chawla 2002). In this technique the plant cells are usually entrapped in calcium alginate or potassium carrageenan or agarose beads. The advantages of using immobilized system are: reuse of expensive biomass, automatic physical separation of product from the cells and stabilization of biocatalyst, etc.

The production of secondary metabolites in plant cell suspension cultures has been reported from various medicinal plants: solasodine from calli of *Solanum eleagnifolium*, pyrrolizidine from root cultures of *Senecio*sp., cephaelin and emetine from callus cultures of *Cephaelis ipecacuanha*, quinoline alkaloids from cell suspension cultures of *Cinchona ledgeriana*, indole alkaloid biosynthesis in the suspension culture of *Catharanthus roseus,* diosgenin from callus cultures of *Dioscorea deltoidea*, cardenolides from hairy root cultures of *Digitalis lanata*, azadirachtin and nimbin from cultured shoots and roots of *Azadirachta indica*, lepidine from *Lepidium sativum*, etc.

Some examples efficient protocols for isolated cell cultures and a large-scale bioreactor system are as follows: production of sanguinarine in cell suspension cultures of *Papaver somniferum*, gisenoside from adventitious root culture and transformed hairy roots of *Panax ginseng*, shikonin from *Lithospermum erythrorhizon*, etc.

Biotransformation involves plant cell cultures not for direct production of secondary metabolites, but for precursor transformation. Biotransformation is the chemical modification (or modifications) made by an organism on a chemical compound. It is the transformation of readily available and inexpensive precursors into a more valuable product. This may involve reduction, oxidation, hydroxylation, glycosylation, esterification, methylation, demethylation, isomerisation or epoxidation of the precursor molecules by the enzymes of plant cells. The most celebrated example is the biotransformation of digitoxin obtained from *Digitalis lanata* into the cardiovascular steroid digoxin by using undifferentiated *D. lanata* cell culture. Other examples of biotransformation by plant cells are: codeinone to codein by *Papaver somniferum*; hydroquinone to arbutinby *Datura* cells; steviol to steviocide and steviobiocide by *Stevia rebaudiana* and *Digitalis purpurea*; and so on.

Germplasm Storage

Most medicinal plants are not cultivated; rather they are collected from wild. In past, quantities needed to meet demand were relatively low; however, increasing commercial demand is fast outpacing supply. Currently about 4,000 to 10,000 medicinal plants are on endangered list and this number is expected to increase (Gomez-Galera 2007). *In vitro* conservation of traditional medicinal and aromatic plant germplasm is important to support chemical analysis and pharmacological and genetic transformation studies.

Conservation of medicinal and aromatic plant diversity can be exercised in a number of complementary ways. The two conventional approaches are: *in-situ* conservation (in the original or natural place or site) and *ex-situ* conservation(outside the original or natural place or site). However, to conserve the germplasm of important medicinal and aromatic plants, which is difficult to be conserved through conventional methods or where additional protection is required, *in vitro* conservation strategies are also adopted (Pandey 2011).

In vitro conservation refers to maintenance of germplasm in a relatively stable form under more or less defined nutrient conditions in an artificial environment. It is suitable for vegetatively propagated material, species with recalcitrant seeds, and material modified by the techniques of molecular biology. The latter material may have been costly to produce, be limited in quantity and have characteristics whose stability during repeated mitotic divisions and following meiosis are unknown.

The potential advantages of using *in vitro* methods of germplasm conservation are (Bhojwani and Rajdan, 1996; Rajdan and Cocking, 1997; Sarasan *et al.*, 2006): (i) Conservation of germplasm of endangered plants; (ii) Conservation of germplasm of vegetatively propagated plants, and plants with recalcitrant seeds; (iii) Conservation of rare germplasm arising through somatic hybridization, somaclonal and gametoclonal variations and genetic engineering; (iv) Relatively little space and time is needed for the storage of large number of clonally multiplied plants; (v) Maintenance of the plants free from pest, pathogen, virus and natural hazards; (vi) Production of large number of plants rapidly, when required, from the preserved material, which serve as nucleus stock to propagate; and (vii) Being free from known pathogens and viruses the material has easy movement in international trade or germplasm exchange with minimum quarantine requirements.

There are two basic methods that are followed to maintain germplasm collections *in-vitro:* (i) slow growth, and (ii) cryopreservation.

In slow growth method, the *in vitro* cultures grown under modified conditions enabling them to be stored for longer periods before transfer to fresh medium constitute slow growth systems (Rajdan 2002). The advantage of this approach is that cultures can be readily brought back to normal culture conditions to produce plants on demand. However, the need for frequent subculturing may pose a great disadvantage including contamination of cultures as well as imposition of selection pressure with subsequent change in genetic make-up due to somaclonal variation.

In cryopreservation the plant material is frozen and maintained at the temperature of liquid nitrogen (LN), which is around -196°C. At this temperature the cells stay in almost completely inactive state *i.e.* in non-dividing or zero metabolism state (Bhojwani and Rajdan 1996). This method offers the possibility for long-term storage with maximal phenotypic and genotypic stability (Steponkus 1985). This method being relatively convenient and economical, large number of genotypes and variants could be conserved and thus maximize the potential for storage of genetically desirable material.

The materials used for *in vitro* conservation may be (i) protoplasts, (ii) cell suspensions, (iii) callus cultures, (iv) meristem tips, (v) propagules at various stages of development, or (vi) organized plantlets. The materials in an undifferentiated phase (protoplasts, cell suspensions and callus cultures) pose two main problems: (a) genetic instability *in vitro* is very common, (b) totipotency *i.e.* potentiality to produce complete plants is not exhibited in many cases or it is lost after some time. Therefore, they are not the ideal systems for germplasm conservation and organized structures, as such shoot apices, embryos and young plantlets are preferred (Bhojwani and Rajdan 1996).

An outline of the strategy, as given by Grout (1990) for the *in vitro* conservation of plant genetic resources is as follows: (i) Select plant material to be conserved; (ii) Disease indexing, Characterization (morphological and biochemical analysis), and Evaluation (based on yield, biomass production and chemical constituent); (iii) Develop protocols for rapid *in vitro* clonal multiplication; (iv) *in vitro* conservation – (a) For short to medium-term conservation develop strategies to maintain *in vitro* cultures under minimal growth conditions, (b) For long-term conservation develop techniques/protocols for cryopreservation in liquid nitrogen (-196°C); and (v) Demonstration of viability, Regeneration of plants retrieved from *in vitro* cultures/ cryopreservation, their field trials and re-characterization.

It is expected that the development of the full potential of *in vitro* culture storage and associated biochemical techniques could revolutionize the handling of germplasm (Chandel and Pandey 1991; Sarasan *et al.*, 2006). *In vitro* techniques have been found to be useful in the propagation of a large number of threatened plants, including medicinal and aromatic plants (Dhar *et al.*, 2000).

Biotechnology of MAPs in India

Based on the recommendations of the then Scientific Advisory Committee to the Cabinet, a National Biotechnology Board (NBB) was constituted by the Government in 1982 to foster programmes and strengthening indigenous capabilities in this newly emerging discipline. Subsequently, a separate Department of Biotechnology (DBT) was set up in February, 1986. Since then a strong infrastructure for biotechnology research and services has been created in both national laboratories and academic institutions.

India has a rich traditional system of medicine. Ayurveda, Unani, Homeopathy, Siddha and Amchi system are popular; make extensive use of herbs in therapeutic treatments. The historical usage of traditional medicine of over 5000 years provides some level of confidence about its safety and efficacy. There are over 7000 manufacturing units in Indian System of Medicine and Homeopathy (ISM&H). About 5,64,000 ISM&H practitioners with an addition of 8000 new practitioners every year. There is growing interest of natural products as a source of new chemical entities for development of modern drugs, also use of natural products as dietary supplements (nutraceutics), ingredients to food and beverages, phytocosmetics and other herbal products. Global market for herbal products which includes medicine, health supplement, herbal cosmetics is around US$65 billion. According to World Bank report, the international herbal medicine market is expanded to reach US$ 5 trillion in 2050 with an annual growth rate of between 10 to 20 per cent. According to Ayurvedic Drug Manufacturers Association (ADMA) estimates, the current value of Indian trade in Indian System Medicine and Homeopathy (ISM&H) medicines is about '4200 crores or US$1billion.

For sustainable utilization of MAPs and to tap the vast and expanding market Department of Biotechnology (DBT), Government of India, has set the objectives, regarding medicinal and aromatic plants, which are as follows: application of biotechnology for conservation, characterization, micropropagation, cell culture production of secondary metabolites, isolation and characterization of novel bioactive agents, development of standardized and safe herbal formulations and genetic improvement of selected medicinal and aromatic plants.

A brief description of work done under the aegis of DBT (2012) is given below:

(A) Micropropagation

In vitro protocols have been developed for multiplication of selected medicinal and aromatic plants such as *Garcinia indica, Holarrhenaanti dysenterica, Lavendula officinalis, Pterocarpus marsupium,* and *Chlorophytum borivilianum*. Multi-locational field trials to evaluate the performance of tissue culture derived *Pogostemoncablin* (Patchauli) carried out over an area of 32 acres involving five centres. Field evaluation of the performance of tissue-culture raised elite varieties of large cardamom (*Amomum subulatum*) over a total area of 50 acres in Uttarakhand initiated in association with the Spices Board.

(B) Genetic Modification and Improvement

There is impressive progress regarding researches in genomic resources and metabolic pathways. The capsaicin synthase enzyme (key regulatory enzyme for capsaicin biosynthesis) and its gene (*csy 1)* have been characterized from placental tissues of *Capsicum* sp. Patent has been filed for csy1 gene. Five lines of *Catharanthus roseus* that hyperproduce serpentine and its product ajmalcine have been developed through DNA marker assisted pyramidation of the concerned gene loci. Projects have been recently initiated to develop ESTs data base, understand the biosynthetic pathway / regulatory genes involved in the production of artemisinin in *Artemisia annua;* morphine alkaloids in *Papaver somniferum;* santalol in *Santalum album;* picrosides in *Picrorhyza kurrooa* and podophyllotoxin in *Podophyllum hexandrum*.

(C) Product Formation

Efforts initiated towards protocol development of production of important therapeutic agents through cell culture methods such podophyllotoxin from *Podophyllum hexandrum*, hyoscyamine from *Hyoscyamus muticus*, guggulsterones Z and E from *Commiphora wightii* and comptothecin from *Ophiorrhyza* spp. Four fast growing cell-lines of *P.hexandrum* capable of synthesizing podophyllotoxin are devoid of α-peltatinsestablished. A bioreactor facility (15-litre capacity) set up for up-scaling of cell culture production of podophyllotoxin, anthraquinones and other therapeutic agents. Cell suspension cultures raised from leaf and hairy root derived callus of *Hyoscyamus muticus* have been scaled-up in bioreactor (15-litre capacity) towards the production of hyoscyamine.

Up-Scaling of process for extraction of lead therapeutic compounds from plants has also been undertaken. A process for extraction of 10-DAB has been scaled-up in pilot-plant with 100 kg fresh *Taxus wallichiana* needles. An improved process has been scaled-up to 30 kg/batch raw material (twigs and stems of *Nothapodytes foetida*) for extraction of comptothecin. A pilot-scale process for isolation of silymarine from the seeds of *Silybum marianum* has been developed (40 kg seeds/batch level)

Isolation and characterization of new bioactives/therapeutic agents: Under multi-institutional project, after bio-activity based *in vitro* screening of 60 medicinal plants (used in Indian traditional system of medicine), a total of 35 lead molecules identified so far: Anti-cancer – 15, Anti-diabetic – 5, Immunomodulatory – 15. Two anti-cancer lead molecules (from *Aegle marmelos* and *Phyllanthus urinaria*) have been patented. A lead medicinal plant extract exhibiting promising osteogenic (bone forming activity) using several *in vitro* and *in vivo* test systems have been identified. The patent for the above is being filed. The active principle (saturated fatty acid) isolated from *Oxalis corniculata* showing significant anti-proliferative activity against *Entamoeba histolytica* has been further characterized. A lead fraction from *Piper nigrum* having anti-tubercular activity (effective against both sensitive and resistant strains of *Mycobacterium tuberculosis*) has been identified. Efforts are in progress for isolation and characterization of anti-cancer, anti-tubercular, anti-viral, hepatoprotective, and immunomodulatory agents from medicinal plants used in Indian traditional system of medicine.

(D) Germplasm Storage

A Network of four national gene banks on medicinal and aromatic plants at TBGRI, Thiruvananthapuram; CIMAP, Lucknow; NBPGR, New Delhi and RRL, Jammu has been established. A total of about 8500 accessions of prioritized species are conserved in different forms such as in field bank, seed bank, *in vitro* repository, cryobank and DNA bank. A germplasm bank for medicinal plants used in Ayurveda has been set up at AryaVaidyaSala, Kottakkal, Kerala.

A brief description of work done in the Central Institute of Medicinal and Aromatic Plants, CIMAP (2012) is given below:

The Plant Biotechnology Department in the Central Institute of Medicinal and Aromatic Plants (CIMAP) is working on molecular biology and tissue culture of Medicinal and Aromatic Plants (MAPs). The major activities includes expression profiling for metabolic and genetic diversity basis, DNA markers for breeding and identification, genomics for pathway engineering, variety development. Work is also being carried out on artemisinin biosynthesis in *Artemisia annua*, withanolide biosynthesis in *Withania somnifera*, morphine biosynthesis in *Papaver somniferum*. The plant tissue culture group is carrying out research for *in vitro* operations and manipulations in MAPs including micro-cloning, somaclonal breeding, haploid production, protoplast culture and genetic transformation etc. The DNA fingerprinting laboratory of the Genetic Resource Management department of CIMAP is dedicated for molecular characterization of germplasm, development of fingerprints of *Withania somnifera*, *Catharanthus roseus* and *Papaver* germplasms by deploying various molecular markers techniques. A large microsatellite marker resource is being generated at the department specifically for *Catharanthus roseus* and *Withania somnifera*. Genetic improvement work of *Artemisia annua, Sylibum marianam, Mentha species, Withania somnifera, Catharanthus roseus, Papaver somniferum, Vetiveriaz izinoides* and *Rosa damacena* is also being carried out by the department.

References

Bhojwani S S and Rajdan M K 1996. Plant tissue culture: theory and practices, a revised edition. *Elsevier*, Amsterdam.

Birch R G 1997. Plant transformation: Problems and strategies for practical application *Ann Rev Plant Physiol Plant Mol Biol* **48**: 297-326.

Chandel K P S and Pandey R 1991. Plant Genetic Resources Conservation: Recent Approaches. *In:* Paroda RS and Arora RK (Eds.), Plant Genetic Resources Conservation and Management Concepts and Approaches. International Board for Plant Genetic Resources, New Delhi.

Chaturvedi H C, Jain M and Kidwai N R 2007. Cloning of medicinal plants through tissue culture – A review. *Ind J Exp Biol* **45**: 937-948.

Chawla H S 2002. Introduction to Plant Biotechnology. Oxford and IBH, New Delhi.

CIMAP 2012. Overview of Divisions.http://www.cimap.res.in/, retrieved on 07February 2012.

DBT 2012. Research and Development: Biotechnology for Medicinal and Aromatic Plants. http://dbtindia.nic.in/, retrieved on 07 February 2012.

Debergh P C and Maene L J 1981. A scheme for commercial propagation of ornamental plants by tissue culture. *Sci Hortic* **14**: 335-345.

Dhar U, Upreti J, Bhatt I D 2000.Micropropagation of *Pittosporumnapaulensis* (DC.) Rehder& Wilson – a rare, endemic Himalayan medicinal tree. *Plant Cell Tiss Organ Cult* **63**: 231-235.

Finer J J 2010. Plant nuclear transformation. *In:* Kempken F and Jung C (eds.), Genetic modification of plants: Biotechnology in Agriculture and Forestry. *Springer-Verlag*, Berlin.

Gantait S, Mandal N and Nandy S 2011. Advances in Micropropagation of Selected Aromatic Plants: A Review on Vanilla and Strawberry. *Am J Biochem and Mol Biol* **1**: 1-19.

George E F 1993. Plant propagation by tissue culture, Vol. I. Exegetics Ltd., Edington, England.

George E F 1996. Plant propagation by tissue culture, Vol. II. Exegetics Ltd. Edington, England.

Gomez-Galera S, Pelacho A M, Gene A, Capell T and Christou P 2007. The genetic manipulation of medicinal and aromatic plants. *Plant Cell Rep* **26**: 1689-1715.

Grout B W W 1990. *In vitro* conservation of germplasm. *In:* Bhojwani S S (ed.), Plant Tissue Culture: Application and Limitations. *Elsevier Sci Publ*, Amsterdam.

Murashige T 1974. Plant propagation through tissue culture. *Ann Rev Plant Physiol* **25**: 135-166.

Naina N S, Gupta P K and Mascarenhas A F 1989. Genetic transformation and regeneration of transgenic neem (*Azadirachta indica*) plants using *Agrobacterium tumefaciens*. *Curr Sci* **58**: 184-87.

Nalawade S M, Sagare A P, Lee C, Kao C and Tsay H 2003. Studies on tissue culture of Chinese medicinal plant resources in Taiwan and their sustainable utilization. *Bot Bull Acad Sin* **44:** 79-98.

Pandey B N 2009.Principles, applications and limitations of plant tissue culture with particular reference to micropropagation. *In:* Vimla Y (Ed), Flower: Prospects and Retrospects. *New Scientific Publ*, New Delhi, 203-225.

Pandey B N 2011. Conservation of Phytodiversity through *in vitro* strategies. *In:* Kumar M and Gupta R K (eds.) Biodiversity: An Overview. IK Internat. Publ House Pvt. Ltd., New Delhi, p. 391-404.

Razdan M K 2002. Introduction to Plant Tissue Culture 2nd edn. Oxford and IBH, New Delhi.

Razdan M K and Cocking E C 1997. Conservation of Plant Genetic Resources *in vitro*: General Aspect. Vol. I Sci Publ Enfield, New Hampshire.

Rout G R, Samantaray S and Das P 2002. *In vitro* manipulation and propagation of medicinal plants. *Biotechnology Advances* **18**: 91-120.

Sarasan V, Cripps R, Ramsay M M, Atherton C, Mcmichen M, Prendergast G, and Rowntree J K 2006 Conservation *in vitro* of threatened plants – progress in the past decade. *In vitro Cell Dev Biol Plant* **42**: 206-214.

Sharma S, Rathi N, Kamal B, Pundir D, Kaur B and Arya S 2010. Conservation of biodiversity of highly important medicinal plants of India through tissue culture technology- a review. *Agric Biol J N Am* **1:** 827-833.

Shibata D and Liu Y G 2000. Agrobacterium-mediated plant transformation with large DNA fragments. *Trends Plant Sci* **5**: 354-7.

Steponkus P L 1985. Cryobiology of isolated protoplasts: Applications to plant cell cryopreservation. *In:* Kartah KK (Ed), Cryopreservation of plant cells and organs. CRC Press, Boca Raton, Florida.

Thorpe T A 1990. The current status of plant tissue culture. In: Bhojwani S S (Ed.) Plant tissue culture: applications and limitations *Elsevier*, Amsterdam.

Tripathi L and Tripathi J N 2003. Role of biotechnology in medicinal plants. *Trop J Pharma Res* **2:** 243-253

Yun D J, Hashimoto T and Yamada Y 1992. Metabolic engineering of medicinal plants: transgenic *Atropa belladonna* with an improved alkaloid composition. *Proc Nat Acad* Sci USA **89:** 11799-803.

Medicinal Plants: Aspects and Prospects (2014) ***Pages* 37-48**
***Editors:* Mukesh Kumar, Anjali Khare and C.P. Shukla**
ISBN: 978-81-7622-309-6
***Published by:* BIOTECH BOOKS, NEW DELHI**

Chapter 3

Hepatoprotective Herbal Plants

Siddhartha Singh, Subhadip Hajra, Archana Mehta and Pradeep Mehta

Department of Botany,
Dr. H.S. Gour University, Sagar – 270 001, M.P., India

ABSTRACT

Use of herbal drugs in the treatment of liver diseases has a long tradition, especially in Eastern medicine. Different solvent extracts of plants such as aqueous extract of the leaves of *Andrographis paniculata,* methanolic extract of *Apium graveolens, Hygrophila auriculata and Enicostema axillare,* aquous extract of *Cleome viscosa and Silybum marianum,* ethanolic extract of bark stem of *Pterocarpus santalinus, Bacopa monnieri and Ziziphus mauritiana,* acetone sub-fraction of total alcoholic extract of the aerial part of *Pergularia daemia,* aqueous seed extract are sufficiently active against certain liver diseases and liver damages. *A. paniculata* is also likely to be active against Hepatitis B virus while Silymarin also stabilize the cell membrane and exert antioxidant properties. The potential of the active components present in these plant extracts has been proved by CCl_4 induced liver damage models in mice.

Keywords: Herbal drugs, Extracts and Carbon tetrachloride (CCl_4)

Introduction

Medicinal plants play a key role in the human health care. About 80 per cent of the world populations rely on the use of traditional medicine, which are predominantly based on plant materials. Liver, the key organ of metabolism and excretion, is constantly endowed with the task of detoxification of xenobiotics, environmental pollutants and chemotherapeutic agents. Thus, disorders associated with this organ are numerous and varied. While a curative agent has not yet been found in modern medicine, the current usage of corticosteroids and immunosuppressive agents only brought about symptomatic relief (Handa *et al.*, 1986). Furthermore, their usage is associated with risk of relapses and danger of side effects. On the other hand, Ayurveda, an indigenous system of medicine in India, has a long tradition of treating liver disorders with plant drugs (Mitra *et al.*, 1988). Enhanced lipid peroxidation produced during the liver microsomal metabolism of ethanol may result in hepatitis and cirrhosis (Smuckler *et al.*, 1998). To focus on a more practical and systematic approach towards the development of standardized medicine, many plant derived drugs used in modern medicine are developed by ethno medical leads and subsequent ethno pharmacological studies. There are more than 100 drugs of known structure that are extracted from higher plants and used in allopathic medicine (Fransworth *et al.*, 1990). Three fundamental roles of the liver are vascular function, including formation of lymph and the hepatic phagocytic system, metabolic achievements in control of synthesis and utilization of carbohydrates, lipids and proteins, Secretory and excretory functions, particularly with respect to the synthesis. Liver diseases remain one of the serious health problems. It is well known that free radicals cause cell damage through mechanisms of covalent binding and lipid peroxidation with subsequent tissue injury. Tumors in the human body or in experimental animals are known to affect many functions of the vital organs, especially the liver, even if the site of the tumor does not interfere directly with that organ's functions. Higher levels of free radicals and malon di aldehyde (MDA), the end products of lipid peroxidation, were reported in cancer tissues than in non-diseased organs.

Assessment of Liver Activity

To assess the liver functions, biologists have chosen three basic biochemical parameters:

- ☆ Serum glutamic oxaloacetic transaminase (SGOT),
- ☆ Serum glutamate pyruvate transaminase (SGPT),
- ☆ Alanine pyruvatetransferase (ALP) and Serum bilirubin.

Serum Glutamic Oxaloacetic Transaminase (SGOT) or Aspartate Amino Transferase (AST)

This enzyme is found in heart and liver cells and released into blood when the liver or heart is damaged. Normal value of this enzyme is 8-20 µ/L at 30°C.

Syrum Glutamate Pyruvate Transaminase (SGPT) or Alanine Amino Transferase (ALT)

ALT is necessary for energy production and found in a number of tissues including the liver, heart and skeletal muscles, but in the liver it is found in higher concentration. Normal value of this enzyme in Serum is 40 U/L at 37°C.

Damaged cells contain both SGOT and SGPT at higher concentration in the blood; the cell walls become permeable and allow the enzyme to escape into the blood stream and the levels of both the enzyme are elevated. In heart attack and muscle disorders the level of SGOT is increased but the SGPT remains normal.

Bilirubin

Bilirubin is secreted from the bile and its levels are elevated in certain diseases. It is responsible for the yellow color of bruises and the yellow discoloration in jaundice. If biliary drainage is blocked, some of the conjugated bilirubin leaks out of the hepatocytes and appears in the urine, turning it dark amber leading to liver's disfunction.

Alkaline Phosphatase (ALP)

Alkaline phosphatase refers to a family of enzymes that catalyze hydrolysis of phosphate esters at alkaline pH. It is mostly synthesized in the liver, bone, intestines, kidneys and the placenta of a pregnant woman. An elevated level of ALP and aspartate amino transferase (AST) or alanine amino transferase (ALT) indicates liver damage while an associated with calcium and phosphate level indicates bone damage. In some forms of liver disease, such as hepatitis, ALP is usually much less elevated than AST and ALT. When the bile ducts are blocked by gallstones, scars from previous gallstones or surgery or by cancers ALP and bilirubin may be increased much more than AST or ALT.

Medicinal Plants and Hepatoprotective Activity

Alcoholism leads to liver diseases such as fatty liver, alcoholic hepatitis and cirrhosis (Bouneva *et al.*, 2003). Increased level in the reduced form of nicotinamide adenine dinucleotide (NADH) causes fat accumulation (Zimmerman 1999), free radicals inducing oxidative stress leads to peroxidation and inflammatory response (Jarvelainen 2000) and ethanol-induced elevation of endotoxin causes liver damage. Endotoxin stimulates Kupffer cells to produce free radicals and pro-inflammatory cytokines such as TNF- α and IL-1 which are the mediators of inflammation and cell death (Boelsterli 2003, Hoek *et al.*, 2002 and Wheeler *et al.*, 2001). There are several herbal plants which show significant hepatoprotective activities.

Phyllanthus amarus (Bhui Amla)

It belongs to the family *Euphorbiaceae,* native of Thailand. It contains phenolic compounds like lignans (phyllanthin and hypophyllanthin), flavonoids (quercetin and astragalin) and ellagitannins such as amarinic acid, amarin and phyllanthisiin-D (Kassuya *et al.*, 2005). It has been reported as potent hepatoprotective drug against paracetamol (Udomuksorn *et al.*, 2000), carbon tetrachloride (Harish *et al.*, 2006) and galactosamine (Khatoon *et al.*, 2006). It may enhance hepatic recovery after ethanol-induced liver injury (Umarani *et al.*, 1985).

Enicostema axillare (Naame/Chhota Chirayata)

It is a seasonal plant, belongs to Gentianaceae family, found all over India. The aerial and root parts have been traditionally used for the treatment of Jaundice, especially the tribes of Western Ghats. Methanolic extract of leaves showed strong hepatoprotective and antioxidant activity in wistar albino rats. These extracts produced significant (p<0.05) hepatoprotective effect by decreasing the activity of serum enzymes, bilirubin, uric acid and lipid peroxidation and significantly (p < 0.05) increased the levels of SOD, CAT, GSH and protein in a dose dependent manner (Gite *et al.*, 2007).

Eupatorium ayapana Vent. (Visalyakarani)

It is an aromatic shrub belongs to the family Asteraceae, native of Brazil and also found in India. Leaves contain 7- ethoxy coumarin [ayapanin], 6, 7-dimethoxy coumarin [ayapin]; carotene, vitamin-C and stigmasterol in addition to coumarins, *viz.* daphnetin, daphnetin dimethyl ether, hydrangetin, daphnetin –7-methyl ether and umbelliferone have also been isolated. Methanolic extract (MEEA) showed protection against carbon tetrachloride intoxicated rats when assessed by measuring the levels of marker enzymes, SGOT, SGPT and ALP, the results were comparable to that of Silymarin (Bose *et al.*, 2007).

Oldenlandia umbellate Lamk (Imbural)

This plant widely grows in costal regions of Tamilnadu, Orissa and West Bengal, belongs to the family Rubiaceae. It is also known as *Hedyotis hispida* and *Hedyotis indica*. Gupta *et al.* (2007) reported the effect of methanolic extract of aerial parts (EBM) against Carbon tetrachloride and paracetamol Induced Hepatotoxicity in Wistar Rats. It showed Hepatoprotective effect by decreasing activities of serum enzymes, bilirubin and total cholesterol.

Ziziphus mauritiana Lam (Jujube)

Belongs to the family *Ramnaceae*. Ethanol extract of the leaf was studied on Carbon tetrachloride induced liver damage. Pretreatment of leaf extract with 200 and 300 mg/kg body wt protected rats by significantly lowering marker enzymes of the serum (Dahiru *et al.*, 2005).

Bacopa monnieri Linn (Water Hyssop/Brahmi)

It belongs to the *Schrophulariaceae family*, distributed throughout India. This is a creeping, glabrous, succulent herb. The plant contains tetracyclic triterpenoid, sapoonins, bacosides A and B, hersaponin, alkaloids (herpestine and brahmine) and flavonoids. The ethanol extract (EBM) at the dose of 300 mg/kg/day and silymarin at 25 mg/kg/day administered to the paracetamol treated rats for seven days. EBM and silymarin produced significant ($p < 0.05$) hepatoprotective effect by decreasing the activity of serum enzymes, bilirubin, and total cholesterol. (Ghosh *et al.*, 2007).

Vernonia amygdalina Del (Bitter Leaf)

Belongs to the Family *Compositae* and used in Nigerian folk medicine as a tonic and remedy against constipation, fever, high blood pressure and many infectious diseases. Aqueous extract of leaves showed significant decrease in the levels of bilirubin, liver enzymes and total protein level against acetaminophen induced liver damage in rats (Iwalokun *et al.*, 2006).

Sarcostemma brevistigma Wight (Somlata)

This is the member of Asclepiadaceae family and grows throughout India and other tropical regions of the world. Phytochemical studies reveal the presence of bergenin, brevine, brevinine, sarcogenin, sarcobiose and flavonoids in it. Hepatoprotective activity was evaluated using CCl_4-induced liver damage. It was reported that CCl_4 intoxicated enlarged liver was normal in extract-treated groups (Nirmal *et al.*, 2003).

***Equisetum arvense* L (Horsetail)**

The aerial parts of this plant *(Family: Equisetaceae)* have been used for the treatment of hemorrhage, urethritis, jaundice and hepatitis in oriental traditional medicine. Phytochemistry of the plant includes onitin, onitin-9-O-glucoside, apigenin, luteolin, kaempferol-3-O-glucoside and quercetin-3-O-glucoside. The ethanolic extract exhibited distinctive hepatoprotective activity at 400 g/ml. Among all, luteolin showed strong hepatoprotective affects on tacrine-induced cytotoxicity in human liver-derived Hep G2 cells (Hyuncheol *et al.*, 1997).

***Acanthus ilicifolius* L (Sea Holly/Hargoza)**

Belongs to the family *Zacanthaceae* and habitated in the tropical regions of the world (Lakshmi *et al.*, 1997). The leaves are used in ethno medical practices to treat rheumatism, snake bite, paralysis and asthma. Analgesic and anti-inflammatory properties of this plant have already been reported. 2-Benzo-xazolinone isolated from this plant, showed leishmanicidal activity (Kapil *et al.*, 1994). The alcoholic extract of leaves inhibited the formation of oxygen derived free radicals (ODFR) *in vitro*. The oral administration of the extract 250 and 500 mg/kg body wt. significantly reduced CCl_4 induced hepatotoxicity in rats (Babu *et al.*, 2001).

***Careya arborea* Roxb. (Slow Match Tree/Kumbhi)**

It is a medium sized deciduous tree known as Padmaka in the Ayurvedic system of medicine (Family: *Barringtoniaceae*). The stem bark is used in the treatment of tumors, urinary discharges, piles, leucoderma, skin diseases, epileptic fits and antidote for snake venom (Kirtikar *et al.*, 1987). Two triterpenoids: lupeol and betulin were isolated from its bark (Row *et al.*, 1964). Methanol extract of bark showed strong *in vivo* anticancer property associated with hepatoprotective and antioxidant activities against Dalton's lymphoma - ascitic and solid tumor models and its *in vitro* cytotoxic properties (Natesan *et al.*, 2007). Potent antitumor properties were also evident in an Ehrlich Ascites Carcinoma (EAC) model in mice.

***Momordica dioica* (Spiny Gourd/Kantola)**

It is a climbing creeper and also known as Kakora. Phytochemical study showed the presence of alkaloids, carbohydrates, proteins, amino acids, phenolic compounds, glycosides and flavonoids in it. The roots are used in head trouble, urinary calculi and in jaundice. This plant possesses hypoglycemic gastro protective and ulcer healing activities (Kushawa *et al.*, 2005), The reduction levels of AST and ALT by the extracts is an indication of stabilization of plasma membrane as well as repair of hepatic tissue damage caused by CCl_4 (Thabrew *et al.*, 1987). The ethanolic extract induced suppression of the increased SALP activity with the concurrent depletion of raised bilirubin and stabilize biliary dysfunction in rat liver during hepatic injury with CCl_4 (Jain *et al.*, 2008).

***Boehmeria nivea* var. *nivea* and *B. nivea* var. *tenacissima* (Ramie, China Grass)**

Belongs to the *Urticaceae* family and are widely distributed in China and Taiwan. In China it is used as a remedy for hepatitis (Kan 1986). Several components such as boehnic acid, palmatic acid, stearic acid, ursolic acid, 19a-hydroxyursolic acid, b-sitosterol, b-sitosteryl-b-Dglucoside have been isolated and identified (Matsuura *et al.*, 1973). The aqueous leaf extracts exhibit liver protective effect against CCl_4 induced hepatotoxicity, anti-lipid peroxidative and free radical scavenging activities(Chun *et al.*, 1998).

Solanum fastigiatum (Jurubeba)

Belongs to the family Solanaceae and commonly known as "jurubeba", widely distributed in the southern states of Brazil. The leaves and roots of the plant are used in tonic for fevers, anemia, erysipelas, liver diseases, hepatitis, spleen disorders, uterine tumors, irritable bowel syndrome and chronic gastritis. Phytochemical analysis has shown the presence of flavonoids and glycosides in the leaves of the plant (Higa *et al.*, 2007). Aqueous extract significantly reduce ($p < 0.05$) $Fe^{2}+$ and SNP-induced TBARS production in a concentration dependent manner. It has important hepatoprotective effect against paracetamol induced liver injury (Sabira *et al.*, 2008).

Apium graveolens Linn (Wild Celery/Bari Ajmod)

It belongs to family Apiaceae and grows widely in the foothills of North-Western Himalayas and the hills of Punjab, Himachal Pradesh and Uttar Pradesh. It is commonly known as 'Ajmod' and the fruits are popularly known as Celery seeds. This plant is used in bronchitis, asthma, liver and spleen diseases (Singh and Handa 1995). The phytoconstituent of the plant are flavonoids, glycosides, steroid and triterpenoids. Administration of CCl_4 led to increase the serum enzymes level by 2–3-folds. Treatment of rats with different extracts at dose of 250-mg/kg showed maximum hepatoprotective activity (Ahmed *et al.*, 2002).

Croton oblongifolius (Nagdanti)

Popularly known as 'Chucka' is middle-sized tree belonging to the family *Euphorbiaceae* and native of India. Bark is used in reducing chronic enlargement of the liver. It is applied externally to the hepatic region in chronic hepatitis. Main phytochemicals are flavonoids, glycosides, steroid and triterpenoids. The methanolic extract of the aerial part showed maximum hepatotoxic activity (Aiyar *et al.*, 1970).

Silybum marianum (Milk Thistle)

Commonly known as 'Milk thistle' and belongs to family 'Compositae'. It occurs widely in Europe, Canada and South America. In India it is indigenous to Kashmir. Silymarin is able to neutralize the hepatotoxicity of several agents, including *Amanita phalloides*, ethanol, paracetamol (acetaminophen) and carbon tetrachloride in animal models. The protection against *A. phalloides* is inversely proportional to the time that has elapsed since administration of the toxin. Silymarin protects against its toxic principle a-amanitin by preventing its uptake through hepatocyte membranes and inhibiting the effects of tumor necrosis factor, which exacerbates lipid peroxidation. The active substances from *S. marianum* also stabilize the cell membrane, exert antioxidant properties and exhibit intrinsic toxicity when tested on primary culture of human hepatocytes (Fraschini *et al.*, 2002).

Andrographis paniculata **(Green Chirayata/Kalmegh)**

It is an annual herb and commonly called as 'Kalmegh', belongs to family 'Acanthaceae'. It is widely distributed in Sri Lanka and throughout India especially in Maharashtra, Utter Pradesh, Madhya Pradesh, Tamil Nadu and Andhra Pradesh. The aqueous leaves extract delaying the hepatic turnerogenic condition by lowering the SGPT, SGOT and alkaline phosphatase level in BHC treated mice (Trivedi and Rawal 2000).

Hygrophila auriculata **(Kokilaksh/Gokulakarita)**

It is a biennial plant found in the regions of Himalaya, Punjab and North Africa. It belongs to family 'Umbelliferae'. Pretreatment of rats with 200 mg/kg body weight of the methanolic extracts of the seeds exhibited a significant reduction in the paracetamol induced levels of different SGOT, SGPT and serum bilirubin (Singh and Handa 1995).

Ginkgo biloba **(Maidenhair Tree)**

It is one of the oldest known trees and is widely distributed all over the world, also known as 'Maiden Hair Tree' and belongs to family 'Gingkoaceae'. It is reported that CCl_4 induction significantly increases the serum hepatic enzyme levels and decreases the total protein and albumin levels, which was reversed with *G. biloba* (Ashok 2001).

Pterocarpus santalinus **(Red Sanders)**

P. santalinus belongs to 'Leguminoseae' family and also known as 'Bijasal' and 'Indian Kino Tree'. It is mainly found in hilly regions of Gujarat, Madhya Pradesh, Uttar Pradesh and Bihar. Animals treated with CCl_4 causes increase in serum levels of bilirubin, alanine transaminase, aspartate transaminase and alkaline phosphatase with a decrease in total protein level. Aqueous and ethanol leaf extracts decrease the serum levels of the markers and significant increase in total protein, indicating the recovery of hepatic cells. Histological study of aqueous extract treated group exhibited moderate accumulation of fatty lobules and cellular necrosis where as ethanol extract treated animals revealed normal hepatic cords without any cellular necrosis and fatty infiltration (Manjunatha 2006).

Pergularia daemia **Forsk. (Utaran or Akasan)**

This plant belong to family Asclepiadaceae and is commonly known as "Dustapu teega" in Telugu, "Uttaravaruni" in Sanskrit and "Utranajutuka" in Hindi. It is a perennial twining herb, grows wildly road sides throughout Andhra Pradesh state. Acetone sub fraction of total aerial part showed significant protective effect by lowering serum levels of various biochemical parameters. Silymarin was used as positive control. The results justify the use of *P. daemia* as a hepatoprotective agent (Kumar and Mishra 1996).

Cleome viscosa **(Asian Spider Flower/Bagra)**

It is an annual, sticky herb belonging to the family Capparaceae. The plant distributed throughout the plains in India. Research suggest that the aqueous seed extract of *C. viscosa* has hepatoprotective activity against carbon tetrachloride (CCl_4) induced liver damage in wistar rats. The aqueous seed extract of *C. viscosa* (200 mg/kg) was administered orally to the animals with hepatotoxicity induced by CCl_4. Silymarin (200mg/kg) was given as reference standard. There was significant reduction in serum enzyme aspartate aminotransferase (AST), alanine aminotransferase (ALT), alkaline phosphatase (ALP), Y-glutamyl transpeptidase and lipid peroxidase and increase in reduced glutathione (GSH). Various pathological changes like centrilobular necrosis and vacuolization observed in CCl_4 treated rats, which were prevented to a moderate extent in groups, treated with *C. viscosa* and Silymarin. It was concluded from the study that aqueous seed extract of *C. viscosa* possesses hepatoprotective activity against CCl_4 induced hepatotoxicity in rats (Sengottuvelu *et al.*, 1998).

Picrorhiza kurrooa (Kutki)

Commonly called as 'Indian Gentian' and 'Kutki'. This perennial herb of family 'Scrophulariaceae' is distributed in upper Himalayas and China and grows naturally in Kashmir and Himachal Pradesh. The hepatoprotective study of *P. kurrooa* was evaluated against alcohol-CCl_4, induced liver damage in rat. It showed a significant hepatoprotective activity by lowering of elevated levels of glutamate oxaloacetate transaminase, glutamate pyruvate transaminase, acid phosphatase, alkaline phosphatase, glutamate dehydrogenase and bilirubin (Tripathi *et al.*, 1991).

Discussion

Medicinal plants are known to have wide therapeutic applications in folk medicine, and scientific advancement through technology has provided substantial evidence to support most of its medicinal claims. *In vivo* studies have further demonstrated the hepatoprotective potential of several medicinal plants. In most of the studies, hepatocellular damage induced in mice is by CCl_4 and the data interpretation is based on significant elevations in SGOT and SGPT activities. There is no doubt that certain herbal products contain chemically defined components that can protect the liver from oxidative injury, promote virus elimination, block fibrogenesis, or inhibit tumor growth. Biologically active molecules derived from herbal extracts can serve as suitable primary compounds for effective and targeted hepato-drugs. Furthermore, the goal of these studies should not be restricted to find new pure compounds as drugs but also the active extracts, fractions or mixture of fractions or extracts which may act in combination to prove an effective and efficient drug. Plant drugs (combinations or individual drug) for liver diseases should possess sufficient efficacy to cure severe liver diseases caused by toxic chemicals, viruses, excess alcohol intake, etc. A single drug cannot be effective against all types of severe liver diseases and therefore, effective combination and formulations have to be developed.

References

Ahmed B, Alam T, Varshney M and Khan S A 2002. Hepatoprotective activity of two plants belonging to the *Apiaceae* and the *Euphorbiaceae* family. *J Ethnopharmaco*, **79**: 313–316.

Aiyar V N and Seshadri T R 1970. Components of *Croton oblongifolius*-III Constitution of oblongifolic acid. *Tetrahedron* **26:** 5275–5279.

Babu B H, Shylesh B S and Padikkala J 2001. Antioxidant and hepatoprotective effect of *Acanthus ilicifolius. Fitoterapia* **72:**.272- 277.

Boelsterli U A 2003. Mechanistic Toxicology: The Molecular Basis of How Chemical Disrupts Biological Targets. Taylor and Francis, London.

Bose P, Gupta M, Mazumder U K, Kumar R S, Thangavel S and Kumar R S 2007. Hepatoprotective and Antioxidant Effects of *Eupatorium ayapana* against Carbon Tetrachloride Induced Hepatotoxicity in Rats. *IJPT*, **6:** 27-33.

Bouneva I, Abou-Assi S, Heuman D M and Mihas A A 2003. Alcohol liver disease. Hospital Physician. 31–38.

Chun-Ching L, Ming-Hong Y, Tsae-shiuan L and Jer-Min L 1998. Evaluation of the hepatoprotective and antioxidant activity of *Boehmeria nivea* var. *nivea* and *B. nivea* var. *tenacissima. J Ethnopharmac* **60:** 9–17.

Dahiru D, William E T and Nadro M S 2005. Protective effect of *Ziziphus mauritiana* leaf extract on carbon tetrachloride-induced liver injury. *African J Biotech* **4:** 1177-1179.

Dvorjak Z, Kosina P, Walterova D, Vilym Sjimanek, Petr Bachleda, and Jitka Ulrichova 2003. Primary cultures of human hepatocytes as a tool in cytotoxicity studies: cell protection against model toxins by flavonolignans obtained from *Silybum marianum. Toxic Lett* **137:** 201-212

Fernandopulle B M R and Karunanyake E H 1994. Oral hypoglycemic effect of *MDR* in rat. *Med Sci Res* **22:** 137–139.

Fernandopulle B M R and Ratnasooriya W D 1996. Evaluation of two cucurbits (Genus: *Momordica*) for gastroprotective and ulcer healing activity in rats. *Med Sci Res* **24:** 85–88.

Fransworth N R 1990. The role of ethnopharmacology in drug development. In: Bioactive compounds from plants. Ciba Foundation Symposium, Chichester, *John Wiley and Sons* **154:** 2-21.

Fraschini F, Demartini G and Esposti D 2002. Pharmacology of Silymarin. Clin Drug Invest. **22:** 51-65

Ghosh T, Maity T K, Das M, Bose A and Dash D K 2007. In *Vitro* Antioxidant and Hepatoprotective Activity of Ethanolic Extract of *Bacopa monnieri* Linn. Aerial Parts. *IJPT* **6:**77-85.

Gite V N, Deshmukh R D, Sane R T, Takate S B and Pokharkar R D 2007. Hepatoprotective Activity of *Enicostema Axillare* Against CCl4 – Induced Hepatic Injury in Rats. *Pharmac online* **1:** 25-30.

Gupta M, Mazumrer U K, Thamilselvan V, Manikandan L, Senthilkumar G P, Suresh R and Kakotti B K 2007. Potential Hepatoprotective Effect and Antioxidant Role of Methanol Extract of *Oldenlandia umbellata* in Carbon Tetrachloride Induced Hepatotoxicity in Wistar Rats. *IJPT* **6:** 5-9.

Handa S, Sharma A and Chakraborthi K K 1986. Natural products and plants as liver protecting agents. *Fitoterapia* **57:** 307–351.

Harish R and Shivanandappa T 2006. Antioxidant activity and hepatoprotective potential of *Phyllanthus niruri*. *Food Chem* **95:** 180–185.

Higa K C, Lopes L, Schild M X, Ana L and Haraguchi M 2007. Flavonóides glicosídicos nas folhas de *Solanum fastigiatum*. *J Brasileira de Fitomedicina* **5:** 174–1174.

Hoek J B and Pastorino J K 2002. Ethanol, oxidative stress and cytokine-induced liver cell injury. *Alcohol* **27:** 63–68.

Hyuncheol O H, Kimb D H, Jung H C and Kim U C 2004. Hepatoprotective and free radical scavenging activities of phenolic petrosins and flavonoids isolated from Equisetum arvense. *J Ethnopharmac* **95:** 421–424.

Iwalokun B A, Efedede B U, Alabi-Sofunde J A, Oduala T, Magbagbeola O A and Akinwande A I 2006. Hepatoprotective and Antioxidant Activities of *Vernonia amygdalina* on Acetaminophen-Induced Hepatic Damage in Mice. *J. Med Food* **9:** 524-530.

Jarvelainen H 2000. Inflammatory Responses in Alcoholic Liver Disease. National Public Health Institute, Helsinki 1–75.

Kassuya C A, Leite D F, deMela, L V, Rehder V L and Calixto J B 2005. Anti-inflammatory properties of extracts, fractions and lignans isolated from *Phyllanthus amarus*. *Planta Medica* **71:** 721–726.

Kapil A, Sharma S, Wahidulla S 1994. *Planta Med* **60:** 187.

Khatoon S, Rai V, Rawat A K S and Mehrotra S 2006. Comparative pharmacognostic studies of three *Phyllanthus* species. *J Ethnopharmacol* **104:** 79–86.

Kirtikar K R and Basu B D 1987. Indian medicinal plants. Bishen Singh Mahendra Pal Singh, Dehradun. **2:** 1061–1063.

Kushawa S K, Jain A, Jain A, Gupta V B, Patel J R and Dubey P K 2005. Hepatoprotective activity of fruits of *Mormordica dioica* Roxb. *Pl Arch* **5:** 613–616.

Jain A, Soni M, Deb L, Jain A, Rout S P, Gupta V P and Krishna K L 2008. Antioxidant and hepatoprotective activity of ethanolic and aqueous extracts of *Momordica dioica* Roxb. Leaves. *J Ethnopharmacol* **115:** 61–66.

Kan W S 1986. National research institute of Chinese medicine. *Pharmaceutical Bot* 184–185.

Ashok K 2001. Hepatoprotective Effects of *Ginkgo biloba* against Carbon Tetrachloride Induced Hepatic injury in Rats. *Ind J Pharmacol* **33:** 260-266

Lakshmi M, Rajalakshmi S, Parani M, Anuratha C S, Parida A 1997. Theoretical Appl Genetics **94:** 1121.

Matsuura S, Lee L T and Iinuma M 1973. Studies on the components of *Boehmeria* sp. I.i) Studies on the components of the roots of *Boehmeria platanifolia* Franch. et Sav. Yakugaku Zasshi **93:** 1682-1684.

Manjunatha BK 2006. Hepatoprotective activity of *Pterocarpus santalinus* L.f., an endangered medicinal plant. *Ind J Pharmacol* **38:** 25-8

Mitra S K, Venkataranganna M V, Sundaram R and Gopumadhavan S 1998. Protective effect of HD-03, a herbal formulation, against various hepatotoxic agents in rats. *J Ethnopharmacol* **63:** 181–186

Natesan S, Badami S, Cherian M M and Hariharapura R 2007. Potent in vitro cytotoxic and antioxidant activity of *Careya arborea* bark extracts. *Phytother Res* **21:** 492–495.

Nirmal K N. Yogendra N S and Mamta M 2003. Hepatoprotective activity of *Sarcostemma brevistigma* against carbon tetrachloride-induced hepatic damage in rats. *Curr Sci* **84:** 1186-1187.

Row L R and Sastry C S P 1964. Chemical examination of *Careya arborea* Roxb. *Indian J Chem* **2:** 510–514.

Sabira S M and Rocha J B T 2008. Antioxidant and hepatoprotective activity of aqueous extract of *Solanum fastigiatum* (false "Jurubeba") against paracetamol-induced liver damage in mice. *J Ethnopharmacol* **120:** 226–232.

Sengottuvelu S, Duraisamy R, Nandhakumar J and Sivakumar T 1998. Hepatoprotective activity of Cleome viscosa against Carbon tetrachloride induced hepatotoxity in rats. *Pharmacogno Magazine* ISSN: 0973-1296

Singh A and Handa S S 1995. Hepatoprotective activity of *Apium graveolens* and *Hygrophila auriculata* against paracetamol and thioacetamide intoxication in rats. *J Ethnopharmacol* **49:** 119–126.

Singh A and Handa S S 1995. Hepatoprotective activity of *Apium graveolens* and *Hygrophila auriculata* against paracetamol and thioacetamide intoxication in Rats. *J Ethnopharmacol* **49:** 119-126

Smuckler E A 1998. Alcoholic drink: Its production and effects. *Fed Proe* **34:** 2038-44.

Kumar S V and Mishra S H 1996. Hepatoprotective activity of extracts from *Pergularia daemia* Forsk. Against carbon tetrachloride-induced toxicity in rats. *Pharmacogno Magazine.*

Thabrew M I, Joice P D T M and Rajatissa W A 1987. Comparative study of efficacy of *Paetta indica* and *Osbeckia octandra* in the treatment of liver dysfunction. *Planta Medica* **53:** 239–241.

Tripathi S C, Patnaik G K and Dhawan B N 1991. Hepatoprotective activity of Picroliv against Alcohol-Carbon tetrachloride induced damage in Rats. *Ind J Pharmac* **23:** 43-49

Trivedi N and Rawal U M. 1998. Effect of Aqueous Extract of *Andrographis Paniculata* on Liver Tumor. *Ind J Pharmacol* **30:** 318-322

Udomuksorn W, Phurthi-Punlai S, Wongnawa M, Nitiruangjaruj A, Wanapong N and Yunyium N 2000. Effects of Phyllanthus amarus Schum. et. Thonn on paracetamol induced hepatotoxicity in rats. *J Mel Sci* **42:** 119–132.

Umarani D, Devaki T, Govindaraju P and Shanmugasundaram K R 1985. Ethanol induced metabolic alteration and effect of Phyllanthus niruri in their reversal. *Ancient Sci Life* **4:** 174–180.

Wheeler M D, Kono H, Yim M, Nakagami M, Uesugi T, Arteel G E, Gabele E, Rusyn I, Yamashina S, Froh M, Adachi Y, Iimuro Y, Bradford B V U, Smutney O M, Conner H D, Mason R P, Goyert S M., Peter J M, Gonzalez F J, Samulski R J and Thurman R G 2001. Serial review: alcohol, oxidative stress and cell injury: the role of Kupffer cell oxidant production in early ethanol-induced liver disease. *Free Radical Biol Med* **31:** 1544–1549.

Zimmerman H J 1999. Hepatotoxicity: the adverse effects of drug and other chemicals on the liver. Lippincott Williams and Wilkins, *Philadelphia* 147–175.

Medicinal Plants: Aspects and Prospects (2014) ***Pages* 49–64**
***Editors:* Mukesh Kumar, Anjali Khare and C.P. Shukla**
ISBN: 978-81-7622-309-6
***Published by:* BIOTECH BOOKS, NEW DELHI**

Chapter 4

Indigenous Medicinal Plants of Garhwal Himalayas Used in the Preparation of Ayurvedic Medicines in Traditional Health Care System

Rakhi Rawat and D.P. Vashistha

Department of Botany and Microbiology
H.N.B Garhwal (Central) University,
Srinagar-Garhwal – 246 174, Uttarakhand

ABSTRACT

The drug plants have been a primary resource for a number of necessary commodities of the tribals living in and around them. About 100 plants are available in Pauri District of Uttrakhand which are used for common Ayurvedic drug formulations. The most notable among them are *Acorus calamus, Adhatoda zeylenica, Asparagus racemosus, Aleo barbadensis, Andrographis paniculata, Berberis aristata, Boerhavia diffusa, Callicarpa macrophylla, Celastrus paniculata, Datura stramonium, Eclipta prostrata, Evolvulus alsinoides, Gloriosa superba, Oroxylum indicum, Fumaria indica, Melia azadirach, Phyllanthus fraternus, Plumbago zeylanica, Ricinus communis, Punica granatum, Terminalia chebula, T. ballerica, T. arjuna, Tinospora cordifolia, Woodfordia fruiticosa, Withania somnifera* etc. The listed plants are given with complete description *viz.* distribution, habit, branded drug, traditionally used and status in Uttarakhand etc.

Keywords: Ayurvedic drug, Traditional use, Habit.

Introduction

India has unique position in the world, where a number of traditional systems of medicines *i.e.* Ayurveda, Sidha, Unani, and Homoeopathy are practiced in health care system. All the systems predominately depend upon medicinal plants (Sharma *et al.*, 2004, Rewani *et al.*, 2005). Medicinal plants provide materials for pharmaceutical industry as well as traditional forms of medicines, generate income and employment and have implications for the conservation of biodiversity and traditional knowledge. Worldwide the number of species used for medicinal purposes is estimated to be more than 50,000 which covers about 13 per cent of all flowering plants (Schipmann *et al.*, 2002).

However, Ayurvedic system of medicine which caters a major portion of health care system, accounts nearly 84 per cent of internal herbal market (Chauhan 2003, Pandey *et al.*, 2006). It currently comprises about 1500 single drugs and 800 compound formulations. About 1500 plant species are used in Ayurved, out of them 603 are available in Uttrakhand out of which 350 plants frequently used in Ayurvedic preparation.

The tremendous growth of pharmaceutical industry considerably affected the popularity of herbal medicines, but the traditional herbal system of medicines still remains intact as cultural heritage in the different regions of the world and continues to cater the needs of 80 per cent of the population of the world mainly in poor countries (Farnsworth and Soejarto 1991, Chauhan 1996). The herbal drugs are low priced, harmless and easily available to common man. These are available even to those people where modern health care system is yet to develop.

The Study Area

Uttarakhand is a newly carved state of India and is well known for its rich repository of biodiversity. The district Garhwal, also known as Pauri, among 13 others is located between 29°202- 29°752 N latitude and 78°102- 78°802 E longitude, covering about 5329 km^2 (Figure 4.1). The valleys and gentle slopes have considerable soil depth developed from colluviums and alluvium and represents thin surface horizon and medium to coarse texture. The South facing slopes are dry and less fertile while North facing slopes support deep, moist and fertile soil. During summer months temperature ranges 35° to 40 °C. December and January are coldest months, the temperature range 1°C. Relative humidity is highest during rainy season 80-97 per cent and 20-40 per cent during winter. Climatically the place links greater Himalayas to Gangetic plains and have varied climatic and altitudinal variations. Due to such variations, it encompasses significant land and water resources and consequently great range of biodiversity. It has great potential to play a significant role in medicinal plant cultivation and industry.

Objectives

The main objectives of present study is to assess the structure and composition of the floristic communities present in and around Garhwal Himalayas, Uttarakhand under the following heads-

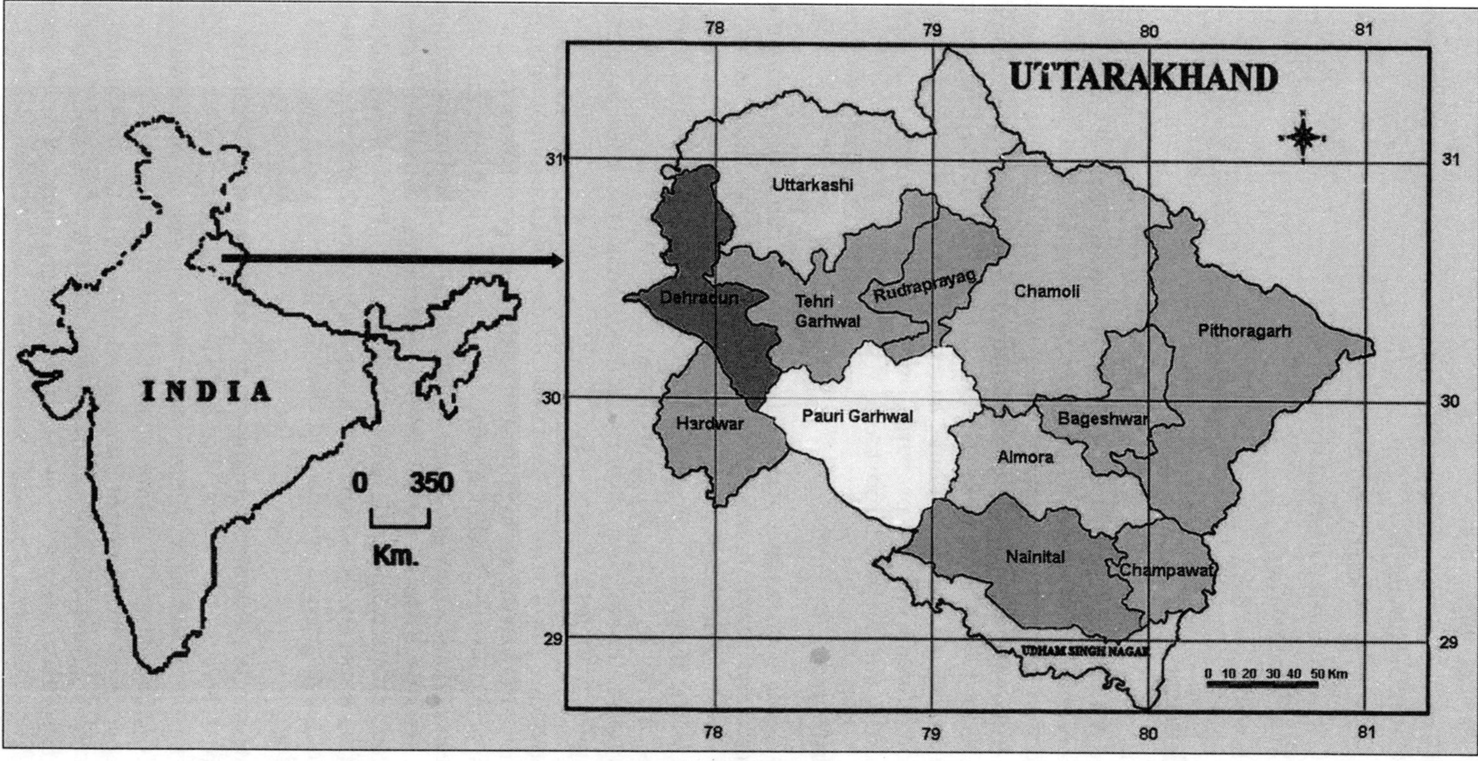

Figure 4.1: Map showing the study area

1. Floristic composition
2. Linkage between plant wealth and its medicinal values in the context of Garhwal Himalaya
3. Association among various species
4. Conservational pattern of species
5. Regeneration rate
6. Relevance of flora

Methodology

Extensive survey of different drug stores selling herbal product in Uttarakhand was carried out and comprehensive list of the herbal component of various Ayurvedic drugs was prepared. Each herbal plant noted as component of particular branded drug prescribed for particular remedy, was given complete taxonomic identification and habit and distribution details by consulting taxonomical literature and flora of the region (Gaur 1999, Naithani, 1984-85, Osmastone 1927). Every plant was provided with details such as botanical name, vernacular name, locality and altitudinal distribution. Notes were also prepared on the medicinal parts of the plants and Ayurvedic medicine prepared from them.

Observations

It has been observed that about 100 plant species currently being used in the reputed herbal industry (Charak, Vaidhnath and Himalaya), are commonly found in district Garhwal (Tables 4.1–4.4). Most of the commercial Ayurvedic medicines cover treatment for the ailments ranging from fever, dysentery, carminative, colic, asthma, epilepsy, skin disease, diarrhoea, aphrodisiac, constipation, ulcer, eczema, jaundice, kidney stone, rheumatic pain, urinary troubles, mental disorder, bronchial cough, hepatitis, piles, diabetes and cancer etc.

The growing popularity of Ayurveda, Unani and Homoeopathy systems of medicine is putting an increasing pressure on the plant population. About hundreds of plants species used for medicinal purposes in India, which once grew abundantly in our forests and elsewhere are at risk as they often are picked indiscriminately from the wild, leaving little scope for their regeneration. Another reason for the disappearance of many species is ignorance with regard to knowledge of the identity and use of such species. In our ignorance, we are invariable treating these otherwise useful species as useless weeds and destroy them, leaving no scope for their regeneration.

It is strikingly evident that many of the herbal plants, aboriginal to the valley region, are *Asparagus racemosus, Aleo barbadensis, Andrographis paniculata, Bacopa monnieri, Boerhavia diffusa, Callicarpa macrophylla, Celastrus paniculatus, Evolvulus alsinoides, Gloriosa superba Oroxylum indicum, Plumbago zeylanica, Tinospora cordifolia, Withania somnifera* etc. survey of literature clearly indicates, that about 31 species such as *Bergenia ligulata, Acorus calamus, Bacopa monnieri, Boerhavia diffusa, Dioscorea bulbifera, Evolvulus alsinoides, Gloriosa superba, Hedychium spicatum, Polygonatum*

Table 4.1: The herbs used in Ayurvedic formulations, their availability and status in District Pauri Garhwal.

Botanical Names/ Local Names/Families	*Parts Use*	*Distribution*	*Traditional Medicinal Use of the Plants*	*Ayurvedic Formulations*
Acorus calamus L.*/ Vacha/Araceae	Rhizomes	Up to 2000 msl	Carminative, colic, dysentery, fever, asthma, epilepsy, bronchital	Chandraprabha vati, Sanjivani vati, Saraswat churn, Ashwagandharista, Laghu vishgarbha tel, Abana tebles, Sumento, Pigmento, Mentat, Alarsin, M_2 -Tone.
Aleo barbadensis Miller/ Kumari/Liliaceae	Leaves	Sub-tropical region	Purgative, constipation	Livomyn, Pigmento, Livosyp, Diabecon, Herbolax, Reostra (tablets and syrup).
Allium cepa L/Piaze/ Amaryllidaceac	Whole plant	Cultivated up to 1800 msl	Digestive and skin disease	Neo, Gerifort
*Amaranthus polygamus auct. non.*L/Chuali/ Amaranthaceae	Leaves	Cultivated plant	—	Femiplex
Andrographis paniculata Wall.ex.Nees/Kalmegh/ Acanthaceae	Whole plant	Up to 600 msl	Fever, stomach disease	Livomyn, Pigmento, Hepatogard
Anethum sowa Roxb.ex Fleming/Sowa/Apiaceae	Leaves	Up to 1000msl	—	Vomiteb, Mental, Himacospaze, Bonnisan, Abana (tablets and syrup).
Bacopa monnieri (L) Pennell.*/Jal birahmi/ Scrophulariaceae	Whole plant	Up to 1200 msl	Nervine tonic, diuretic blood purifier	Bharmi churn, Bharmi ghrit, Brahmi rasayan
Bergenia ligulata (wall) Engle.*/Silpara/ Saxifragaceae	Rhizomes	1000-3000 msl	Fever, diarrhoea, kidney stone, swelling	Ashmeriher kwath, Pashanbhedadi kwath, Pushyanug churn, Vatvidhv ansak ras. Cystone tablets.
Boerhavia diffusa L*/ Punarnava/Nyctaginaceae	Roots	Up to 2000 msl.	Jaundice, asthma, bronchitis, eye problems	Punarnava kwath, Punarnava -sova, Punarnavadi mandoor, Punarnavadi guggulu, Abana, Crush, Lukol, Geriforte, Livomyne, (tablets, and syrups)
Cassia absus L./ Cheaksu/Caesalpiniaceae	Seeds	Up to 1000 msl,	Opthalmia, habitual constipation	Parpaundrik kwath, Khadiradi vati and Surma.

Contd...

Table 4.1–*Contd...*

Botanical Names/ Local Names/Families	*Parts Use*	*Distribution*	*Traditional Medicinal Use of the Plants*	*Ayurvedic Formulations*
Centella asiatica (L.) Urban/Brahmi/Apiaceae (Umbelliferae)	Whole plant	Up to 1500 msl	Mantal disease, skin disease, blood purifier, diuretic	Brahmi vati, Brahmi ghrit, Brahmi tel, Saraswatarista, Geriforte, Mental (tablets and syrup).
Coriandrum sativum L/ Dhaniya/Apiaceae	Fruits	Cultivated species	Urinary disorder, digestive troubles, carminative, diuretic tonic, vomiting	Livomyn
Cuminum cyminum L./ Zeera/Apiaceae	Fruits	Up to 1000 msl.	Cold	Ojus, Femiplex, M_2- Tone, Apimore
Cyperus rotundus L./ Motha/Cyperaceae	Tuberous roots	Up to 1500 msl	Malaria fever, diaphoretic, jaundice, stomach disorder, dyspepsia, wounds, ulcer	Pigmento
Datura stramonium L./ Dhatura/Solanaceae	Leaves, flowering tops and seeds	Up to 1500 msl	Control motion, sickness, asthma, rheumatism, jaundice	Kankasava, Maha laxmi vilas ras, Asthma relief Alarex, Spasmolin (tablets and syrup).
Dioscorea bulbifera L.*/ Genthi/Dioscoreaceae	Roots	Up to 1500 msl	Bronchial cough, antiseptic burn, wound	Chayavanpras, Mahamash tel, Narsingh tel, Grabbdharin vati
Eclipta prostrata (L.) L./ Bhringraja/Asteraceae (Compositeae)	Whole plants	Up to 1000 msl	Intestinal parasites anemia, hepatitis, asthma, hair tonic	Bhringrajasava, Sutsekherras Anand bhairas, Neo, Ojus, Acidocid syrup, Geriforte Mahabhringraj tel, Gandhak rasayana, Livomyn, M_2- tone, Hepatogard.
Evolvulus alsinoides L.*/ Sankhpuspi/ Convolulaceae	Whole plant	Up to 1500 msl	Fever, dysentery, cough, cold, bronchitis, asthma, brain tonic	Sumento
Fumaria indica (Haussk.) Pugsley/Pit papra/ Fumariaceae	Whole plant	Up to 1400 msl	Fever, influenza, diuretic, diaphoretic, wound, urinary troubles	Arvindasava,Chandrakala ras, Chandanasava, Mahasudorshan kwath Livomyn drops (tablets and syrup)
Gloriosa superba L.*/ Kalihari/Liliaceae	Rhizomes and seeds	Up to 1500 msl	Painful delivery, suppressed treatment of gout and cancer	Mahavishgarbh tel, Nirundi tel, Rumalaya cream, Langali rasayan

Contd...

Table 4.1–*Contd...*

Botanical Names/ Local Names/Families	*Parts Use*	*Distribution*	*Traditional Medicinal Use of the Plants*	*Ayurvedic Formulations*
Hedychium spicatum Ham.ex smith*/Banhaldi/ Zingiberaceae	Rhizomes	1200-2000 msl	Stomachic, carminative, expectorant bowel complaints, vomiting	Shatyadi chrun, Dasamoolarista, Kankayan gutika, Vomiteb, Chikara, Lahmina,Jivahakalpa, Himanshu tel.
Hygrophila spinosa T/ Talimkhana/Acanthaceae	Leaves, seeds and roots	Outer Himalayan belt	Liver trouble	Femiplex, Crush
Nelumbo nucifera Gaertn/ Kamal/Nelumbonaceae	Rhizome, flowers and seed	Up to 1500 msl	—	Arvindasava, Pushyanug churn, Sarivadi vati, drakshavleha.
Phyllanthus fraternus Webster/Jarmala/ Euphorbiaceae	Whole plant	Up to 1000 msl	Gonorrhoea, genitourinary, jaundice, dyspepsia	Chyavranpras, Swanshar kwath, Kashhar kwath, Satpatradi churn, Tonali Livomyn, Livomap (tablets and syrup).
Picrorhiza kurrooa Royle ex Benth.*/Kutaki/ Scrophulariaceae	Rhizomes and roots	3000-4200 msl	Cathartic, stomachic, fever, dyspepsia, purgative	Arogyavardhi vati, Laxmi narayan ras, Mahayogray guggulu, Amritarista, Cruminillsyrup, Livertone, Pigmeto, Carminex, Livomyn, Hepatogard, Calcury Apimore (tablets and tonic)
Polygonatum verticillatum (L) All*/Kantula/Liliaceae	Rhizomes	2000-3000 msl	Gastric complain, wounds	Chyavanpras, Triphala ghrit, Bramh rasayan, Mahanarayan tel
Rheum australe D. Don*/ Dolu/Polygonaceae	Rhizomes and roots	3000 to 5200 msl	Headache, fever, piles, ulcer, asthma, bronchitis, dysentery, jaundice, eye disease, internal injury	Lavanbhaskar churn, Mahajwarankush ras, Gurukulbal ghutti, Livertone, Uteronol
Sida cordifolia L./Balu/ Malvaceae	Leaves and roots	Up to 1000 msl	Urinary troubles,	M_2 -Tone
Spinacia oleracea/ Palak/Chenopodiaceae	Leaves and fruits	Cultivated species	—	Manoll
Solanum nigrum L./Makoi/ Solanaceae	Whole plant	Up to 2800 msl	Spleen, diarrhoea, eye ailments, piles	Pigmento

Contd...

Table 4.1–*Contd...*

Botanical Names/ Local Names/Families	*Parts Use*	*Distribution*	*Traditional Medicinal Use of the Plants*	*Ayurvedic Formulations*
Solanum surattense Burm. F./Kantakari/ Solanaceae	Whole plants	Up to 1500 msl	Cough, asthma, chest pain, fever, diuretic	Kantakariavleha, Kantakari ghrit, Vyaghiharitaki, Chyavanpras, Dasamoolarist, Dasamool kwath
Swertia chirayita Roxb. ex. Flem*/Chiraita/ Gentianaceae	Whole plant	1200 to 3000 msl	Fever, malaria, bronchitis, blood purifier	Pigmento and Diabacon
Tagetes minuta L./Jangali genda/Asteraceae	Flowering tops	Up to 2000 msl	Antiviral activity	Essential oil
Tephrosia purpurea Pers./Pal/Fabaceae	Whole plant	Submontane Himalaya and plain	Asthma, hepatic disorder, cough, dyspepsia, diarrhoea, bronchitis, piles, rheumatism	Livomyn, Pigmento
Thalictrum foliolosum DC.*/Kirmuri (Mamira)/ Ranunculaceae	Roots	1200-2500 msl	Dyspepsia, purgative, febrifuge, convalescence	Surma (Yogi Pharmacy)
Tribulus terrestris L./ Gukhru/Zygophyllaceae	Seeds	Up to 800 msl	Asthma, cough	Manoll, Calcury
Valeriana jatamansi Jones.*/Sumaya/ Valeriancaceae	Rhizomes and roots	Up to 2000-3000 msl	Carminative, antispasmodic, hysteria, epilepsy, cholera, neurosis	Tagardi kwath, Chandana lodhrasava, Sarivadyasava, Chandrasekhar ras, Newrocardine, Nervoplex Mondo-valerian, Neocardial liquid, Sumenta (tablets and syrup).
Vernonia anthelmintica (L) Willd./Kalijiri/Asteraceae	Whole plants	Up to 1200 msl	skin disease, fever, leuco-derma, asthma, kidney troubles, colic inflammations, itching of eyes, snake bite, cough and flatulency	Obenyl, Divya kayakalp vati.
Viola pilosa Blume*/ Vanfsa/Violaceae	Whole plants	Up to 1500 msl	Antipyretic febrifuge, bile and lungs troubles	Kafsina, Kofteb, Bans cough syrap, Cheston, Mehasudarsan ark.
Zingiber officinale Roscoe/Aadarak/ Zingiberaceae	Rhizome	Cultivated species	Cold, cough, fever, bronchitis, dysentery	Livomyn, Vomiteb

Table 4.2: The shrub used in Ayurvedic formulation, their availability and status in District Garhwal

Botanical Names/ Local Names/Families	*Parts Use*	*Distribution*	*Traditional Medicinal Use of the Plants*	*Ayurvedic Formulations*
Adhatoda zeylanica Medic./Vasaka/ Acanthaceae	Leaves, roots	Up to 1500 msl	Wooping cough, skin disease, dysentery	Vasakamadu, Vasavaleha, Diakof, Vasakasva, Kasni, Kankasava Livomyn, M_2-Tone, Styplon, Geriforte (tablets and syrup).
Asparagus racemosus Willd/Satawari (Jhirni)/ Liliaceae	Roots	Up to 1500 msl	Sexual debility, urinary disorders	Femiplex, Neo, M_{2}.Tone, Ojus, Himaplasia, Tentex Geriforte, Lukol, Renalka.
Berberis aristata (D.C.)*/ Kingore/Berberidaceae	Roots, root bark, stem	Srinagar, Pauri	Ophthalmia, fever, jaundice	Femiplex, Livomyn
Berberis lycium Royle*/ Daruharidara/ Berberidaceae	Roots, roots bark and stem	800-2500 msl	Diabetes	Pushyanug churn, Aswagantharista, Khadirarista, Patrangasava
Callicarpa macrophylla Vahl./Daiya/Verbenaceae	Flower buds	Up to 1400 msk.	Rheumatic pain, blood purifier, dysentery	Priyangwadi tel, Ashwagandharista, Dasamoolarist, Chandana sava, Draksharista, and Eladi churns.
Celastrus paniculatus Willd*/Kaunya/ Celastraceae	Seeds and seeds oil	Up to 1200 msl	rheumatism	Jyotismasti tel, Himcotin cream, Sexotex cream Gerifort, Remem, Sumenta (tablets and syrup).
Cissampelos pareira L. Var. Hirsuata/Pahre/ Menispermaceae	Roots	Up to 2000 msl	Antiperiodic, diuretic, purgative, dyspepsia, urinary trouble	Pusyanug churn, Pathadi kwath, Mahayograj guggulu Agnimukh churn.
Ephedra gerardiana Wall ex. Stopf.*/ Tutgautha/Ephedraceae	Stem	Up to 3000-4200 msl	Asthamatic cardiac, circulatory, stimulants, rheumatism, hepatic diseases	Som kalp, Swankalp.
Holarrhena antidysenterica Wall. ex.A.DC/ Karu/Apocynaceae	Barks and seeds	Up to 1250 msl	Dysentery, dropsy, astringent, fever, diarrhoca, intestinal worm	Kutajarista, Kutajahan vati, Kutajawteha ras, Karpur ras
Lawsonia alba Lam. Mehandi/Lythraceae	Leaves	Up to 1000 msl.	Blood dysentery, wounds, cut	Femiplex

Contd...

Table 4.2–*Contd...*

Botanical Names/ Local Names/Families	*Parts Use*	*Distribution*	*Traditional Medicinal Use of the Plants*	*Ayurvedic Formulations*
Plumbago zeylanica L.*/ Chitrak/Plumbaginaceae	Roots	Up to 2500 msl	Dyspepsia, piles, skin disease	Chitrakadi vati, Chitrak haritake, Chitrak ghrit Mahasankh vati, Mahamritaunjay ras, Pigmento, Liv-52, Alarsin, Vigiroll (tablets and syrup).
Pueraria tuberosa (Roxb. Ex.Willd)*/Siralu/Fabaceae	Tuber	Up to 1200 msl	Fever, sexual excitement, demulcent, swelling, rheumatism	Vidari churn, Laxamivilasras, Narsing churn, Ashwagandha churn, Saraswatista, Chyvanpras.
Punica granatum L./ Dadim/Punicaceae	Fruits	Up to 1500 msl	Diarrhoea, dysentery, pimples, boil, cold, diuretic, asthma, anthelmintic, stomachic, vomiting, bronchitis	Dadimastak churn, Dadimadi ghrit, Lavanbhaskar churn Eladi churn.
Ricinus communis L./ Arandi/Euphorbiaceae	Roots, root bark, leaves and seeds	Up to 1400 msl	Asthma, bronchitis, skin disease, jaundice, nervous disease	Erand pak, Erand tel Balarista, Pradrantak lauh, Bishgorbh tel, Cibolic capsules.
Solanum anguivii Lam./ Kandyari/Solanaceae	Whole plant	Up to 1500 msl	Jaundice, cough, asthma, fever, colic, skin disease	Brahtyadi kwath, Dasamoolarista, Dasamool kwath, Chyavanpras, Mahasudarshan churn, Mahanarayan tel
Symplocos racemosa Roxb./Ladha/ Symplocaceae	Bark and leaves	Up to 1200 msl	—	M_2- Tone
Vitex negundo L./ Siwain/Verbenaceae	Leaves roots and fruits	Up to 2000 msl	Rheumatism, arthritic, fever, ulcer, cuts, wound, skin disease	Nirgundi ghrit, Nirgundi kwath, Nirgundi tel, Bishagarbh tel, Rumalagy cream, Himcolin cream.
Withania somnifera (L.) Dunal/Aswagandha/ Solanaceae	Roots	Up to 1200 msl	Urinary disorders, fever, cold, rheumatism	Asbwagandha churn, Ashwa gnadha rasayan, Ashwagan dharista, Kamdev ghrit, Chyaranpras, Phalaghrit Mahamash tel, Manoll, M_2 tone (tablets and syrup).

Contd...

Table 4.2–*Contd...*

Botanical Names/ Local Names/Families	*Parts Use*	*Distribution*	*Traditional Medicinal Use of the Plants*	*Ayurvedic Formulations*
Woodfordia fruiticosa (L.)/Kurz/Luthraceae	Flowers	Up to 1500 msl	Haemorrhoids, febrifuge, vaginitis	Gangadher churn, Pushyanug churn, Abhyarista, Ashokarista, Arvidiasava, Femiplex.
Zanthoxylum armatum DC*/Timroo/Rutaceae	Whole plants	Up to 2000 msl	Toothache, toothdecay	Bharmi Vati, Panchtikta guggula, Madusnhi rasayan, Dantmanjan.

Table 4.3: The trees used in Ayurvedic formulation, their availability and status in District Pauri Garhwal

Botanical Names/ Local Names/Families	*Parts Use*	*Distribution*	*Traditional Medicinal Use of the Plants*	*Ayurvedic Formulations*
Abies pindrow Royle/ Raga/Pinaceae	Bark	Montane region	Asthma, bronchitis	Vigroll
Acacia catechu (L.F.) Willd/Kher/Mimosaceae	Wood bark	Tarai – bhabar tracts	Diarrhoea, dysentery, bronchitis	Pigmento
Aegle marmelos (L.) Correa/Bael/Rutaceae	Fruit, roots bark	Up to 1200 msl	Diarrhoea, diabetes	Bilvadi kwath, Bilava churn, Bilva-majjadi yog, Dasamoolarist, Tentex forte, Lukol, Diarex, Sumento (Tablets).
Azadirachta indica A.H.L Juss/Neem/Meliaceae	Leaves, wood, seeds, flowers	Up to 1000 msl	Skin disease, ulcer, eczema, diabetes	Livosyp, Hepatogard, Pilex
Bauhinia variegata L. Kanchnar/ Caesalpiniaceae	Stem bark	Up to 1200 msl	Cough and tumour	Kanchanar gugulu, Kanchnar kwath, Ushirasava, Chandanasava
Bombax malabaricum DC/Semal/Bombaceae	Bark, fruits and seeds	Up to 1200 msl	Demulcent tonic, stimulant, expectorant, diuretic	Femiplex, M2-tone (tablets and syrup).)
Butea monosperma (Lamk.) */Tabu Plash, DhakFabaceae	Gum, seeds and flowers	Submontan tracth.	Diarrhoea, dysentery	Krmighna churn, Krimimudgar ras, Krmikutharras, Plashbeejj churn. Plash ghrit, Crush Lukol (ablets and syrup)
Cassia fistula L. Amaltas/Caesalpiniaceae	Fruits and roots	Upto 1400 msl	Fever, heart disease, urinary trouble	Aragvadhadirsta, Aragvadhadi ghrita Aragvadhadi tel, Agni- kumar churn, Kankayan gutika.
Cedrus deodara (Royale ex.D.Don)/Devdara/ Pinaceae	Wood oil and bark	2000-3000 msl	Diarrhoea, dysentery, bilious fever, ulcer, skin disease	Devdaryadi kwath, Mahamash tel, Khadirarista Dasant lep, M_2 Tone
Cinnamomum tamala (Ham.) Nees ex.Eberm.*/ Tejpatta/Lauraceae	Leaves and bark	500-2200 msl	Stimulate, carminative, rheumatism, colic, diarrhoea, scorpion sting	Lodhrasava, Vasasava, Sudarshan churh, Chandraprabha vati, Yograj guggulu, Dasamoolarista, Aswagandharisata
Cinnamomum zeylanicum (Breyn)/Dalchini/Lauraceae	Bark	7000-2000 Feet	Carminative, rheumatism, colic, diarrhoea	Vomiteb

Contd...

Table 4.3–*Contd...*

Botanical Names/ Local Names/Families	*Parts Use*	*Distribution*	*Traditional Medicinal Use of the Plants*	*Ayurvedic Formulations*
Emblica officinalis Gaertn. Amla/Euphorbiaceae	Fruits	Up to 5000 feet	Stomach problem	Amlaki churn, Triphala chrn, Triphala ghrit, Brahmi amla tel, Fresh fruit pulp (Chavanprash)
Eugenia jambolana Lam/ Jamun/Myrtaceae	Bark and seed	Up to 900 msl	Diabetes	M_2-Tone
Ficus glomerata Roxb., Pl. Corom/Gular/Moraceae	Fruit, Bark and roots	Up to 1200 msl	Diabetes, piles, dysentery, smallpox, throat pain, skin disease, leprosy, urinary troubles, heel cracks	Femiplex
Ficus benghalensis (Linn) Bargad/Moraceae	Fruits, barks and leaves	Up to 900 msl	Rheumatism, dysentery, diabetes	M_2-Tone
Juglans regia L.*/ Akhrot/Juglandaceae	Leaves, stem bark and fruits	700-2500 msl	Eczema, scropula, anthelminitic	Amritpras ghrit, Dantasodhak, Yavih
Mangifera indicia L./ Aam/Acanthaceae	Fruits, seeds and Bark	Up to 1000 msl	Jaundice, rheumatism, diarrhoea, scorpion sting, uterus, lungs, intestinal disease, asthma, check nasal bleeding	M_2-tone
Mallotus philippensis (Lamk.)/Kmbhal/ Euphorbiaceae	Ripe fruits	Up to 1200 msl,	Tapeworm killer, skin disease, snakebite, ulcer, constipation, purgative, external parasites	Krimikuthar ras, Krimighatini vati, Kshisadi ghrit, Herbinol cream
Melia azedarach L. Daikan/Meliaceae	Leaves barks and fruits	Up to 1400 msl	Anthelmintic, diuretic, leprosy, rheumatism, skin disease, cuts, wounds	Pigmento
Myrica esculenta Ham. ex.D.Dun.*/Kaphal/ Myricaceae	Barks and fruits	Up to 900-2100 msl	Fever, asthma, diarrhoea, dysentery,	Katphaladi churn, Purhyanug churn, Khadiradi gutika, Irimedadi tel, Brihatphal ghrit.

Contd...

Table 4.3–*Contd...*

Botanical Names/ Local Names/Families	*Parts Use*	*Distribution*	*Traditional Medicinal Use of the Plants*	*Ayurvedic Formulations*
Oroxylum indicum (L.) Vent*/Tantia/Bignoniaceae	Root bark	Up to 1500 msl	Diaphoretic, stomachic, anodyne, dysentery, diarrhoea	Dasomoolarista, Dasamooll kwath, Pushyanug churn, Amritarista, Mahanarayan tel Chayavanpras etc.
Pistacia khinjuk Socks./ Kakarsingi/Anacardiaceae	Leaves, petiole and branches	Up to 2500 msl	Asthma, cough, dysentery	Shringyandi churn, Dashmularista, Kantakari Avaleha, Chyavanpras Nokufcough, Kasni.
Sapindus mukorossi Gaertn*/Ritha/Sapindaceae	Fruits	Up to 1800 msl	Snakebite, hair tonic, fever, epilepsy, chlorosis	Maha laxadi tel
Shorea robusta Roxb. ex./Sal/Dipterocarpaceae	Leaves and bark	Up to 1000 msl	Dysentery, diarrhoea, ulcer, bacterial infection	Arshonyte forte.
Taxus baccata auct. (Non L.)*/Thuner/ Taxaceae	Leaves and bark	Up to 2400 msl	Sedative antiseptic, asthma, chronic genitor, urinary infection liquors	Talisadi churn, Lavanbhaskar churn, Sudarshan churn, Kanksava
Terminalia bellirica (Gaerth.) Roxb.*/Bahera/ Combretaceae	Fruits	Up to 1200 msl	Stomach problem, fever, cold, heart disease, dropsy, diabetes, gastric problem, liver problem, neural pain, piles, urinary disorder	Triphala churn, Triphala kwath, Sanjivani vati, Kutajavaeleha, Herbotone Hepatogard
Terminalia chebula (Gertn.) Retx/Harrd/ Combretaceae	Fruits	Up to 1600 msl	Stomach problem, cough, cold, fever, constipation, dysentery, spleen problems, and ulceration	Haritaki churn, Triphala churn, Agastgy haritaki aveleha, Abhyarista, Triphala ghrti, Chitrak, Haritaki, M_2 tone, Hepatogard, Ojus, Mentat, Liv-52, Herbolax Pilex, Menosan, Gerforte, Abana (tablets and syrup).
Terminalia arjuna (Roxb) ex. DC/Arjuna/ Combretaceae	Bark	Cultivated in Dehradun	Jaundice, sores, dysentery, pneumonia, diabetes, hair tonic	Sumenta, Mental, Menosan, Geriforte, Reosta, Liv-52 Mentat.

Table 4.4: The climbers used in Ayurvedic formulation, their availability and status in District Garhwal

Botanical Names/ Local Names/Families	*Parts Use*	*Distribution*	*Traditional Medicinal Use of the Plants*	*Ayurvedic Formulations*
Abrus precatorious L.*/ Ratti/Fabaceae	Seed and roots	Up to 1000 msl	Skin disease, ulcer, nervous disease	Tranquil
Mucun pruriens (L.) DC.*/ Kauch/Fabaceae	Roots and hairs on pods	Up to 1300 msl	Nervine tonic, cholera, eczema, diabetes, eye disease	Kawanch pak, Kawanch churn, Kamdev ghrit, Ashwagandha churn, Musli pak, Dhatupaustic churn, Agstya haritaka, Badam pak, Strenex, Tentax forte, Speman tab.
Rubia cordifolia L./ Majethi/Rubiaceae	Roots	Up to 2500 msl	Paralysis, jaundice, fever, menstrual skin disease	Mahamanjisthadyarista, Mahamanjisthadi ark, Manjisthadi tel, Mahanarayan tel, Ashwagandharista.
Tinspora cordifolia (Willd.) Hook Fr*/ Giloy/Menispermaceae	Stem	Up to 1600 msl	Debility, leprosy, urinary trouble, malaria	Giloy satva, Amritasava, Amritarista, Ashwaganda churn, Chandra, prabha Vati, Guduchi tel, Livonyn, Calcury, Pigmento Livosyp Dibecon, Mentat,Rumalaga Gasex, Geriforte, Liv- 52. Diakof, Manoll, Apimare
Vitis vinifera L./ Angoor/Vitaceae	Whole plant	Cultivated plant	Urinary disorder	Ojus

* Threatened medicinal plants of Uttarakhand (Pandey *et al.*, 2006).

verticillatum, *Rheum australe*, *Swertia chirayita*, *Picrorhiza kurrooa*, *Thalictrum foliolosum*, *Viola pilosa*, and *Valeriana jatamansi* etc. are listed threatened medicinal plants of Uttarakhand (Pandey *et al.*, 2006), are the main ingredients of the certain Ayurvedic formulation. Therefore, it is worthwhile to note that special attention is utmost necessary to conserve it for sustainable utilization.

Conclusion

Present study clearly shows that out of total plant being used, about 30 per cent plants belong to high altitude and rest 70 per cent from valley region and plains out of which 40 are the herbs, 20 shrubs, 29 trees and 5 climbers. The production, processing and use of medicinal plants have great potential for employment generation particularly at village level. Most of the herbal plants used for Ayurvedic medicine are either native to Uttarakhand or can be successfully introduced in the state. However, to preserve the biodiversity, the promotion of herbal exploitation will have to be paralleled with the systematic introduction of herbal cultivation at large scale, which will avoid over indiscriminate exploitation and ensure constant and authentic supply of Ayurvedic drug preparation industry will decidedly guide the herbal trade and related activities in the State.

Acknowledgements

This work is a part of University Grant commission, New Delhi funded project. The authors thankfully acknowledge the financial assistance of the agency.

References

Chauhan N S 1996. Business potential of medicinal and aromatic plants. *Sci. Tech: Enttp*. **2:** 34-38.

Chauhan N S 2003. Important medicinal and aromatic plants of Himachal Pradesh *Indian Forester* 129.

Fransworth N R and Soejarto D D 1991. Global importance of medicinal plants *In:* Akeral D, Heywood V and Synoge H (eds) Conservation of Medicinal Plants Cambridge University Press. Cambridge, UK.

Gaur R D 1999. Flora of the District Garhwal, North-West Himalaya (With Ethonobotanical Notes) *Transmedia* Srinagar Garhwal, Uttarakhand.

Naithani B D 1984-1985. Flora of Chamoli, BSI Howrah.

Osmoston A E 1927. A forest flora of Kumaon, Government Press, Allahabad.

Medicinal Plants: Aspects and Prospects (2014) ***Pages* 65–80**
***Editors:* Mukesh Kumar, Anjali Khare and C.P. Shukla**
ISBN: 978-81-7622-309-6
***Published by:* BIOTECH BOOKS, NEW DELHI**

Chapter 5

Manual for the Propagation and Cultivation of Medicinal Plants at Rohilkhand Region (U.P.)

***Beena Kumari*[1] *and Mukesh Kumar*[2]**

[1]*Department of Botany, Hindu College, Moradabad – 244 001*
[2]*Department of Botany, Sahu Jain P.G. College, Najibabad – 246 763*

ABSTRACT

Man depends on plants for his existence and this relationship must be sustainable. Herbs have saved man's life on innumerable occasions and for this reason their over-use concerns everyone.In U.P. medicinal plants form the raw materials for new and emerging industries especially the manufacturers of 'bitters' or alcohol-based medicines. The plants required for these industries are mostly harvested from the wild. With this increasing harvesting pressure, the plant supplies will, in time, dwindle and eventually become extinct. To ensure that medicinal plants are always available, we must learn to grow our own.

This manual has been prepared to help the gardening man and woman to propagate and grow these plants. It provides helpful and practical advice on how to propagate and cultivate a wide range of plants and how to maximise success from those efforts.

Keywords: Medicinal plants, Propagation.

Introduction

Medicinal plants have always been a part of man's life on earth and there is a close relationship between plants and human beings. Man depends on plants for his existence and this relationship must be sustainable. Herbs have saved man's life on innumerable occasions and for this reason their over-use concerns everyone. It is important to note that medicinal plants in U.P. remain wild and undomesticated. In U.P. medicinal plants form the raw materials for new and emerging industries especially the manufacturers of 'bitters' or alcohol-based medicines. The plants required for these industries are mostly harvested from the wild. With this increasing harvesting pressure, the plant supplies will, in time, dwindle and eventually become extinct. To ensure that medicinal plants are always available we must learn to grow our own. Therefore, this manual has been written to help the gardening man and woman to propagate and grow these plants. It is true that most plants can be propagated from seeds but the success of this is dependent on factors such as the type of soil, the rainfall and pests. This manual concentrates on seed propagation and stem and root cuttings. It looks at the collection of seedlings, vines, suckers, rhizomes and corms. I hope that this manual will be of some use to anyone who is considering cultivating his or her own medicinal plants. It provides helpful and practical advice on how to propagate and cultivate a wide range of plants and how to maximise success from those efforts.

Things to consider when growing your medicinal plants:

1. Chose suitable sources for your plants
2. Plan the nursery well
3. Avoid unnecessary stress on the plants that you are propagating
4. Prepare the planting ground well
5. Look after newly planted plants until they are well established.

Setting Up your Nursery

Site Selection for your Nursery

The site should be flat and situated near a permanent source of water and must not be too far from the house. The area needs some natural shade but should not be overshadowed with dense trees. Ensure that you have rights to use the land for some time into the future.

Establish Stock Beds

The purpose of a stock bed is to provide material for you to propagate plants from. You may want to collect seed from these plants or use young new shoots to take cuttings (Figure 5.1). Your stock plants should be the kind of plants that you have selected as both useful and manageable within the compound or farm. Grow your stock plants in a place that is easy to get to but does not take up too much space on your land. Keep these plants well-pruned and free from pests and disease. Water them in the dry season to encourage the growth of new add mulch and manure at the beginning of the growing season. You may want to consider inter-planting with other plants to provide natural shading or with nitrogen fixing plants to enrich the soil. You may choose to keep your stock plants in pots if for example, you are likely to move from the land that you are currently on. The biggest danger with stock plants in pots is that they may dry out. To address this problem, water regularly and check the pots periodically to see if the roots are growing out of the base of the pot. A plant that

Figure 5.1: Stock plants provide fresh plant material

Figure 5.2: Young shoots from a tree stump

dries out too often may need re-potting. Alternatively, you can cut the roots back when they start to grow through the base of the pot and keep the plant well-pruned.

Stump Coppicing

You may have the opportunity to make use of a tree that has been felled nearby. The young shoots that grow from the stump may provide you with useful cutting material for a number of years to come. It is important that you leave some of the shoot on the stump so that the remaining young leaves can keep the stump alive (Figure 5.2). Try to keep the stump partially shaded and prune it to ensure that no large shoots start to form.

Seedling Beds

Make a bed 120cm wide and as long as is convenient for your nursery. The bed should be raised 12cm high to ensure that heavy rain drains away at the sides of the beds but does not wash the seedlings away. It is important that the bed is not wider than twice the length of your arm, allowing you to sow and weed without standing on the soil. A path should also be left between the beds to allow easy access. The soil in the bed should be finely graded, without stones or debris that will make sowing and subsequent germination uneven. The beds should be in level to avoid the seeds washing down into gullies and dips on the surface.

Direct Sticking Beds

These are beds that are prepared for hardwood cuttings and leafy cuttings. Like the seedling beds, these need to be twice the length of your arm in width so that you can reach to the middle of the bed from both sides and can be as long need or have space for. The cuttings, once prepared are inserted directly into the beds, in rows and labelled.

Standing Down Area

When the seeds have developed their first set of true leaves or the cuttings have roots and shows signs of leafy new growth, they are ready for potting-up or planting-out. The standing down area is an area of flat, shaded ground, where you place the newly potted-up plants or plant the new plants temporarily to encourage them to grow larger. This gives them the additional space they need to start growing. The plants stay in this area until they are ready to be planted in a permanent site.

Shading

The intense heat of the sun may cause over-heating in your seedlings or cuttings. If there is not enough light the plant will not be able to manufacture food for new roots and the weakened plant will be vulnerable to disease and fungal attack. Young seedlings should be protected from direct sunlight using palm fronds. Palm fronds will also protect the tender young plants from heavy droplets of rain that may wash-away the seedlings or disturb the young roots of the cuttings (Figure 5.3).

Figure 5.3: Protection of young plants from rain and sun

Water

The best way to keep the cuttings from drying-out is to:

1. Shade the propagating area
2. Trim the leaves

The most vulnerable times for the young plants are at the time of seed germination, before the roots have developed in the new cuttings and immediately after you have transplanted them from the seed or direct sticking bed.

Compost

You will achieve the best results if you provide a good rooting medium for the cuttings. Ordinary soil is not ideal because it does not have the properties required for cutting compost. These include high aeration, good moisture retention and drainage. You can also ensure that your own compost is free of pests by sterilising it. Heat the compost in tin drums over a fire for a few minutes. This will kill all insects and their eggs. The ingredients for a good compost mix are sharp sand, grit or gravel and old weathered sawdust or coconut palm husks. Ideally use one part of each in your mix (Figure 5.4).

Figure 5.4: Making your own compost

Propagation Unit

The propagation unit is used to create an atmosphere of high humidity for leafy cuttings (Figure 5.5). This prevents wilting and promotes root growth. The unit can be of any size suitable for the volume of plants that you hope to propagate. The basic materials required are timber, nails, timber preservative and polythene.The unit is wrapped in transparent polythene allowing the light in, but restricting the evaporation. Note that the propagation unit is well drained with a thick layer of sand beneath to allow the water to flow out.

Beware of allowing the heat in the propagator to rise too high. The temperature should not rise above 28-33°C. There is a removable lid that allows adjustment hot days when ventilation is needed to keep the unit cool.When the roots have formed the new plant is transplanted to a fresh site on the nursery where it is grown on until it is strong enough to be planted in its final position where it can grow independently of the nursery.

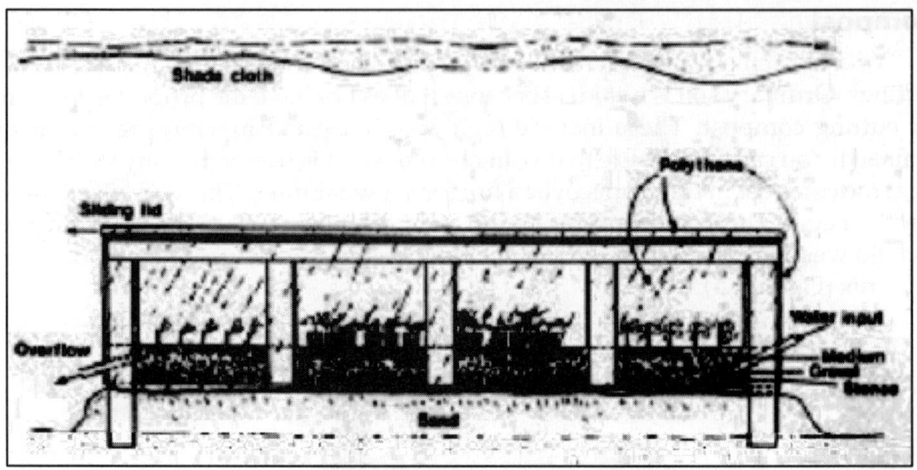

Figure 5.5: Propagation unit

What Causes Stress and Failure in your Plants?

Water Stress

- ☆ Too little water
- ☆ Too much water
- ☆ Too dry an atmosphere on young cuttings
- ☆ Too little root system to support the leaves

Light Levels

- ☆ Too sudden removal of shade
- ☆ Keeping the plants in dim light for too long

Temperature Stress

- ☆ Too hot or too cold

Nutrient Stress

- ☆ Lack of main elements
- ☆ Lack of trace elements

Others

- ☆ Wind, torrential rain, pest and diseases

Nursery Management

Arrangement of Seedlings

Black polythene bags measuring 15cm to 18cm wide and 19.5-20cm high should be filled with nursery compost or topsoil enriched with well-decomposed manure at a ratio of 3:1. The bags should be arranged in rows, leaving lanes of about 45-60cm to allow for watering, weeding and chemical spraying. For direct seed sowing or the direct sticking of cuttings, the beds should be raised and the soil should be shifted to remove large stones and other debris. Firm down the soil and rake the surface before sowing. Ensure that you sow the seeds in neat rows. This will help you to distinguish between the emerging weeds and the plants you want to hand weeding. Water with a spray of droplets after sowing.

Watering

It is important that plants are watered regularly. Periods of dryness weaken the plant and make it susceptible to disease. Regular watering encourages strong root growth and vigorous shoot growth. The seed bed should be watered twice every other day until seedlings emerge. Leafy cuttings require high humidity, as they are unable to draw water up through the soil until their roots have begun to grow. When growing rows of plants be careful not to miss watering those growing on the ends or corners of beds. Potted plants require even moisture throughout the compost and inadequate watering will leave the base of the pot dry. The consequence of poor watering is that the roots will remain at the surface and not reach down to moisture stored lower down in the soil. Transplanting and potting-on will be less successful as a result of poor root formation and they will not be able to withstand periods of drought.

Pest and Diseases

Hygiene on the nursery is important. Sweep away any leaf debris that is on the ground as this will harbour pests. When the nursery is temporarily empty, introduce fowls to the nursery for a few hours. They will eat any small grubs and insect eggs that may infest your plants later on. Do not leave plants in standing water where the roots will become water logged. This will encourage rotting and weaken your plants making them more susceptible to fungal attack.

Figure 5.6: Ventilate your propagation unit

Ventilation

Regularly prop open the lid of the propagation unit with a piece of wood to avoid a build-up of heat. This is particularly important in the day time when, without a breeze, the sun heat will be captured in the unit and kill the young cuttings (Figure 5.6).

Propagation Methods

By Seed

Seeds produce many plants at once and they can be collected and transported very easily. The main problems that you may encounter with seed propagation are failure to germinate and loss through transplanting. Both of these problems can be reduced with careful nursery stock management. Many tropical seeds are best sown fresh and do not survive if they are allowed to dry-out. Always check that your seed is free from pests and disease. Even if an infected seed does germinate it is unlikely to thrive.

Failure to germinate is often a result of

1. A natural chemical inhibitor in the seed or
2. The seed having a physical inhibitor *i.e.* a hard seed coat etc.

These inhibitors are designed to protect the seed from germinating in an unsuitable environment.

To succeed with these kinds of seeds you may need to experiment to find out what kind of inhibitor the seed has, sometimes they have both. The main method of breaking down the chemical inhibitor is by soaking the seeds in water. This can wash away a chemical inhibitor. A physical inhibitor requires scarifying. This is the use of abrasion to break through the hard outer coat of the seed without damaging the seed itself. Hard-coated seeds may also respond to overnight soaking in water. The prepared seeds should be distributed uniformly over the seedbed. After sowing rake the soil over lightly to ensure good seed-soil contact. If the seeds are very large you may choose to plant them individually. The general rule is to plant a seed at a depth equal to the size of the seed itself. The seeded areas are watered carefully. Avoid water-logging and the displacement of seeds with careless watering. When planting seeds into polythene bags the soil in the bags should be moist. The seed should be planted at a depth of 2cm or so. If in doubt about which way upto plant the seed place the seed flat in the soil at a depth that is the same as the thickness of the seed. Alternatively, seeds can be planted by digging a hole and dropping two seeds in with the pointed ends upwards. The holes are then covered up with soil. At the four-leaf stage, remove one seedling, leaving the stronger to grow.

Vegetative Propagation

This is when you take a vegetative part of the plant, stem or piece of root and sow new plants from it directly *i.e.* not seed. By rooting cuttings on your nursery you do not need to rely on the parent plant producing seed and you can overcome of the problems that you may have with trying to get seeds to germinate successfully (Figure 5.7). Vegetative cuttings mean that you can select the exact plant that is best for your needs *i.e.* the individual plant that you know to be particularly potent and valuable for medicine rather than leaving this to chance with a random selection of seedlings.

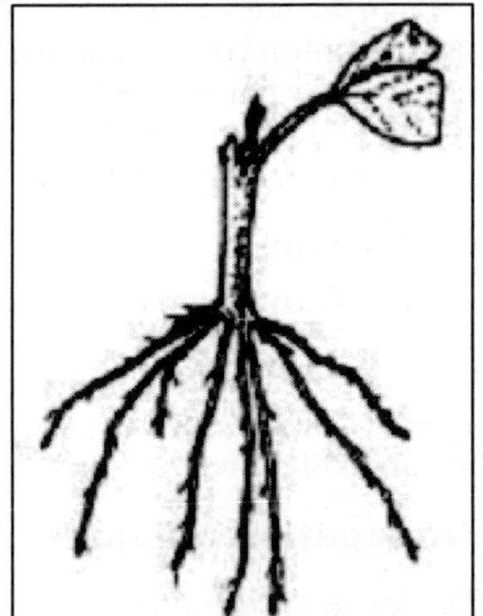

Figure 5.7: Rooted cutting

Stem Cuttings

Commonly, the vegetative part of the plant used for propagation is the stem. Collect small branches of vigorous growth, which shows no sign of pest or disease and makes every effort to prevent this from wilting. When you collect your cutting materials avoid the heat of the day as this is when the plant is losing a lot of its moisture. Do not keep the material in a polythene bag as it will overheat very easily. Once in the nursery, find a shaded area and prepare your cuttings. You may like to have a bucket of cold water standing by to receive the trimmed cuttings. Take a moderately vigorous shoot and cut to a length of between 2.5-12cm long. The length of stem you required depends on the length of internode (the space between the leaves). A node is the point at which the leaf meets the stem and a cutting is measured by the number of nodes it has. For example, a single node cutting is suitable from plants where the internode space is no less than 1.5cm. In this case you cut just above the node and insert that end into the compost. Two-node cuttings are taken from plants where there is a shorter distance between the nodes. Cut just under a node, as this is a good place for roots to be formed. Many noded cuttings are for species that have very short internodes and lots of small leaves. These are best taken at 5-15cm long. If the stem is leafy this is usually known as a softwood cutting. Most tropical trees can be propagated from softwood cuttings. Using a clean knife, remove all but the top two leaves of the cutting. Trim each leaf down by half. It is preferable to dip the stem base into a hormone solution or powder to encourage rooting. However, this is not absolutely necessary. Insert two thirds of the stem into the prepared cuttings bed or a polythene bag of compost and gently press the compost down to firm it in. Water and keep shaded. To avoid drying breezes make sure that the cuttings are sheltered by palm leaves or by some other kind of structure. It is important that the new cuttings are kept cool and do not dry out. The leaves of the cutting should never wilt. The best way to stop the leafy soft wood cuttings drying-out is to build a small propagator. This can be as simple as a frame covered in polythene. The propagator can be as large or as small as you require. The purpose of the propagator is to maintain high air humidity. If you take a cutting from a stem of a plant that is leafless this is usually known as a hardwood cutting. If there are leaves still on the cutting these can be cut off using a sharp knife leaving a very few leaves at the tip. The wood is usually firm and there is less danger of the cutting drying out. These cuttings are usually much larger than softwood cuttings (as much as 1-2 meters) and can usually be planted directly into the ground but will take longer to root. The cuttings should have an inactive bud as it is better for the new shoot growth to begin after rooting.

Root Cuttings

These can be cut from the main plant without destroying it. Cut a piece of stem as long and as thick as your index finger. Do not use any growth hormone, as this will stop the root growing. Place the root cutting horizontally in a polythene bagfull of compost or directly into your nursery beds. Cover with a light layer of soil and firm down. Watering is done as and when necessary. Roots develop between 4-8 weeks after planting.

Tubers

Take these cuttings when the plant is not actively growing *i.e.* when it is dormant. Cut a small piece of the tuber, which should include a bud. Do not use any growth hormone. To allow you to harvest some of the plants tubers later without destroying the whole plant, it will be beneficial to plant this cutting into a raised bed or individual mound, as for yam.

Suckers

Suckers are the shoots that are produced from the roots. Treat these as either unrooted cuttings or separate them from the main plant keeping a piece of root attached. Avoid these suckers drying out by preparing the planting ground in advance and water them as soon as you have planted them.

Offsets

These are the little clumps of new shoots that form clumps or sets of buds at the base of some species *e.g.* banana and plantain. Separate these from the main plant and divide them into individual plants or small clumps. Do not allow these to dry out or be left in the sun. Plant these clumps into the nursery beds or into their final planting position and water them well (Figure 5.8).

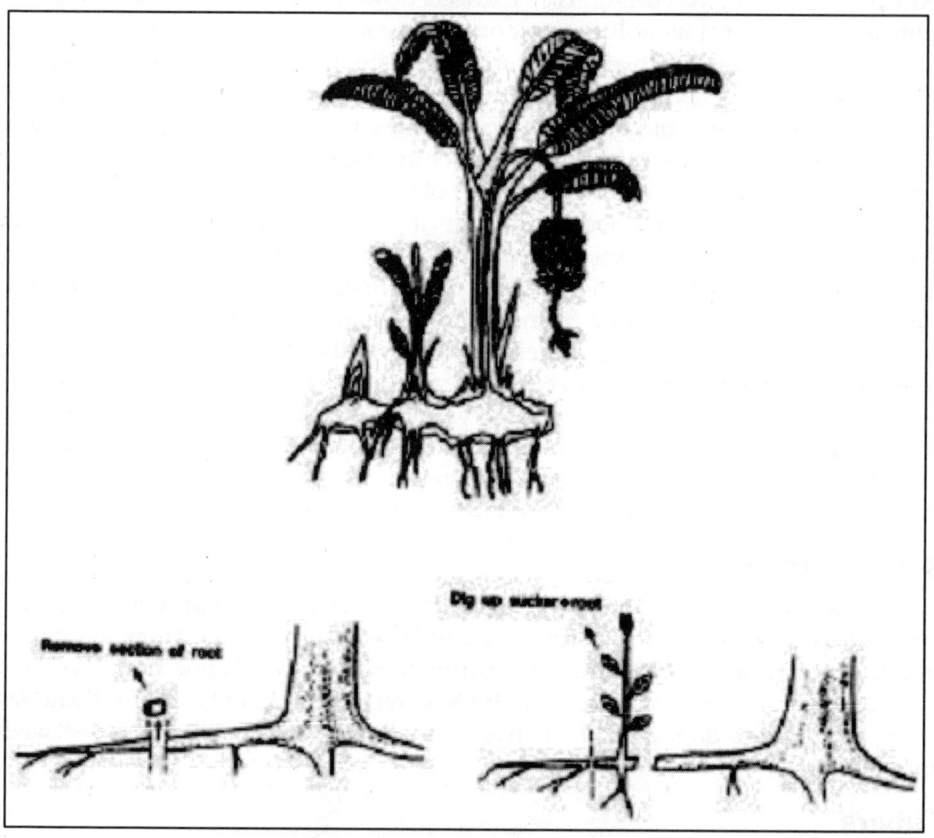

Figure 5.8: Offsets and suckers

Undercutting Seed Beds

Drawing the blade of a machete at ground level, under a raised bed of young plants, is known as 'undercutting'. This is a useful technique when growing young plants in the nursery to help them to develop a very fibrous root system.This results in good plant establishment when they are transplanted to their final planting site.

Undercutting promotes vigorous fibrous roots and prevents the roots from penetrating too far down into the nursery bed (Figure 5.9). Undercutting can only be done on a raised bed that is twice the width of a machete.

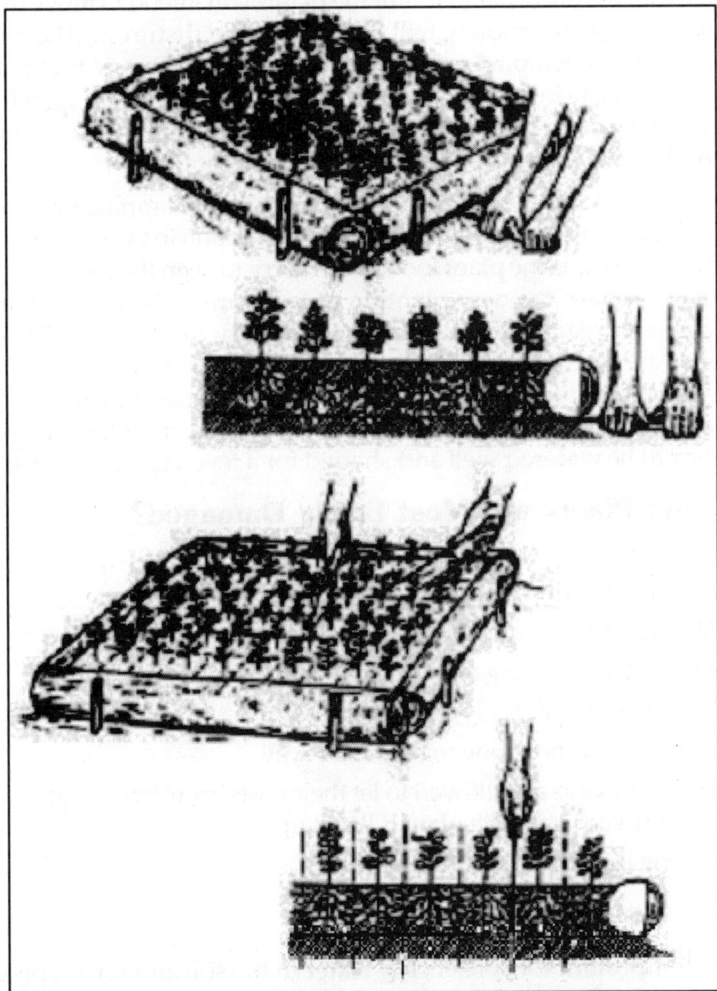

Figure 5.9: Undercutting promotes good root growth

Planting and Aftercare

Transplanting Seedlings or 'Potting on'

Potting-on is what you do when your seeds or cuttings have established an adequate rooting system to sustain them and you need to move them into a bigger pot or to a larger site where they can grow on for a little while longer. It is the point at which they have developed roots and are ready to start growing into young plants. The correct time for potting-on seedlings is when the first set of true leaves have grown. The correct time for cuttings is when the roots are visible through the holes in the bottom of the pot or, if they are directly planted into beds, when they show signs of independent growth in the form of a new flush of leaves. Lifting these delicate young plants and putting them into new pots or a new piece of ground is a most stressful time for the young plant. Water the plants you intend to move the day before the operation. Handle the rooting ball gently, without disturbing the roots and ease the plants out of their compost rather than pulling them. Ensure that you loosen the soil around the plant using a stick or other small utensil before lifting it from the bed. Hold the young seedling by the first two seed leaves not by the stem or roots. If you bruise the neck of the soft stem, the plant is likely to root.

If in pots, gently loosen the root ball in the pot before tapping the whole root ball into your upturned hand. Always carry out his operation in the cool of the day and in the shade. Do not handle the plant too often and try to keep the roots cool throughout the operation. Prepare the new planting area before initiating the transplanting. Have the compost ready mixed or the bed already dug-over and raked level in expectation of the transplanting. Water the newly transplanted plants thoroughly to ensure that the roots are in close contact with the soil around them. The plants must not be allowed to dry out or the roots be unduly damaged or broken. All newly potted-on plants should be watered well and shaded for a few days after potting-on.

When Young Plants are Most Easily Damaged?

- ✰ When seeds are just germinating
- ✰ When leafy cuttings are being handled
- ✰ When plants have just been potted up
- ✰ If unsuitable potting compost has been used
- ✰ If conditions suddenly become stressful
- ✰ If watering is not done regularly or well
- ✰ If potted plants are allowed to let their roots grow into the ground (these are then broken when the plant it lifted up)
- ✰ During the first week after planting

Mulch

Mulch is important for conserving water in the soil and for keeping roots cool and weed free. Mulch can be made from old coconut husks, chopped bark or other organic material to a depth of approximately 6cm (Figure 5.10). Always water the newly transplanted plants well before putting on mulch.

Propagation Guide to Selected Medicinal Plants

1. *Name*: Adrak, Ginger (common name), (***Zingiber officinale*** **Rosc.**)

   Habitat: Cultivated

   Distribution: Originating in Asia, now widely cultivated in tropical zones.

   Description: This plant has rhizomes and two kinds of aerial erect stems; one non flowering which grows to 1-1.5m high with long pointed leaves,

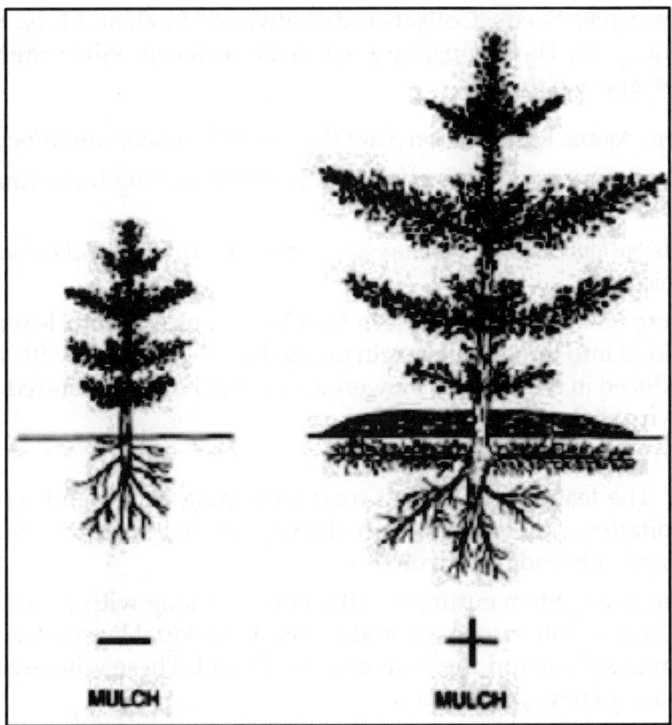

Figure 5.10: Mulching promotes root growth

the others, has flowers and does not exceed 20cm high. Oval shaped flower petal with yellow-greenish centre.

Part used: Rhizome and rhizome-bark

Use: The rhizome is used for many things including the treatment of coughs, haemorrhoids, boils and against whooping cough.

Propagation: Rhizomes: Select and cut rhizomes with buds. Plant directly into beds. Water regularly both morning and evening.

2. *Name*: Neem (English) (***Azadirachta indica* A. Juss.**)

   Habitat: Cultivated

   Distribution: Native of India, widely spread in all tropical regions of the world.

   Description: Small tree up to 10m high, evergreen, leaves compound, uneven leaflets in pairs of 5-8. Flower heads in panicles of small flowers, white fruits, berry-like, yellow at maturity, fragrant.

   Part used: Leaves, root-bark, stem-bark, seed, bark.

   Use: The leaves are used to treat ringworm, boils, fever and hepatitis. The stem-bark is used to treat Malaria. Seeds and root-bark are used to treat intestinal helminthiasis and wounds. The bark is used to treat pharyngitis.

Propagation: Seeds: Collect seeds between March and July. Sow in drills 20cm apart. These must be sown fresh or fewer will germinate. Thin to 15cm apart in the row.

3. *Name*: Ajuba, Resurrection plant (English) (***Bryophyllum calycinum* Salisb.**)

   Habitat: Often found in compounds but also along footpaths, in thickets and in secondary forests.

   Distribution: Native of Malaysia, widely distributed and naturalised in the tropics.

   Description: A perennial herb 60-120cm high, smooth leaves simple or divided into three, leaf margin toothed, and sometimes with young leaves produced in the notches. Flowers are in bunches, greenish-red and hanging down, petals brown.

   Parts used: Leaves, stems and roots.

   Use: The leaves are used to treat whooping cough and whitlow, heart palpitations, ear ache and sore throat. Juice from the root is used to treat epileptic fits and heart troubles.

   Propagation: Stem cuttings: Cut stems 6cm long with 2-3 nodes on each. Insert into polythene bags and water. Individual leaves can be detached from the plant and the base inserted in soil. These will develop roots to produce a new young plant.

4. *Name*: Sausage tree (English) (***Kigelia africana***)

   Habitat: Species spread over all Rohilkhand region from the thick forests in Bareilly.

   Description: Tree up to 15 meters high, 50cm long leaflets in pairs of 3-6. Oblong toothed, smooth flower heads hanging down and up to 60cm in length. Individual flowers reddish purple and up to 5cm long and 15cm in diameter.

   Part used: Fruit

   Use: The fruits of this plant are commonly used as a base ingredient for many different medicines.

   Propagation: Seeds: Collect ripened fruits, remove the seeds and soak in water overnight. These can then be sown in drills on seedbeds. Water until seedlings emerge.

5. *Name*: Barbados nut (English) (***Jatropha curcas* Linn.**)

   Habitat: Frequently cultivated, traditionally used as fencing.

   Description: Shrub 3-4m high with thick smooth small branches, bark greenish, peeling, leaves smooth, plain oval shaped or with 5 lobes; leaf stem long, flowers yellowish green, fruit ellipsoid.

   Part used: Leaves, roots, whole plant and seed.

Use: The roots are used for jaundice and male sexual impotence. Leaves for wounds and convulsions and fever. Roots are used to treat male impotence. Seeds for the treatment of syphilis.

Propagation: Seeds: Seeds are black when ripe.

6. *Name*: (***Mormordica charantia*** **Linn.**)

   Habitat: Commonly found in abandoned cultivation's.

   Distribution: Pantropic species.

   Description: Herbaceous, climber with tendrils; leaves alternate, with lobes roughly toothed. Flowers yellow, spindly, up to 8cm long; fruit oblong with longitudinal ridges.

   Part used: Leaves, stem and roots.

   Use: Leaves are used for hypertension, chicken pox, cut wounds. The stem is used for abnormal pains, dysentery and yaws. Fruits for wound treatment and diabetes.

   Propagation: Seeds: Extract from pods and sun-dry before sowing in drills or in propagation boxes. Water until seedlings emerge.

7. *Name*: Ocimum (Tulsi), Basil (English), (***Ocimum gratissimum*** **Linn.**)

   Habitat: Found commonly around villages.

   Distribution: Pantropical species, probably native of Asia.

   Description: Scented shrub. Leaves opposite, elliptical, with toothed margins about 10cm long. Flowers hang in a long raceme in small groups of 4-6. They are white and the flower base remains after the petals have dropped off.

   Parts used: Leaves, roots and the whole plant.

   Use: Leaves used in the treatment of bacterial infections and diarrhoea. Roots are used with guava and plantain to treat snakebite.

   Propagation: Seeds: Sow in drills or propagation boxes and water until seedlings emerge.

References

Anon 2003. WHO Guidelines on good agricultural and collection practices for medicinal plants. World Health Organization, 20 Avenue Appia, 1211 Geneva 27, Switzerland.

Balick M J 1996. Annals of the Missouri botanical garden. Missouri Botanical Garden. **4:** 57-65.

Dokosi O B 1998. Herbs of Ghana.Council for Scientific and Industrial Research Ghana, Ghana Universities.

Dubey N K, Kumar R, Tripathi P 2004. Global promotion of herbal medicine: India's opportunity. *Curr Sci* **86:**37–41.

Harnischfeger G 2000. Proposed guidelines for commercial collection of medicinal plant material. *J Herbs Spices Med Plants* **7**: 43–50.

Jain S K 2001. Medicinal Plants. National Book Trust, New Delhi.

Leaman D 2002. Medicinal plants. Briefing notes on the impacts of domestication/ cultivation on conservation. Paper for the "Commercial captive propagation and wild species conservation" workshop, 7–9.12.2001, Jacksonville.

Medicinal Plants Cultivation, Conservation and Development Plan–A Status Paper, (May 2002). Forest Department, Uttarakhand.

National Medicinal Plants Board 2003. <http://www.balkanherbs.org/ .MedicinalPlantsand ExtractsNo8.pdf>.

National Medicinal Plants Board 2002. <http://www.nmpb.nic.in>.

Uniyal R C, Uniyal M R and Jain P 2000. Cultivation of medicinal plants in India. A reference book– New Delhi, India, TRAFFIC India and WWF India.(Source of All Figures) (http://www.fao.org/DOCREP/005/Y4586E/y4586e08.htm)

Uniyal M R 2003. Issues in Cultivation of Medicinal Plants.

(http://www.technopreneur.net/timeis/technology/SciechMay03/ cultivation.html)

Medicinal Plants: Aspects and Prospects (2014) ***Pages* 81–97**
***Editors:* Mukesh Kumar, Anjali Khare and C.P. Shukla**
***Published by:* BIOTECH BOOKS, NEW DELHI**

Chapter 6

Ethnobotanical Notes on Alpine Medicinal Herbs of the Kedarnath Range of the Garhwal Himalayas

***Vaibhav Sharma*[1], *B.L. Sharma*[2] *and K.A. Bharati*[3]**

[1] *Department of Botany, M.K.R. Government Degree College, Ghaziabad – 201 009, U.P.*

[2]*Department of Botany, N.R.E.C. College, Khurja – 203 131, U.P.*

[3]*Raw Materials Herbarium and Museum Delhi (RHMD), CSIR-NISCAIR, New Delhi – 110 012*

ABSTRACT

Indigenous methods of medical treatment using plants predominate in the remote villages of the Garhwal Himalayas. This study aims to document ethnobotanical information on plants used for the treatment of various ailments by inhabitants of that region. Information on the use of these plants was collected from March, 2007 to June, 2010 through semi-structured interviews with local people and traditional healers. This study resulted in the documentation of forty-eight plant species belonging to twenty-one families, which were being used to treat forty different ailments. The majority of the plants were used in the various treatment categories: health tonic (10 species), cough and cold (6 species), fever (6 species), wound (6 species), diarrhoea and dysentery (4 species), cuts and bruises (4 species), toothache (4 species), insecticide (3 species), dysmenorrhoea (3 species), and gastric complains (3 species). The need to document this traditional knowledge is urgent since, in many regions, both traditional knowledge and plant diversity are rapidly disappearing.

Keywords:* *Ethnobotany, Kedarnath range, Garhwal Himalayas, Alpine, Medicinal herbs.

Introduction

Indigenous knowledge of medicinal plants is gaining worldwide recognition. The World Health Organization has estimated that more than 80 per cent of the world's population in developing countries depends primarily on herbal medicine for basic healthcare needs (Vines 2004). The traditional use of plants and plant medicines has a long history in the Himalayas and its use is rapidly spreading due to the lack of side effects seen with traditional remedies the low-cost and availability of native plants and the unavailability of other health care sources for the poor (Acharya *et al.*, 2009).

The Garhwal region of the Western Himalayas is one of the most significant hot spots for endemic species diversity and genetic diversity in all of India (Khoshoo 1991, Rau 1975), sharing over 50 per cent of the economically or medically useful vegetation on the Indian subcontinent. The Kedarnath region, the area under study and a verdant jewel of the Garhwal Himalayas, is located in the Rudraprayag district of the State of Uttarakhand. The whole region has been named after the holy Kedarnath shrine, a temple to the Lord Shiva, situated at approximately 33°44'N latitude, 79°62'E longitude, at an elevation of 3581 above sea level in the Kedar valley. In the Kedarnath range, the alpine zone starts at approximately 3200m and ends at 4600 m ASL or higher. The major portion of alpine zone between the tree line and snow line is covered by lush green alpine meadows, commonly known as "Bugyals" in the local language. The *bugyals* are true representatives of alpine vegetation and are chiefly herbaceous in composition. There is an amazing diversity of plants in this alpine ecosystem which serve as life sustaining *Jadibuties* (medicinal herbs) for the hill-dwelling people.

Thomas Hardwicke, a European, was the first person to collect plants from the Himalayas during his visit in 1796 as part of a political mission with the King of Garhwal at Shrinagar in the Alakhnanda Valley (Duthie 1906). The actual work of plant exploration in the western Himalayas and particularly in the Garhwal region began after the establishment of the Botanic Garden in Saharanpur by Hastings in 1820 (Rau 1963 a). Hardwicke was followed by Royle (1833-40), Duthie (1882-1906), Richard Strachey (1906), Babu (1977), and many others. Members of the Kamet expedition of 1931 also published a few historic reports on alpine vegetation. Rau (1963, a, b, 1964, 1968, 1975, 1981), Naithani (1969-84), Raizada and Saxena (1978), Semwal and Gaur (1986), Gaur (1983-99), Uniyal *et al.* (1989) have also made major contributions on alpine vegetation.

Ethnobotanical studies have been done on certain Himalayan plant groups or individual plant species including the studies by Mehra (1967) on Sesame, Shah (1977) on *Acorus calamus*, Chandler *etal.* (1982) on *Achillea millefolium*, Merlin (1984) on Poppy, Bist and Badoni (1990) on family Arecaceae in folk life within Garhwal Himalayas, Gaur and Nautiyal (1993) on fiber yielding plants of Garhwal, and Negi and Gaur (1994) on wild edible plants of Western Himalayas. The work done on ethnomedicinal plants of the Garhwal Himalayas by De (1962), Nautiyal (1981), Gaur *et al.* (1995), Negi *et al.* (1987-88) are of immense importance in this field. But, so far, there has not been a holistic exploration of alpine medicinal plants. With this in mind, an attempt has been made to explore and document traditional knowledge on alpine medicinal plants of the Kedarnath range.

Methodology

The following information was obtained through personal interaction with community elders, forest dwellers and traditional healers and then gathered on datasheets with details such as local names for the plants, plant part(s) used, diseases treated, methods of preparation, methods of administration, doses and duration of treatment. The trekking route has been furnished in Figure 6.1.

The data presented is based on interviews conducted from March 2007 to June 2010 with eighty informants (65 per cent men and 35 per cent women) who are natives of the studied locality. The methods adopted for obtaining information from the locals were the same as those of Jain (1995), Martin (1995) and Pujadas *et al.* (2004). Conversations were held in Garhwali, the local language. The prevalent diseases, their diagnostic details, medicinal plant species and other raw materials used in the treatments were documented. This data has been verified with other informants within the communities. The specimens were dried and preserved in herbaria (Singh and Subramanium 2008). The plant specimens were identified with the help of relevant floras (Hooker 1872-1897, Gupta 1994, Polunin *et al.*, 1984 and Naithani 1984) and with the help of Professor R. D. Gaur, retired Head, Department of Botany, H.N.B.University, Garhwal and Professor Som Deva, retired Head, Department of Botany, D.A.V.College Dehradun. Plant identifications were also compared and confirmed using herbarium specimens from the Forest Research Institute, Dehradun and Botanical Survey of India, Northern Circle, Dehradun. The herbarium specimens were then deposited at the herbaria of the Department of Botany, N.R.E.C. College, Khurja-203131, Bulandshahar, Uttar Pradesh, India.

Enumeration

Plants are listed alphabetically in Table 6.1 with respect to their botanical names, families, local names, flowering periods, distribution and ethnomedicinal uses.

Discussion

A total of forty-eight species were documented, which belong to twenty-one families used to treat approximately forty different ailments (see Table 6.2). Out of the forty-eight plants, the majority were herbs (89 per cent) and shrubs (9 per cent) because the harsh alpine climate does not permit trees, climbers and large shrubs to grow. The largest representation was of the Asteraceae family with eight species present, followed by Ranunculaceae with seven species, Ericaceae with five species, and so on. This finding is probably due to the fact that the Asteraceae family is cosmopolitan whereas Ranunculaceae and Ericaceae are characteristic families of the temperate and boreal regions of the Northern Hemisphere.

The most widely used remedies are derived from roots or rhizomes (62 per cent) followed by leaves, (20 per cent), whole plants (9 per cent), flowers (5.4 per cent) and seeds (3.6 per cent). The use of roots or rhizomes indicates that this part of the plants may have strong medicinal properties, further indicated by the fact that roots can

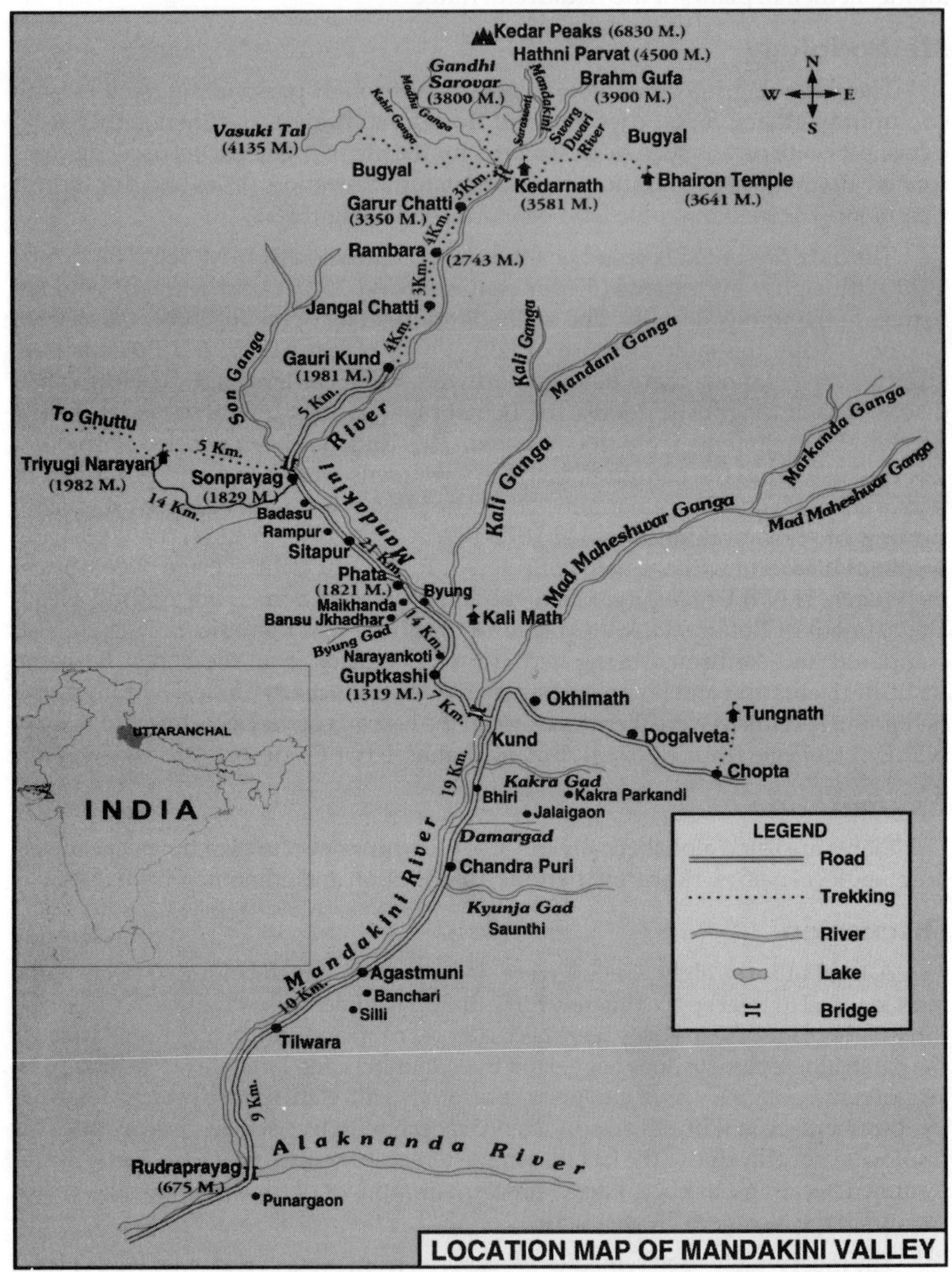

Figure 6.1: Geographical location and trekking route of the study area

Table 6.1: Enumeration of medicinal plants with ethnomedicinal uses

Sl.No.	Botanical Names and Families (in parenthesis)	Local Names	Flowering Periods	Altitude (Meters)	Ethnomedicinal Uses
1.	*Achillea millefolium* Linn. (Asteraceae)	Gandrain	May-October	2800-3300	The leaves and flowering shoots are used for gastric and digestive disorders, as a homeostatic in lung hemorrhage, kidney hemorrhage, and for excessive menstrual flow.
2.	*Aconitum balfourii* Stapf Syn. *A. atrox* (Bruhl) Mukerjee (Ranunculacease)	Meetha bis	July-August	3600-4000	Roots contain highly toxic alkaloids and are used in the treatment of gastritis and tonsillitis.
3.	*Aconitum heterophyllum* Wall. (Ranunculaceae)	Atis	August-September	3000-4500	Roots are extensively used in Ayurvedic, Unani, and homeopathic systems of medicine as a febrifuge, stomachic, bitter tonic, aphrodisiac, and antiperiodic. It is efficaciously used against diarrhoea, dysentery, and chronic enteritis. It is also used to create an excellent tonic for combating debility after fever.
4.	*Aconitum violaceum* Jacquem. ex Stapf. (Ranunculacease)	Dudhia bis	August-September	Above 3000	The tuberous roots are poisonous and are used in the treatment of fever.
5.	*Anemone obtusiloba* D. Don (Ranunculaceae)	Kakrya	June-July	2800-3500	Roots are used for treating menorrhoea. Seeds are purgative and seed oil is used to cure rheumatism
6.	*Anemone rivularis* Buch.-Ham. ex DC. (Ranunculaceae)	Mirchilee, Angeli	May-June	2200-3500	Leaves are used on wounds and sores and for headaches.
7.	*Angelica glauca* Edgew. (Apiaceae/Umbelliferae)	Chora	July-August	2400-3200	The roots are considered a cardioactive and stimulant, carminative, expectorant, diaphoretic and cordial. It is also used for stomach disorders, cough and dyspepsia. Roots yield an essential oil that is used in confectioneries and for the preparation of liquors.
8.	*Arnebia benthamii* (Wall ex G. Don) Johnston (Boraginaceae)	Balchhar	July-August	3200-4000	Leaves and flowers are used in cooling drinks; the whole plant is used to treat diseases of the tongue and throat ulcers.
9.	*Berberis jaeschkeana* Sch. (Berberidaceae)	Kingora	May-June	1800-2200	An extract of the root is astringent, diuretic and a blood purifier. It is also used for jaundice, eye troubles, and skin diseases.

Contd...

Table 6.1–*Contd...*

Sl.No.	*Botanical Names and Families (in parenthesis)*	*Local Names*	*Flowering Periods*	*Altitude (Meters)*	*Ethnomedicinal Uses*
10.	*Berginia ciliata* (Haw.) Sternb. (Saxifragaceae)	Silpara, Silfora	April-May	2000-3500	Rhizomes and roots are used in medicines for the treatment of kidney stones, piles, diabetes, and heart ailments. A paste of the fresh rhizome is very effective in the treatment of swelling, especially in livestock.
11.	*Caltha palustris* Linn. (Ranunculacease)	Kushnya, Maniru	May-June	2600-3800	Roots are reported to be poisonous, and horses often die from eating the herb. The leaf paste is used to treat abscess and boils.
12.	*Cirsium verutum* Spreng. (Asteraceae)	Kandaru	June-July	3000-3600	Plants are used as a diaphoretic and roots are used to cure fever.
13.	*Cotoneaster microphyllus* Wall. ex Lindl. (Rosaceae)	Bheda	May-July	2400-4000	Plants are astringent.
14.	*Corydalis cashmeriana* Royle. (Papaveraceae)	Indrajata	May-June	3000-4000	Roots are used as a tonic, diuretic and antiperiodic. The juice of the plant suppresses urination. It is also prescribed for syphilitic and cutaneous afflictions.
15.	*Dactylorhiza hatagirea* (D. Don) Soo. (Orchidaceae)	Hathazari	June-July	2800-4000	Tubers are used in Ayurvedic and Unani systems of medicine as a nervine tonic and aphrodisiac. Tubers are also used as an expectorant, astringent, and to increase virility.
16.	*Delphinium denudatum* Wall. (Ranunculacease)	Nirbishi		3500-4000	Roots are used as an expectorant and a bitter, stimulant tonic.
17.	*Elsholtzia strobilifera* Benth. (Lamiaceae)	Pothi	July	2500-3300	Leaves are a source of essential oil. Plants are astringent, carminative and diuretic.
18.	*Ephedra gerardiana* Wall. (Ephedraceae)	Tutgantha	May-July	3000-4200	Extract of the plants is used in bronchitis, asthma, and for relieving bronchial spasms. It is also used as a heart stimulant. It is an important medicine used in Ayurvedic, Unani and allopathic preparations.
19.	*Fritillaria roylei* Hook. (Liliaceae)	Kakoli	July-August	3000-3800	Bulbs are used as an antipyretic, expectorant and galactagogue.

Contd...

Table 6.1–*Contd...*

Sl.No.	Botanical Names and Families (in parenthesis)	Local Names	Flowering Periods	Altitude (Meters)	Ethnomedicinal Uses
20.	*Gaultheria trichophylla* Royle (Ericaceae)	Bhuinla	June-July	3800-4500	Oils from the leaves are aromatic, stimulant, carminative, antiseptic, and used to treat neuralgia. Fruits are a good invigorant for the heart.
21.	*Geranium wallichianum* D.Don (Geraniaceae)	Lal Zari, Neenai	July-September	2400-3000	Roots of the plants are astringent and are used to treat eye diseases and toothache.
22.	*Heracleum candicans* Wall. ex DC. (Apiaceae/Umbelliferae)	Kakriya	June-August	2500-3500	Roots are aromatic and used in the treatment of Leucoderma. They are used to prepare lotions to re-pigment white patches of skin.
23.	*Hyoscyamus niger* Linn. (Solonaceae)	Bisajain	August-September	3000-3800	Leaves are expectorant; used for pain in the liver, gouty swellings, and inflammation of the breasts and the testes.
24.	*Inula racemosa* Hook.f. (Asteraceae)	Poshkar	July-September	2500-3000	Roots are antiseptic and expectorant.
25.	*Iris kumaonensis* Auct. (Eridaceae)	Balupu-chhya, Pyzya	May-July	3000-3800	Roots and leaves are given to treat fevers and are also used as an insecticide.
26.	*Jurinea dolomaiea* Boiss. (Asteraceae)	Dhoop	August-September	3000-3800	A decoction of the roots is made into a cordial and given to treat colic and puerperal fever. The roots are also a commercial source of *Dhoop*, which is sold in local markets for its aromatic smell.
27.	*Lilium polyphyllum* D.Don (Liliaceae)	Kalihari	June-July	2800-3500	The bulbs are used as a general tonic. They are sold in market with local trade name *Khirkakoli.*
28.	*Malaxis muscifera* (Lindley) Kuntze (Orchidaceae)	Zeevak bhed	July-August	2500-3500	The rhizome is used as general tonic for health and is sold in the market as one of the *Ashtavarga* medicines by the name of *Rishibhak.*
29.	*Meconopsis aculeata* Royle (Papavaraceae)	Kalyari	July-August	3000-4200	Roots are considered narcotic and poisonous.

Contd...

Table 6.1–*Contd...*

Sl.No.	*Botanical Names and Families (in parenthesis)*	*Local Names*	*Flowering Periods*	*Altitude (Meters)*	*Ethnomedicinal Uses*
30.	*Morina longifolia* Wall. ex DC. (Morinaceae)	Biskandara	July-August	3000-4000	The roots are aromatic and are used to cure wounds and boils.
31.	*Nardostachys jatamansi* DC. (Valerianaceae)	Jatamansi	July-August	3000-4500	*Jatamansi* has been held in high regard in Hindu medicine since ancient times as an antispasmodic tonic and nervous stimulant and has been employed in the treatment of nerve-related ailments, especially palpitations of the heart, chorea, flatulence, and so on. It is also used with much success in treating epilepsy and similar nervous and convulsive ailments. It has also been recommended as a diuretic, emmenagogue and treatment for many diseases of the digestive and respiratory organs. It is also used in treating jaundice and to promote the growth and blackness of hair.
32.	*Origanum vulgare* Linn. (Lamiaceae)	Sathra	August-October	2500-3500	The whole herb is medicinal and contains a volatile oil that is stimulant and tonic and also used in treating diarrhoea, hysteria, headache, and paralysis. The leaves are good for earaches, bronchitis and asthma. Flowers are used to treat hemicrania.
33.	*Picrorhiza kurroa* Royle ex. Benth (Scrophulariaceae)	Kutki	June-July	3200-4000	Plant is effective in treating chronic dysentery, asthma, dyspepsia, hepatic derangement, and jaundice. The drug is bitter, diuretic, antipyretic, anti-stress, antiinflammatory, an analgesic, a muscle relaxant, and a CNS depressant.
34.	*Pleurospermum candollei* Benth. ex C.B.Clarke (Apiaceae/Umbelliferae)	Dhoop	August-September	3600-4500	The whole plant is used to make an aromatic *Dhoop*, a type of incense.
35.	*Podophyllum hexandrum* Royle (Berberidaceae)	Ban Kakri	August-September	2400-4500	The rhizome contains the active components podophyllin and podophyllotoxin. Podophyllin powder and resin are powerful purgatives and hepatic stimulants. They are also supposed to be able to cure cancer.
36.	*Polygonatum verticillatum* All. (Convallariaceae)	Deoringal	May-June	2000-3500	Rhizomes are used as a tonic and aphrodisiac.

Contd...

Table 6.1–*Contd...*

Sl.No.	Botanical Names and Families (in parenthesis)	Local Names	Flowering Periods	Altitude (Meters)	Ethnomedicinal Uses
37.	*Polygonum viviparum* Linn. (Polygonaceae)	Amjabat	June-July	3000-3500	Roots are useful in afflictions of the chest and lungs, piles, chronic diarrhoea, vomiting, rhinitis, chronic bronchitis and wounds.
38.	*Rheum australe* D. Don (Polygonaceae)	Archa, Dolu	July-August	3000-3600	Rhizomes are purgative, astringent and tonic. Root powder is sprinkled on ulcers and wounds for quick healing. The drug is suitable for children and elderly people.
39.	*Rhododendron anthopogon* D. Don (Ericaceae)	Semuru	June-July	3500-4200	Leaves are aromatic and astringent; they are used in the treatment of insomnia and as a sedative for anxiety.
40.	*Rhododendron campanulatum* D. Don (Ericaceae)	Semuru	May-June	3000-4000	Leaves are used to treat rheumatism, cold and cough as well as being a good remedy for sciatica and syphilis. Leaves are poisonous to livestock.
41.	*Roscoea alpina* Royle (Zingiberaceae)	Kakoli	July-August	2500-3800	Tuberous rhizomes are used as general tonic and also used as *Ashtvarga* in Ayurvedic preparations.
42.	*Saussurea costus* (Falc.) Lipsch. (Asteraceae)	Kuth	June-July	2500-4000	Roots are used as an insect repellent and smoked as a substitute for opium. Root oil is used to cure arthritis. Root powder is used to cure asthma and bronchitis.
43.	*Saussurea gossypiphora* Wall. (Asteraceae)	Phen Kamal	July-September	3800-4200	Local people use the whole plant to make a medicine for epilepsy.
44.	*Saussurea obvallata* (DC.) Sch.Bip.(Asteraceae)	Brahm Kamal	July-September	3600-4500	Flower heads are used to cure hydrocele. To prepare the medicine, the bracts are removed and the flower heads are roasted with ghee. One to two teaspoonfuls are given to the patient in the morning for three to six days. Root paste is used to cure leucoderma. In remote villages of Garhwal, people believe that the flowers of the *Brahm Kamal* plant protect their families from evil sprits.

Contd...

Table 6.1–*Contd...*

Sl.No.	*Botanical Names and Families (in parenthesis)*	*Local Names*	*Flowering Periods*	*Altitude (Meters)*	*Ethnomedicinal Uses*
45.	*Selinum vaginatum* C.B. Clarke (Apiaceae/Umbelliferae)	Bhoot Keshi, Moor	July-August	2500-4000	A decoction of roots mixed with ginger and black pepper is used to treat dysmenorrhoea. The fragrance of the roots is used to relieve bouts of hysteria. Afterwards, the patient is given a decoction of the root powder mixed with honey as a permanent cure.
46.	*Tanacetum longifolium* Wall. ex DC. (Asteraceae)	Guggulu	July-August	3000-4000	Peoples of Garhwal use the leaf juice as a vermifuge.
47.	*Valeriana jatamansi* Jones. (Valerianaceae)	Balchhari, Samewa	April-June	2000-3300	Locally, the rhizomes and roots are used as a tonic. Local people add rhizome extract to beverages for the aroma. Dried root powder is also used as insecticide and kept in clothes and fabric to save them from insects.
48.	*Viola biflora* Linn. (Violaceae)	Vanafsa, Kauru	May-August	2700-3600	Locally, the leaves are used as a laxative and also to cure fever, cold, and cough. Roots are used as an emetic and the flowers are used as an emollient, diaphoretic and antispasmodic.

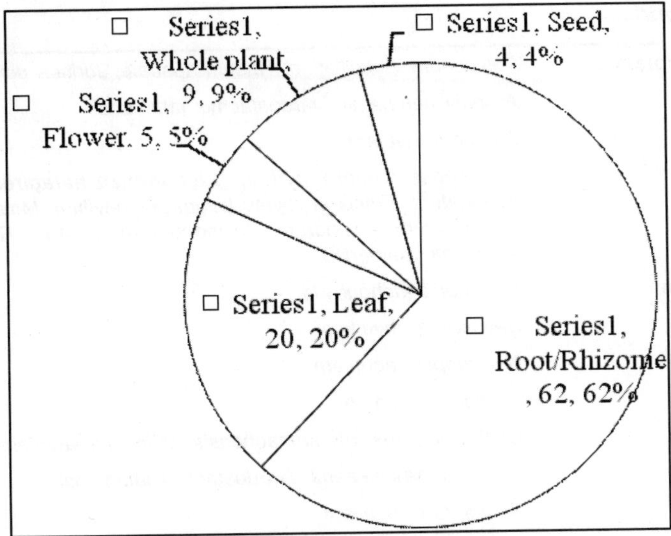

Figure 6.2: Pie chart showing the varying percentages of plant parts used

Table 6.2: A listing of folk medicinal claims arranged by diseases treated

Abdominal pain	*Delphinium denudatum, Picrorhiza kurroa*
Abscesses	*Caltha palustris, Polygonum viviparum*
Antispasmodic	*Viola biflora*
Anxiety	*Rhododendron anthopogon*
Aphrodisiac	*Aconitum bafourii, Lilium polyphyllum*
Astringent	*Polygonum viviparum*
Blood purifier	*Berberis jaeschkeana, Meconopsis aculeata*
Bone fractures	*Polygonatum verticillatum*
Bronchial asthma	*Fritillaria roylei, Saussurea costus*
Carminative	*Achillea millefolium*
Cough and cold	*Achillea millefolium, Geranium wallichianum, Rhododendron anthopogon, R. campanulatum, Hyoscymus niger, Viola biflora*
Cuts and bruises	*Arnebia benthamii, Circium verutum, Geranium wallichianum, Morina longifolia*
Diarrhoea and dysentery	*Circium verutum, Cotoneaster microphyllus, Geranium wallichianum, Picrorhiza kurroa*
Dysmenorrhoea	*Inula racemosa, Origanum vulgare, Selinum vaginatum*
Eczema and itches	*Hearcleum candicans*
Eye troubles	*Berberis jachkeana, Geranium wallichianum*
Fever	*Aconitum heterophyllum, A. balfourii, Circium verutum, Iris kumaonensis, Picrorhiza kurroa, Viola biflora*

Contd...

Table 6.2–*Contd...*

Gastric complaints	*Polygonum verticillatum, Rheum australe, Jurinea dolomiea*
Hair tonic	*Arnebia benthamii, Nardostachys jatamansi*
Headache	*Anemone rivularis*
Health tonic	*Cotoneaster microphyllus, Dactylorhiza hatagirea, Delphinium denudatum, Fritillaria roylei, Lilium polyphyllum, Malaxis muscifera, Podophyllum hexandrum, Polygonum verticillatum, Roscoea alpina, Valeriana jatamansi*
Heart trouble	*Gaultheria trichophylla*
Hydrocele	*Saussurea obvallata*
Hypertension	*Meconopsis aculeata*
Hysteria	*Selinum vaginatum*
Insecticide	*Caltha palustris, Iris kumaonensis, Valeriana jatamansi*
Jaundice	*Berberis jaeschkeana, Nardostachys jatamansi*
Leucoderma	*Saussurea obvallata*
Leucorrhoea	*Polygonum viviparum*
Muscular pain	*Rheum australe*
Rheumatism	*Origanum vulgare*
Sexual dysfunction	*Dactylorhiza hatagirea, Polygonum verticillatum*
Skin diseases	*Heracleum candicans*
Snake bite	*Morina longifolia*
Stomach disorders	*Aconitum heterophyllum, Delphinium denudatum*
Toothache	*Achillea millefolium, Aconitum heterophyllum, Hyoscymus niger, Origanum vulgare*
Medicinal teas	Leaves of *Gaultheria trichophylla* and *Rhododendron anthopogon*
Ulcers	*Delphinium denudatum, Polygonum viviparum*
Vermifuge	*Tanacetum longifolium, Morina longifolia*
Wounds	*Anemone rivularis, Arnebia benthamii, Berginia ciliata, Circium verutum, Cotoneaster microphyllus, Elsholtzia strobilifera*

contain high concentrations of bioactive compounds (Robinson, 1974).The majority of the plants were used in the following treatment categories: health tonic (10 species), cough and cold (6 species), fever (6 species), wound (6 species), diarrhoea and dysentery (4 species), cuts and bruises (4 species), toothache (4 species), insecticide (3 species), dysmenorrhoea (3 species), and gastric complaints (3 species), etc (Table 6.2). According to the village residents, fever, coughs, and colds are common problems in the area and had used indigenous medicines to treat this diseases, Six species: *Aconitum heterophyllum, A. balfourii, Circium verutum, Iris kumaonensis, Picrorhiza kurroa* and *Viola biflora* are used to treat fever, cough and cold are being treated by *Achillea millefolium, Geranium wallichianum, Rhododendron anthopogon, R. campanulatum, Hyoscymus niger* and *Viola biflora* (Table 6.2).

***Aconitum heterophyllum* Wall.**

***Anemone obtusiloba* D.Don**

***Pleurospermum brunonis* Benth. ex C.B.Clarke**

Jurinea dolomiaea **Boiss.**

Saussurea obvallata **(DC.) Sch. Bip.**

Meconopsis aculeate **Royle**

A few species like *Nardostachys jatamansi*, *Picrorhiza kurroa*, *Podophyllum hexandrum*, *Rheum australe*, *Saussurea costus*, *Saussurea gossypiphora*, Saussurea *obvallata*, and others are endangered yet local people continue to harvest them from the wild due to their strong medicinal properties. Over-exploitation has put a question mark on the survival of these species. As soon as possible, an integrated approach should be taken for the conservation of rare and endangered species by the creation and implementation of protective laws.

During our study, it was found that elderly people have greater knowledge of the use of medicinal plants in comparison to the younger generation. The reasons that were documented are as: allopathic medicine provides quick relief, there has been a loss of biodiversity, and people have fewer interactions with nature due to the influences of the Western lifestyle. Hence, the survival of indigenous knowledge is at risk and should be documented before it becomes extinct.

Acknowledgements

The authors are highly indebted to Professor R.D. Gaur, Former Head, Department of Botany, H.N.B. Garhwal University, Srinagar, Professor Som Deva, Former Head, Department of Botany, D.A.V. College Dehradun, the Forest Research Institute, Dehradun and the Botanical Survey of India, Northern Circle, Dehradun for their invaluable help in the identification of plants and other references. We express our gratefulness to Julia M.S. Mudrick of Provo, Utah, United States for her valuable help in language editing of paper. Bharati K Avinash would also like to thank the Council for Scientific and Industrial Research (CSIR), India for providing financial assistance.

References

Acharya K P, Chaudhary R P and Vetaas O R 2009. Medicinal plants of Nepal: Distribution pattern along an elevational gradient and effectiveness of existing protected areas for their conservation. *Banko Jankari* **19:** 16-22.

Babu C R 1977. *Herbaceous Flora of Dehradun* Publication and Information Division, CSIR, New Delhi.

Bist M K and Badoni A K 1990. Araceae in the Folk Life of Tribal Populace in Garhwal Himalaya. *J Eco Bot Phytochem* **1:** 21-24.

Chandler R F, Harper S N and Harney M J 1982. Ethnobotany and phytochemsitry of Yarrow *Achillea millefolium* Compositae. *Econ Bot* **36:** 203-223.

De A C 1962. Folklore of Medicinal Plants of Bhagirathi Valley (Himalaya) Colombo. Govt Press, Ceylon.

Duthie J F 1882. Plants Found in Kumaon Garhwal and the Adjacent Pats of Tibet by Capt Richard Strachey and J E Winterbottom *In*: Atkinson's E T (Ed.) *Gazetteer of North Western Provinces* **10:** 403-670.

Duthie J F 1906a. Catalogue of the Plants of Kumaon and of the Adjacent Portions of Garhwal and Tibet based on the collections made by Strachey and Winterbottom during the years 1846-1849, London.

Gaur R D 1999. *Flora of the Dist Garhwal: North-west Himalaya (With Ethnobotanical Notes)*. Transmedia, Srinagar (Garhwal).

Gaur R D and Nautiyal S 1993. A Survey of Fiber Yielding Plants of Garhwal Himalaya. *Higher Plants of Indian Subcontinent* **2:** 193-208.

Gaur R D, Rawat D S and Dangwal L R 1995. A Contribution to the Flora of Kuaripass-Dalisera Alpine Zones in Garhwal Himalaya. *J Eco Taxo Bot* **19:** 9-26.

Gaur R D and Semwal J K 1983. Exploration and Threat to Survival of Some High Altitude Plant of Garhwal Himalaya. *In*: Jain S K and Rao R R (eds.) 37-39. *An Assessment of Threatened Plants of India*, Howrah

Gupta R K 1994. Arcto-Alpine and Boreal Elements in the High Altitude Flora of North West Himalayas. *In*: Pangtey and Rawal (Ed.) 11-32. *High Altitudes of the Himalaya (Biogeography Ecology and Conservation)*. Gyanodaya Prakashan, Nainital.

Hooker J D 1872-97. *The Flora of British India* Vol 1-7, London.

Jain S K 1995. *A Manual of Ethnobotany*, 2nd ed, Scientific Publishers, Jodhpur.

Khoshoo T N 1991. Conservation of Biodiversity in Biosphere. *In*: Khoshoo and Sharma (Ed.) 178-233. *Indian Geosphere-Biosphere* Vikas Publications, New Delhi.

Martin J G 1995. *Ethnobotany*. Methods Manual, Chapman and Hall, London.

Mehra K L 1967. History of Sesame in India and it Cultural Significance. *Vishveshwaranand Indological J* **5:** 93-107.

Merlin M D 1984. *On the Trial of the Ancient Opium Poppy*, London.

Naithani B D 1969. Plant with the Kedarnath Parbat Expedition. *Bull Bot Survey India* **11:** 224-233

Naithani B D 1978. New Plant Records from Northern India from Garhwal Himalaya. *Ind J For* **1:** 244-246

Naithani B D 1984-85. *Flora of* Chamoli, Vol 1-2, Botanical Survey of India, Howrah.

Nautiyal S 1981. Some Medicinal Plants of Garhwal Hills: A Traditional Use. *J Sci Res Pl Med* **2:** 12-17

Negi K S, Tiwari J K and Gaur R D 1987-88. A Contribution to the Flora of Khatling Galcier in the Garhwal Himalaya (Distt. Tehri). *J Bombay Nat History Society* **84:** 585-598.

Polunin O and Staninton A 1984. Flower of the Himalaya. Oxford Press, New Delhi.

Pujadas J J, Coma D and Roca J 2004. *Etnografia*. Universitat Oberta De Catalunya Barcelona

Raizada M B and Saxena H O 1978. *Flora of Mussoorie*. Vol I, Bishen Singh Mahendra Pal Singh, Dehradun.

Rau M A 1963a. Illustration of West Himalayan Flowering Plants, *Bull Bot Surveyof India*, Calcutta.

Rau M A 1963b. The Vegetation around Jumnotri in Tehri Garhwal UP. *Bull Bot Survey India*. **5:** 287-280.

Rau M A 1964. A Visit to the Valley of Flowers and Lake Hemkund in North Garhwal UP. *Bull Bot Survey India* **6:** 169-171.

Rau M A 1968. Flora of the Upper Gangetic Plain and the Adjacent Siwalik and Sub Himalayan Tracts. *Bull Bot Survey India* **10:** 1-87.

Rau M A 1975. *High Altitude Flowering Plants of West Himalaya*. Botanical Survey of India, Howrah.

Robinson T 1974. Metabolism and function of alkaloids in plants. *Sci* **84:** 430-435.

Royle J F 1833-40. Illustrations of the Botany and Other Branches of the Natural History of the Himalayan Mountains and of the Flora of Kashmir, London.

Semwal J K and Gaur R D 1986. Addition to the Flora of Tungnath in Garhwal Himalaya. *J Bombay Nat Hist Soc* **83:** 267-271.

Shah N C 1977. Ethnobotany of *Acorus calamus. Indian J Pharmacol* **39:** 8-11.

Singh H B and Subramanium B 2008. *Field Manual on Herbarium Techniques,* NISCAIR, New Delhi.

Uniyal M R 1989. *Medicinal Flora of Garhwal Himalayas.* Vaidyanath Ayurved Bhawan Pvt Ltd, Nagpur.

Vines G 2004. Herbal harvests with a future: towards sustainable sources for medicinal plants, Plant life International; www.plantlife.org.uk

Medicinal Plants: Aspects and Prospects (2014) ***Pages* 98–110**
***Editors:* Mukesh Kumar, Anjali Khare and C.P. Shukla**
ISBN: 978-81-7622-309-6
***Published by:* BIOTECH BOOKS, NEW DELHI**

Chapter- 7

Spices and Condiments Used as Folk-Medicines by the Tribes of Uttarakhand

***Deepak Kumar*[1], *Deepika*[1] *and Mukesh Kumar*[2]**

[1]*Department of Botany, R,S,M, Degree College, Rampur, Ghoghar, Moradabad – 244 002*
[2]*Department of Botany, Sahu Jain P.G. College, Najibabad, Bijnor – 246 763*

ABSTRACT

Medicinal plants are community of living trees and associated organisms covering a considerable area, utilizing earthy materials to attain maturity and reproduce it and are capable of furnishing mankind with indispensable products and services. Ethnobotanical surveys were conducted for more than two years in Uttarakhand state. The tribal and other communities depend on the folk medicines for the treatment of the diseases and ailments. Firsthand information on ethnomedicinal use of plants, plant part and mode of administration have been collected from the local traditional healers and other knowledgeable men and women. This Chapter deals with medicinal plants used by different tribal or communities for the treatment of different diseases with the help of spices and condiments.

Key words: Folk medicines, Diseases, Tribal, Ailments.

Introduction

The ethno-medicinal applications of the plant species used primarily as spices and condiments among the people of Uttarakhand were experimented. A total of 25 species belonging to 12 families were recognized to have varying applications in ethnobotany and ethnomedicines. The studies indicate that the indigenous people have also developed different methods for collecting, processing, using and conserving these valuable plants and/or their products. The contributions of this study towards the understanding, documentation and safeguarding of indigenous knowledge and use of plants has been studied. Most aspects of the ethnomedicine of Uttarakhand are deeply rooted in spiritualism reflecting the dual character of culture. Spices and condiments are substances of plants, which are mostly used for seasoning, flavouring and enhancing the taste of foods, beverages and drugs (Manandhar 1995). The knowledge and use of plants as spices and condiments has been told in the history of mankind (Garland 1972). Plants used as spices and condiments are usually aromatic and pungent. Spices and condiments constitute a huge component of trans-boundary trade in areas such as India, Ceylon, China, Indonesia East and West Africa and West Indies (Parry 1969). The authors have reported that the use of spices and condiments has widened to include pickles, chutney, sausages, cakes, bread and alcoholic drinks. In some regions of Uttarakhand, many of spices and condiments are collected from the forest. These spices and their herbs are used generally to prepare pepper soups which may be taken hot or cold and especially during the cold and rainy seasons. Achinewu *et al.* (1995) concluded that these spices are particularly very important in the diets of postpartum women as an aid to the contraction of the uterus. Literature on ethno-botany and ethno-medicine of plants in the Uttarakhand is very scanty. Few taxonomic listings carried out in the area fail to incorporate indigenous knowledge and utilization of the plants. Information on ethno-medicinal applications, of the plant species used as spices and condiments are in adequate or completely lacking. Attempt has also been made to provide the most acceptable scientific, common and local names for various species. The present information is further intended to contribute in the documentation and provision of accurate record of indigenous knowledge, use and conservation of these plants, and their subsequent integration in the efforts towards the development of natural products and indigenous health care management processes. The specimens used for present study had been collected from different parts of Kumaon and Garhwal regions. The studies involved field trips and surveys. Information was obtained through oral interviews and questionnaires administered to local herbalists, older household heads and women.

Ethno-Medicinal Plants of Uttarakhand

The investigations have revealed a total of 16 species of angiosperm families used as spices and condiments in Uttarakhand. These species have been found to posses varying therapeutic applications. The data on the correct identification including common, English and local names, families, plant parts used and the ailments treated are summarized in Table 7.1. There are ample evidence that increasing numbers of people across the world depend on traditional herbal remedies for their health care. The local uses of plants and products in health care are even much higher, particularly in the areas with little or no access to modern health services (Sayeed *et al.*, 2004). Spices have been extensively used in the history for flavouring and seasoning foods, beverages and medicines (Stethberger *et al.*, 1996). The present studies however show that apart from the use of these plants as spices and condiments, they have several other wide applications in the local treatment and management of many diseases. Infact, on many occasions, the study team observed that the indigenous people value the plants more for their ethno-medicinal uses than for spicy foods. For instance, ginger is more valued for the treatment of coughs, asthma, colds and hypertension than as condiment. The indigenous people have therefore developed various ways of harvesting, processing and administering preparations of these plants in the cure of the different ailments. Trade and commercial utilization of the plants, though informal, constitute dominant enterprise of the local people in Uttarakhand. Uncontrolled exploitation, due to increasing population and its attendance pressure on resources and new wave of emphasis on natural products is threatening to most of the species investigated. In addition to these, the loss of habitats due to pollution and environmental degradation harbours much of UK flourishing petroleum business further states the treats to these species. The need to inventory, collect, describes and document that these plants will certainly from the basis of articulate programmes on their conservation. The increasing emphasis on the need to document customary knowledge and use of plant genetic resources (Cunningham 1994) provided the basis for attempt to capture these data especially the local names in this present study. This work, moreover, is a part of an on-going effort at the gradual build up of strong databank and knowledge on the development of ethno-medicine and its eventual integration into the formal health care system. The investigations revealed that a total of 16 species distributed genera and unrelated angiosperm families are used as spices and condiments around the Uttarakhand. Their uses in ethno-medicine include acting as stimulants, antiseptic carminatives, expectorants, laxatives, purgatives, anticonvulsant, antihelmintic and sedatives to the treatment of diarrhoea, malaria, rheumatism, asthma and bronchitis. The data on the correct identification including common, English and local names, families, plant parts used and the ailments treated are summarized in Table 7.1. The details of the ailments cured the methods of preparation and treatment is further described.

Medicinal plants are an important aspect of daily lives of many people and an important part of the South African heritage. The first good doctors were wise plant gatherers, with a superior knowledge of practical botany. Depending upon the species and its chemical components, they knew which plants caused hallucinations, heart palpitations, poisoning or successful healing.

Valuable Aspects on Medicinal Plants

Developing a medicinal plant sector, across the various states of the Indian Himalaya has become an important issue. Different stake-holders in the medicinal plants sector have projected Uttarakhand, one of the Himalayan states, as an Herbal state. This notions has made medicinal plants as a commodity of high value across the Uttarakhand. There are many aspects of research associated with the medicinal plants sector. We have identified the following five aspects that have not been studied properly or lesser to carry out research work.

1. The majority of current research programmes on medicinal plants conservation are being shifted from ecosystem to the species level. There is an urgent need for identification and notification of areas for conservation of medicinal plants on a priority basis.

Table 7.1: Summary of data on the species used for spices and condiments

Sl.No.	*Scientific Names*	*Family Names*	*Common Names (English)*	*Parts Used*	*Ethnomedicinal Applications*
1.	*Ocimum americanus* L.	Lamiaceae	Scent leaf	Whole plant and leaves	Anticonvulsant, diaphoretic and carminative. It cures cough, catarrh, cold, fever, chest pains and diarrhoea. Others are earache, ringworm, nasal bleeding, antispasmolytic and relief of pains of the colon.
2.	*Ocimum basilicum* L.	Lamiaceae	Sweet Basil Harry Basil	Whole plant and leaves	Diaphoretic, stimulant and carminative. Juice of the leaves is antihelmintic.
3.	*Ocimum canum* Sims	Lamiaceae	Scent leaf	Whole plant and leaves	Headache, cough, gouts, catarrh conditions and gonorrhoea.
4.	*Ocimum gratissimum* L.	Lamiaceae	Tea, bush	Whole plant and leaves	Diaphoretic, stimulant and carminative. Juice of the leaves is antihelminic.
5.	*Ocimum guineesnse* Schum et.Thonn	Lamiaceae	Scent leaves	Whole plant and leaves	Diaphoretic, stimulant and carminative. Juice of the leaves is antihelmintic.
6.	*Ocimum viride* Willd	Lamiaceae	Scent leaf	Whole plant and leaves	Anticonvulsant to stop diarrhoea, treatment of cold, fever chest pains and treatment of catarrh and bronchitis.
7.	*Thymus vulgaris* L.	Lamiaceae	Thyme	Leaves	Antiseptic antihelmintic, expectorant, carminative, diurotic emenagogic and sedative.
8.	*Tetrapleura tetraptera* Taub	Fabaceae	Unknown	Stem, bark and fruit pod	Flatulence, fever, convulsions, bone fractures, rheumatism, gonorrhoea.
9.	*Allium cepa* L.	Liliaceae	Onion	Bulb and leaves	Asthma, ulcers, cough, cold and skin infections.
10.	*Allium sativum* L.	Liliaceae	Garlic	Bulb	Fevers, cough, constipation, asthma, nervous disorders, hypertension, ulcers and skin diseases, antihelmintic.
11.	*Myristica fragrans* Houtt	Myristicaceae	Nutmeg, mace	Seeds	Diarrhoea, rheumatic pains,

Contd...

Table 7.1–*Contd...*

Sl.No.	Scientific Names	Family Names	Common Names (English)	Parts Used	Ethnomedicinal Applications
12.	*Murraya koenigii* Spreng.	Rutaceae	Curry leaf	Stem, bark, roots and leaves	Diarrhoea, dysentery, vomiting, fevers, herpes and bruises, post-partum pains.
13.	*Capsicum annuum* L.	Solanaceace	Chili, Red Pepper	Fruits and seeds	Cold, fever dysentery, malaria and gonorrhoea.
14.	*Capsicum frutescens* L.	Solanaceace	Red Pepper Tartashi	Fruits and seeds	Cold, fever, dysentery, malaria and gonorrhoea; additives as flavours in many medicines.
15.	*Capsicum minimum* Roxb	Solanaceace	African Pepper	Fruit and seeds	Cold, fever, dysentery, malaria and gonorrhoea
16.	*Zingiber officinale* Rose	Zingiberaceae	Ginger	Rhizome	Toothache, congested nostrils, cough, colds, influenza and flu, asthma, stomach problems, rheumatism, piles, hepatitis and liver problems.

2. Over all collection of medicinal plants has identified as one of the severe threats to medicinal plant population. Habitat alteration and specificity, narrow range of distribution, overstocking and overgrazing of areas by domestic animals are some of the threats endangering the existing population of medicinal plants.
3. There is an urgent need to carry out detailed investigations on the habitat utilization patterns, feeding ecology, geographical distribution patterns and impact of herbivores on medicinal plant population.
4. Most of the documentation and research on indigenous uses of medicinal plants is focused on the human aspects.
5. Animal husbandry is backbone of economy in the Himalayan region and maintaining a healthy animal population will obviously benefit various ethnic and non-ethnic communities prevalent in the Himalayas.

There are some aspects of medicinal plants, which are basically ignored but they could be important from the conservation point of view. On many occasions, collection of plant material, especially of rare and endangered medicinal plant species from natural habitats for various experimental purposes by researchers also consists a threat on their natural population in the wild. The researches must aware the people on the germination potential, seedlings and rhizomes survival strategies of the desired species collected from the wild for scientific experiments.

Conceptual Framework of Indigenous Medicines

Medicine is the system that a group of humans has developed over a long period of time, in order to ensure its survival and good health, which are pre-conditions for the individual and collective progress of the society at any point in time and space. Progress in this regard, is physical, mental and spiritual. Therefore medicine is tied to the apron strings of culture and environment. Culture includes a complex of knowledge, belief, art, law, morals, custom and many other capabilities and habits acquired by man, as a member of society. However, since this definition was given, many others have followed. One important insight we have got in recent times is that culture is clearly divisible into broad two parts as follows:

1. The actual behaviour of a group of people.
2. The abstract values, beliefs and perceptions of the world underlying behaviour.

This means that culture is an intricate combination of the seen and unseen phenomena. The observable or empirical dimension of behaviour or culture necessarily derives from the abstract component in a mutually understandable manner. One of the problematic of cultural researches around the World emanates from the fact that, while members of a given society are able to appreciate and appropriate this relationship and its centrality to their corporate existence and progress, the outsider hardly understands. This arises basically from the pre-conceived, but wrong notion of what should be the standard way of behaving by all human groups, regardless of the diversity of socio-cultural and environmental circumstances. This mode of thinking creates more epistemological gaps in the data generated by the researcher. However, we recognize the fact that, every culture is unique and dynamic, thus making it possible to adapt to new socio-cultural and ecological circumstances. But adaptation in this context has to be gradual and systematic so that it becomes beneficial to the affected people. It is not to be forced or imposed on a people by another group. This is to ensure some amount of continuity and orderliness, with respect to the corporate survival and progress of a people. A lack of appreciation and appropriation of the fundamentals of culture makes most 'modern' development efforts to assist the technologically weaker peoples around the World unworkable. Indigenous medical practices are as old as the time of the emergency of the earliest man. These practices involved experimenting with different plants and to a lesser degree, animals with a view to determining whether or not they had therapeutic values. This would sere as the basis for ensuring good health for a community. Much information about ethno-medicine is documented orally. Every environmental set-up, had numerous plants with chemotherapeutic values that mankind can use to treat a wide range of disease including illnesses at any point in time (Haviland 1999). But in practical terms, this is not always a simple and straight forward enterprise, and herein lies the gulf between Yoruba ethnomedicine and Western health care solely embedded in the domain of the biomedical germ theory.

Medicinal Plants Challenges and Opportunities

Information related to medicinal plants and traditional medicine can be found in documents and database aimed at readers in a wide range of disciplines including botany, ecology, chemistry, medicine, veterinary science, etc. However these are few publications reporting current work or reviewing and analyzing recent advances. Access to relevant information by the public, decision makers and local communities is still very limited. The convention of biological diversity, an international treaty that has been signed by more than 160 member states of the United Nations provides an international legal framework for the conservation of biological diversity including access to and exchange of genetic materials. While many different approaches are being tried to minimize the loss of biodiversity, the reduction of habitat loss and its accompanying loss of bio-cultural diversity are still unfortunately some way off. The need for internationally agreed methodologies for giving effect to the equity provisions of the convention on biological Diversity (CBD) is now widely recognized. The issue of benefit sharing has received considerable attention during the last decade. Many developing countries are behind the rest of the World in the development of national policies with respect to access to genetic resources and trade in medicinal plants. Appropriate strategies to increase awareness or policy makers and donors about the need for sustainable use and conservation medicinal plants and traditional medicine, can complement research efforts aimed to achieve this objective. Traditional medicine, in the estimate of the World Health Organization is used by up to 80 per cent of the population of most developing countries. These plant-based medicines are used for primary health care needs. Between 25-50 per cent of modern drugs are derived from plants. Demand for medicinal plants is increasing in both developing and developed countries. At the same time, the bulk of the material traded still derives from wild-harvesting. Only a very small number of species are cultivated. Herbal medicines and traditional healers are receiving attention from mainstream health officials and international medical research and training institutions as governments confront the high cost and inefficiencies of official health programmes. There is growing recognition of the need for increased efforts to produce medicines from plants. A number of international organizations now support projects and programmes in this area. There is still a need for substantial support at the national level and regionally to promote medicinal plants, traditional medicine and ethno-pharmacology, and to assure that biological resources are being harvested at a sustainable level.

Indigenous Knowledge, Bioprospecting and Benefit Sharing

The absence of an internationally agreed methodology for sharing economic benefits from the commercial exploitation of biodiversity with the primary conservers and holders of traditional knowledge and information is leading to a growing number of accusations of biopiracy committed by business and industry in developing countries. Biodiversity in both developing and developed countries has been accessed for a long time, for various purposes, by outside researchers, private companies as well as local communities, with little or no returns to conservation activities. Bio-prospecting has been practiced for many years in different forms but in more recent times in particular with the development of CBD, the issue of sharing of benefits arising from bio-prospecting has attained significance. However, certain critical issues remain unresolved, particularly in relation to how to go about legalizing and formalizing the bio-prospecting prospecting process in a way which ensures that there is full and prior informed consent for fair and equitable benefit sharing with the originator of the knowledge and resource that enable the bio-prospecting. On the other hand, traditionally, bio-prospecting in developing countries had been the preserve of field researchers in universities and botanical gardens.

Indeed, until recently, most bio-prospectors in developing countries have been individual professors or collectors who collected samples on contract with foreign companies or sold samples left over from research expeditions. These small-scale activities added little value to the biodiversity resource in any case are now likely to be discouraged by national legislation implementing the biodiversity convention. There has been a recent growth of interest in traditional medicine from the international pharmaceutical industry, as well as from the national product industry in Europe and America. Traditional medicine has become to be viewed by the pharmaceutical industry as a source of 'qualified leads' in the identification of bioactive agents for use in the production of synthetic modern drugs. Bio-prospectors express optimism that they can help to implement the 1992 convention on Biological Diversity by encouraging biodiversity, conservation and stimulating capacity building in developing countries. Those concerned with the development of bio-resources for human health recognize that when local custodians of biodiversity benefit from their sustainable use by theirs, conservation opportunities increase. The CBD codifies this benefit-sharing principle, but the absence of applicable instruments to equitably compensate all stakeholders within a country leaves it largely untested. Currently this aspect of the debate on access and benefit sharing has not received much attention, as the focus has been on the development and establishment of policies and legislation.

It is clear that many local and indigenous communities will not be able to go through this process alone and would need assistance and capacity development. Many legal and practical problems relating to protection of IPR remain yet to be fully understood and addressed: the collective ownership/custodianship of traditional medicine; The problem of ownership and exercise of right in traditional medicinal knowledge which exists across different countries in a region; practical means for the exercise and management of rights; mechanisms for application of customary law to protection of traditional medicine; and the need for comprehensive documentation standards, for traditional medicine.

Future Prospects

The commercial viability of biotechnology and gene technology in medicinal plant research is strongly influenced by the common perception of both, the plant and biotechnology. As outlined above, genetically-by the public as unsafe and dangerous. The crop industry learned its lesson following the rejection in Europe of genetically modified crops that were introduced into the food chain. Clearly, companies producing herbal compounds would face the same problems in obtaining permission to conduct farm-scale trails, to document the safety of the final product, and to overcome the in-principle resistance of the consumer as a strong and perhaps immovable barrier (Canter *et al.*, 2005). Within an open scientific environment the discovery and development of botanical therapeutics and medicinal plant biotechnology must be accepted, as its expansion is very unlikely to cease.

Constraints in Developing Traditional Medicine

A large portion of the population in number of developing countries still relies mainly on traditional practitioners, including traditional birth attendants, herbalists and bone-setters, and local medicinal plants to satisfy their primary health care needs. Practices involving use of traditional medicine vary greatly from country to country and from region to region as they are influenced by factors such as culture, mentality and philosophy. Despite its existence over many centuries and its expansive use during the last decade, in most countries, traditional medicine, including herbal medicines has not yet been officially recognized, and in most countries the regulations and registration of herbal medicines have not been well established. Furthermore research and training activities for traditional medicine has not received due support and attention. As a result, the quantity and quality of safety and efficacy data are far from sufficient to meet the demands for the use of traditional medicine in the World. Safety and efficacy date exit only in respect of much smaller number of plants and their extracts and active ingredients, as well as preparations containing them. Reasons for the lack of research data involve not only policy problems, but also the research methodology for evaluating traditional medicine. There is literature and data on the research of traditional medicine in various countries, but all scientists may not accept them. There is a need for validation and standardization of phyto-medicines and traditional medical practices so that this sector can be accorded its rightful place in the health care system. As the characteristics and applications of traditional medicine are quite different from Western medicine, how to evaluate traditional medicine and what kind of academic research approaches and methods may be used to evaluate the safety and efficacy of traditional medicine are new challenges which have emerged in recent years. Along with increased interest in medicine is an increased interest in the safety aspects of the practice of herbal medicine. Private sector involved in the business of herbal drugs should take responsibility and ensure the safety and efficacy of the preparations that they put on the market.

Social Dimensions

Gender issues in medicinal plant use, conservation and cultivation. The role of spiritual beliefs in traditional healing system:

- ☆ The place and contribution of traditional medicine in primary health care, with particular attention to priority diseases such as malaria, HIV/AIDS, TB; *e.g.*, gaps between the finality of current research with regard to public health concerns and local communities needs and priorities.
- ☆ The global context (bio-prospecting, IPR, TRIPs, CBD, etc):Implications for promotion of the sector *e.g.*, protection of traditional knowledge and practices relating to the uses of medicinal plants and traditional medicines.
- ☆ Sustainable use and conservation of medicinal plants: The contribution of forests in community health care.
- ☆ Propagation and domestication of medicinal plants, over harvesting and controlled harvesting, resource management.
- ☆ Economic and policy incentives and legal tools for conservation and sustainable use.
- ☆ Strategies to enhance income generation and benefit sharing from medicinal plants and traditional medicine.
- ☆ Integration of traditional medicine and public health programmes and systems.
- ☆ Spiritual and cultural values that is supportive to medicinal plants conservation, traditional healing and practices.
- ☆ Tools, mechanisms and strategies to enhance information exchange, cooperation and collaboration.
- ☆ Benefit sharing and development of and collaboration with indigenous peoples and local communities.
- ☆ Options for national policies and legislation to access genetic resources and legal and financial aspects related to benefit sharing.
- ☆ Prospects for private sector participation in biodiversity prospecting the developing countries.
- ☆ Options for legislation, policies and incentives to add value to medicinal plant genetic resources and increase capacity in bio-prospecting.
- ☆ Improvement of the multidisciplinary information on medicinal plants needed for conservation, agriculture, primary health-care and manufacturing activities.
- ☆ The economic significance of traditional medicinal knowledge and systems.

- ☆ The enhancement of medicinal plant usage through intellectual property rights, capacity building and information technology and transfer.
- ☆ Developing and using indigenous and traditional knowledge for promoting bio-prospecting for the benefit of all stakeholders involved.
- ☆ Mechanism, strategies, partnership and co-operation for encouraging bio-prospecting-based business in medicinal plants and phyto-medicines.

Policy, Planning and Legislation

A coherent national programme on medicinal plants should include the following institutions.

- ☆ **Ministry of health:** Formulation of national policy, legislation, regulation and licensing, collection, analysis and dissemination of information on medicinal plants and approval of selected plant remedies for use by health services.
- ☆ **Ministry of agriculture:** Cultivation of medicinal plants, protection of endangered species.
- ☆ **Universities or Pharmacy:** Inventory of indigenous medicinal plants and natural products, identification of constituents of traditional remedies, pharmacological evaluation of medicinal plants and natural products, identification of active substances, their extraction and toxicity testing, dosage and formulation.
- ☆ **Clinical medicine:** Clinical trials and field testing
- ☆ **Public health:** Studies on indigenous remedies and their uses, training of health personnel.
- ☆ **Botany:** Cultivation of medicinal plants, cloning and cell culture, taxonomy identification, studies on ethnobotany.
- ☆ **Pharmaceutical industry:** Pharmaceutical development, processing and pilot production, trial marketing, full-scale production.
- ☆ **Ministry of trade:** Assessment of local trade in medicinal plants, export and imports.

A national strategy on medicinal plants should highlight the following:

- ☆ Assessment of the importance of traditional medicine in the country.
- ☆ Current status of medicinal plant and traditional medicine at the national level.
- ☆ Activities and institutional framework.
- ☆ Coordinating and implementing agencies; support of R&D.
- ☆ Role of NGO's and traditional healers associations.
- ☆ Respective role of public and private sectors.
- ☆ Actions and strategies to promote traditional medicine and medicinal plants list of most common medicinal plants and their uses.

- ✰ Efforts to integrate traditional medicine in official public health system-advantages, disadvantages, implications, obstacles, policy options and key players.
- ✰ Lessons learnt and new perspectives;
- ✰ Limitations and potentials
- ✰ Future directions- Research priorities, Technology transfer and capacity building; Planning and Development, Policy options; Information exchange and management; Co-operative strategies and mechanisms.

Conclusion

In order to achieve better understanding and wider consensus of these issues it is necessary to address basic conceptual problems and test practical solutions to the protection of traditional medicine. There is a need to continue debate with true stakeholders-practitioners of traditional medicine, representatives of the medical community, the pharmaceutical and bio-technology industries, intergovernmental organizations, etc. Lasting solutions can only be found if all stakeholders work together in good faith and bring their specific expertise and experience towards a common understanding and solution of the problems.

It is difficult to predict the future for medicinal plants, but it is likely that herbal drugs, isolated natural products and recombinant low and high-molecular weight products will hold at least the same significance. Plants as a renewable source with low energy consumption that can offer complex biochemical synthesis will be even more compatible in the future.

References

Achinew S C, Anjemna M I and Obomanu F G 1995. Studies on spices on food value in the South-eastern states of Nigeria 1: Antioxidants Properties. *J African Med Pl* **18:** 135-139

Canter P H, Thomas H and Ernst E 2005. Bringing medicinal plants into cultivation: opportunities and challenges for biotechnology. *Trends Biotechnol* **23:** 180-185.

Carland S 1972. The Herbs and Spices Book. Frances Lincoln Publishers, London.

Cunningham A B 1994. Integrating local plant resources and habitat management. *Biodiversity and Conservation* **3:** 104-115

Haviland W 1999. A cultural Anthropology. Fort Worth: Harcourt Brace Wlege publishers.

Manandhar N P 1995. Substitute spice in Nepal. *J Herbs, Spices Med Plants* **3:** 71-77.

Ogundele S D 2006. Our Ancestors Speak: Cultural Heritage, Knowledge and Development crises in contemporary Nigeria. Ibadan: Faculty of Science, Public lecture series, UWV of Ibadan.

Parry J W 1969. Spices-Morpholgy, Histology, Chemistry, Vol. II, Food Trade Press Ltd London.

Saeed M, Arshad M, Ahmad E and Ishaque M 2004. Ethnophytotherapies for the treatment of various diseases by the local people of selected areas of NWP (Pakistan). *Pak J Bot Soc* 7: 1104-1108

Stethberger S Bomme U and Rothenburger W1996. Economics of medicinal and condiment plants. Germuse-Muchen. 32: 117-1118

Medicinal Plants: Aspects and Prospects (2014) ***Pages* 111–125**
***Editors:* Mukesh Kumar, Anjali Khare and C.P. Shukla**
***Published by:* BIOTECH BOOKS, NEW DELHI**

Chapter 8

Therapeutic Potential of Phytochemicals

Jinu John, Pradeep Mehta and Archana Mehta

Plant Biotechnology Lab, Department of Botany,
Faculty of Life Sciences, Dr. Hari Singh Gour University,
Sagar – 470 001, M.P.

ABSTRACT

Nature is blessed with a wide variety of medicinal plants; most of them are in use and phytochemically and pharmacologically evaluated. Herbal medicine, or phytotherapy, is the science of using herbal remedies to treat the sick. The study of the use of medicinal plants is now a scientific subject, a field of medicine in the same way as chemotherapy. Knowledge of medicinal plants and their uses has been recorded from antiquity. The rise of chemistry, development of synthetic drugs and the possibilities opened up by experimental pharmacology has caused herbal knowledge to be neglected. However now there is an increasing popularity for herbal medicine mainly due to the side effects and increasing drug resistance of modern medicine experienced in last decades. Many of potent drugs of modern medicine are the result of such exploration. Multiple and effective therapeutic properties of medicinal plants make it an interesting field of research. Research till to date confirms that plant kingdom has the remedy for all types of health problems either by directly or indirectly turning the immune system or by the removal of pathogens from the body. Scientific and systematic way of approach will help to explore and utilize the immense source of herbal diversity effectively.

Keywords: Herbal medicine, Phytochemicals, Pharmacological evaluation.

Introduction

Nature provides a rich store house of herbal products to cure all ailments of mankind. Human beings have been utilizing plants for basic preventive and curative health care since time immemorial. In recent years, research on medicinal plants has attracted a lot of attention globally and proved the potential of medicinal plants used in various traditional, complementary and alternative systems. Therapeutic agents of plant origin used to cure a wide range of diseases like hypertension, leprosy skin disorders and even for various cancers. Phytochemical and pharmacological evaluation of these drugs help to know the mechanism of action and subsequently several significant phytochemicals have been isolated and today they become the lead molecules in the discovery of several potential allopathic drugs. Allopathic medicine owes a tremendous debt to medicinal plants: About 25 per cent of drugs prescribed worldwide come from plants and more than 121 such active compounds being in current use (Rates 2001). In the present article, the authors have made an attempt to discuss the approach to herbal drug research and the therapeutic potential of different plants.

Plants contain substances that can be used for therapeutic purpose or precursors for the synthesis of useful drugs. Therapeutic properties are due to the phytochemicals, the non-nutritional secondary metabolites. Plants synthesize a wide spectrum of aromatic substances, most of which are phenols or their oxygen-substituted derivatives. At least 12,000 secondary metabolites have been isolated, which are less than 10 per cent of the total (Schultes 1978). The recent advances in the technology increased the momentum and large number of phytochemicals have been characterized. Generally, these substances serve as plant defense mechanisms against number of pathogens. Some of them such as terpenoids give plants their odors; flavonoids and anthraquinons gives pigments. Many compounds are responsible for plant flavor (*e.g.*, the terpenoid capsaicin from chili peppers). The herbs and spices used by humans to season food contains useful medicinal compounds (Marjorie 1999).

Herbal medicines are based on long-term experiences, often hundreds or thousands of years hence it might expect any bioactive compound obtained from such plants to have low human toxicity. In addition to this chemical diversity of plant secondary metabolites that results from plant evolution may be equal or superior to that found in synthetic combinatorial chemical libraries. Moreover, the expense of launching a new synthetic drug in to the market is very high as well as it is time consuming. It was estimated that in 1991 in the United States, for every 10,000 pure synthetic compounds that are biologically evaluated (primarily *in vitro*), 20 would be tested in animal models, and 10 of these would be clinically evaluated, and only one would reach U.S. Food and Drug Administration for approval for marketing. The time required for this process was estimated as 10 years at a cost of $231 million (Vagelos 1991). However, the evaluation of plants as a source of therapeutic agents may helps to isolate bioactive compounds for direct use as drugs (digoxin, digitoxin, morphine, reserpine, taxol, vinblastine, vincristine), to produce bioactive compounds of novel or known structures as lead compounds for semi-synthesis to produce entities of higher activity and/or lower toxicity [*e.g.*, metformin, nabilone, oxycodon (and other narcotic analgesics), taxotere, teniposide, verapamil, and amiodarone], to use agents as pharmacologic tools (*e.g.*, lysergic acid diethylamide, mescaline) and to use the whole plant or part of it as a herbal remedy (*e.g.*, Cranberry, garlic, *Ginkgo biloba* and St. John's wort) (Daniel *et al.*, 2001).

Both quantitative and qualitative parameters of phytochemicals depend on plant species and their abiotic and biotic environments (Facchini *et al.*, 2000). The major significant classes of secondary compounds are the alkaloids, phenylpropaniods, flavonoids, and the terpenoids. Many phytochemicals can act as antioxidants, including carotenoids and flavonoids. Among the flavonoids, isoflavonoids in soya and other legumes have estrogen-like effects. Some phytochemicals, such as isothiocyanates in the cabbage family and organo sulfur compounds in garlic helps the body to dispose off the chemical carcinogens and block their carcinogenic action.

Isolation of phytochemicals in large quantities is one of the major problems faced by the herbal industry. Elucidation of biochemical pathways and application of elicitors in tissue cultures made a tremendous progress in the production of phytochemicals. It may gradually substitute the traditional way of collecting phytochemicals from field grown plants to an extent. Evolution of chromatographic technique along with other chemical methods like partial precipitation, made easy the isolation of phytochemicals in the pure form. Exploration of new plant sources which gives high yield of a particular compound and suitable cultivation technique can also increase the production. India is the home for a number of commercially important medicinal and aromatic crops; its geographical location with salubrious climate is congenial for cultivation of a majority of important medicinal and aromatic crops.

General Approach to Phyto-Pharmacological Evaluation of Herbal Drugs

Traditional uses of plant in primary health care are supported by actual pharmacological effects or merely based on folklore (Visen *et al.*, 1998). Use of herbal medicines can be traced back as far as 2100 B.C. in ancient China (Xia dynasty) and India (Vedic period). The first written reports date back to 600 B.C. with the Charaka Samhita of India and the early notes of the Eastern Zhou dynasty of China that became systematized around 400 B.C. The recipes, once formulated, were usually expanded rather than abandoned during subsequent centuries. Expansion was stimulated by a growing understanding of the natural evolution of frequently encountered diseases and by emerging hypotheses regarding their causes. Herbal drugs are generally used in the form of decoctions, powders, tinctures etc. The experiences derived from the natural practices for ages resulted in the origin of different traditional medicinal systems like Ayurveda. Lesser side effects, efficiency and economy made herbal drugs popular not only in developing but in developed countries also. Improvement in collection methods, new screening techniques and the development of molecular techniques increased the momentum and popularity of herbal drug research.

The selection of plants for phyto-pharmacological evaluation is either random selection or on the basis of traditional use. Traditional medicine is a broad term used to define any non-Western medical practice. Ethnopharmacology is a highly diversified approach to drug discovery involving the observation, description, and experimental investigation of indigenous drugs and their biologic activities. It is based on botany, chemistry, biochemistry and pharmacology that contribute to the discovery of natural products with biologic activity (Rivier and Bruhn 1979). In general the evaluation of ethanobotanical data involves various steps (Figure 8.1) like chemical screening followed by various *in vitro* and *in vivo* analyses. Observation of some of such studies which proves the potential of herbal drugs is the main point of description in this article.

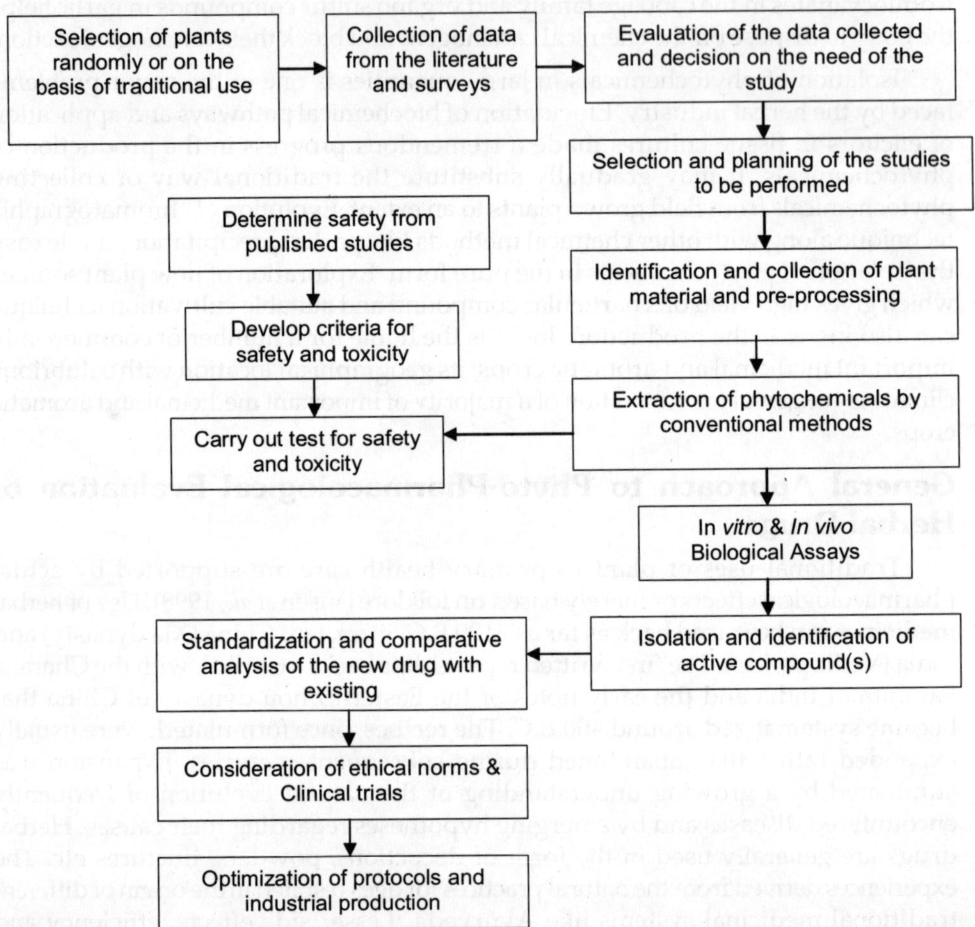

Figure 8.1: Flow chart of sequence for the study of plants used in traditional medicine

Pharmacological Significance of Phytochemicals

Plants are efficient biosynthetic laboratories; produce a wide spectrum of chemicals for their existence/defense and development. Most of such phytochemicals are found to be beneficial in several aspects such as potent drug for several diseases.

Phytochemicals as Antioxidants

Antioxidants are compounds that protect cells against the damaging effects of reactive oxygen species (ROS). Some ROS, such as superoxide and hydrogen peroxide, are normally produced in cells as by-products of biochemical reactions. When there is an imbalance between antioxidants and ROS, it creates some detrimental effects on body. Oxidative stress, which may result in cellular damage, has been linked to cardiovascular disease, diabetes, cancer and other degenerative conditions. Herbal antioxidants, such as ginseng, grape seed, green tea, and *Scutellaria baicalensis* are well known and extensively in use, may protect against the diseases by contributing to the total antioxidant defense system of the human body. The activity of these antioxidant herbs to affect various pathological processes mediated by ROS depends on their ability to access the sites or sub-cellular compartments of biochemical activity, in other words they can play potential role in preventing and treating diseases.

Free radicals are atoms or groups of atoms that have at least one unpaired electron, which makes them highly reactive. Free radicals promote beneficial oxidation that produces energy and kills bacterial invaders. In excess, however, they produce harmful oxidation that can damage cell membranes and cell contents. Antioxidants intercept free radicals and protect cells from the oxidative damage that leads to aging and disease; they donate an electron to the free radical and convert it to a harmless molecule. Antioxidants prevent injury to blood vessel membranes, helping to optimize blood flow to the heart and brain, defend against cancer-causing DNA damage and helps to lower the risk of cardiovascular disease and dementia, including Alzheimer's disease.

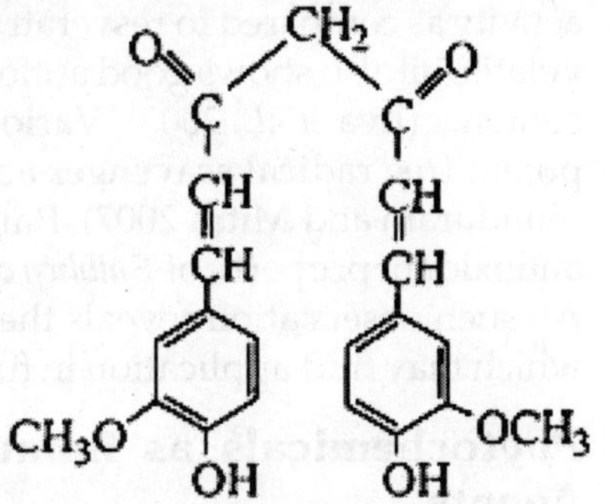

Curcumin

Ascorbic acid

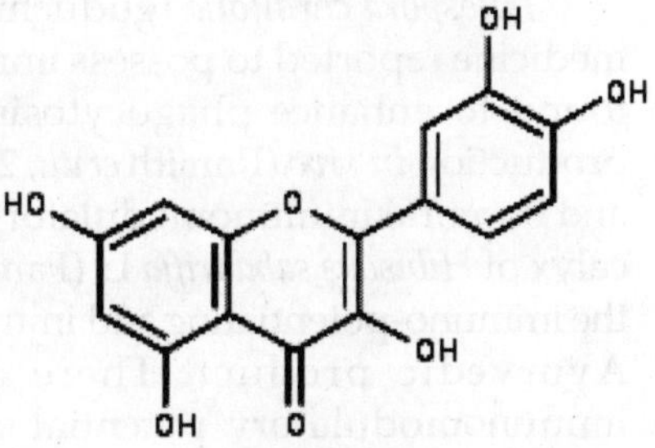

Polyphenols are a broad family of naturally-occurring physiologically-active nutrients. Vegetables and fruits are rich source of phenolic compounds. Curcumin, polyphenol present in turmeric root is a powerful anti-inflammatory and antioxidant agent. Ascorbic acid is another potent naturally occurring antioxidant. Flavonoids are poly-phenolic compounds that give vegetables, fruits, grains, seeds, leaves, flowers, and bark their color. Bio-flavonoids like quercetin act as antioxidants that battle and neutralize a wide variety of free radicals including nitric oxide, the hydroxyl radical, singlet oxygen, the super-oxide radical,

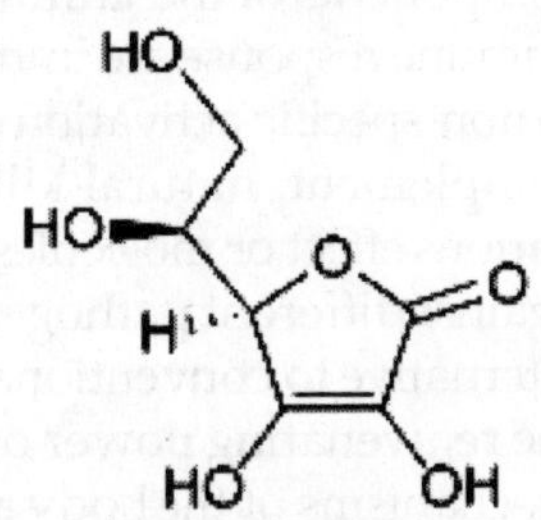

Quercetin

and the super-potent combination of superoxide and nitric oxide called the peroxynitrate radical. Bio-flavonoids help to regulate nitric oxide levels and keep them from becoming excessive. The hydroxyl groups present in the flavonoids are responsible for their radical scavenging effect (Gryglewski *et al.*, 1987).

Evaluation of different plants for their antioxidant properties by different *in vitro* and *in vivo* models like DPPH radical scavenging, nitric oxide (NO.), etc., proves that they are good sources of antioxidants and there is a correlation between phenolic content and antioxidant property. In most of the cases flavonoids are found to be responsible for their activity. Ethyl acetate fraction of *Terminalia chebula*, prepared from crude methanol extract showed maximum activity and phenolic content compared to other extract/fractions (Walia *et al.*, 2007). There are several such results like the ethanolic extract of *Excoecaria indica* leaves (Ahmed *et al.*, 2007), extracts of *Areca catechu*, *Paeonia suffruticosa* and *Cinnamomun cassia* showed better antioxidant activity as compared to resveratrol (Lee *et al.*, 2003). Rosemary (*Rosmarinus officinalis*) volatile oil also shows good antioxidant capacity, which is related to the total phenol contents (Eva *et al.*, 2003). Various extracts and fractions of *Acacia arabica* showed potent free radical scavenger activity in *in vitro* and *in vivo* experimental models (Sundaram and Mitra 2007). Ramasundaram *et al.* (2007) have reported the strong antioxidant property of *Emblica officinalis* during restrain-stress study in albino rats. All such observation reveals the fact that plants are good sources of antioxidants, which may find application in future like *Panax ginseng*.

Phytochemicals as Immunomodulatory and Anti-inflammatory Agents

Biological or synthetic substances, which can suppress or modulate any of the components of the immune system, including both innate and adaptive arms of the immune response are immunomodulators. The concept of immunomodulation relates to non-specific activation of the function and efficiency of macrophages, granulocytes, complement, natural killer cells, and lymphocytes, and also to the production of various effect or molecules generated by activated cells (Para immunity) give protection against different pathogens including bacteria, fungi, viruses, etc., and constitute an alternative to conventional chemotherapy (Atal *et al.*, 1986). Recent researches proved the rejuvenating power of herbal drugs and its combinations to improve the defense mechanisms of the body and enhance longevity due to its immunomodulatory effect, which is mentioned in Ayurveda as rasayana.

Tinospora cordifolia (guduchi) is a widely used shrub in Ayurvedic systems of medicine reported to possess immunomodulatory properties. The aqueous extract found to enhance phagocytosis *in vitro* and induced an increase in antibody production *in vivo* (Ranjith *et al.*, 2008). Fakeye *et al.* (2008) reported the cell mediated and humoral immunomodulatory activity of water and alcohol extracts of the dried calyx of *Hibiscus sabdariffa* L. (Family Malvaceae) in mice. Pallabi *et al.* (1998) studied the immuno-potentiating and immunoprophylactic activity of Immue-21, a polyherbal Ayurvedic product. There are several such studies which prove the immunomodulatory potential of different plants alone and in combination as medicinal preparations.

Gossypin, a bioflavonoid obtained from yellow petals of *Hibiscus vitifolius* L. (Malvaceae) was found to reduce the carrageenin induced paw edema significantly and also found to be effective against the adjuvant and formalin induced chronic arthritis in rat models (Parmar 1978). Betulinic acid, a triterpene isolated from *Nelumbo nucifera* L. demonstrated significant anti-inflammatory activity when tested in carrageenin and 5-hydroxytryptamine induced paw edema models. Premnazole, an isoxazole alkaloid isolated from *Premna integrifolia* L. and *Gmelina arborea* L. (Verbenaceae) found to be effective in reducing cotton pellet-induced granuloma formation in rats. These activities were comparable to that of betamethasone and phenylbutazone (Mukherjee *et al.*, 1997; Barik, 1976). The water-soluble alkaloid, achyranthine isolated from *Achyranthes aspera* L. (Amaranthaceae) showed significant anti-inflammatory activity in different models though they were less active compared to the standard drugs, phenylbutazone and betamethasone (Neogi *et al.*, 1969). Thus phytochemicals are not curative only but can also rejuvenate and prepare the body to face pathogenic threats.

Phytochemicals as Anticancer Agents

Cancer is a major public health burden in both developed and developing countries. Herbal drugs provide a wide range of phytochemicals which are preventive as well as curative for various types of cancers. Natural products discovered from medicinal plants have played an important role in the treatment of cancer. Natural products or derivatives comprised 14 of the top 35 drugs in 2000 based on worldwide sales (Butlet 2004). Exploration of plants can provide potential bioactive compounds for the development of new 'leads' to combat cancer diseases. Several anticancer agents derived from plant including taxol, vinblastine, vincristine, the camptothecin derivatives, topotecan and irinotecan, and etoposide derived from epipodophyllotoxin are in clinical use all over the world. A number of promising agents such as flavopiridol, roscovitine, combretastatin A-4, betulinic acid and silvestrol are in clinical or preclinical development.

There is no successful treatment for cancer till to date. The world-wide search for novel anti-tumor agents from natural sources continues through the collaboration of scientists. Despite major scientific and technological progress in combinatorial chemistry, natural products still make significant contributions to drug discovery today. An analysis of the number of chemotherapeutic agents and their sources indicate that over 60 per cent of approved drugs are derived from natural compounds (Cragg *et al.*, 1997). Scheme of screening involves treatment of plant extract or purified phytochemicals with cancer cell lines and measurement of cell cytotoxicity by different assays like MTT or SRB assay. Mechanism of action can be determined by flow-cytometry or by DNA electrophoresis. *In vivo* studies were conducted by measuring the effect of plant drug on the growth or proliferation of induced tumors on animal models.

The isolation of the vinca alkaloids, vinblastine and vincristine from the Madagascar periwinkle, *Catharanthus roseus* Don (Apocynaceae) introduced a new era on the use of plant material as anticancer agents. They were the first agents to advance into clinical use for the treatment of cancer. Vinblastine and vincristine are primarily used in combination with other chemothe-rapeutic drugs for the treatment of a variety of cancers, including leukemia, lymphomas, advanced testicular cancer, breast and lung cancers, and Kaposi's sarcoma (Cragg and Newman 2005).

Withanolides, the active constituents of *Withania somnifera* Dunal (Solanaceae) are group of pharmacologically active compounds present in roots and leaves. Withanolides are similar to ginsenosides (the active constituents of *Panax ginseng*) in structure and activity. Withaferin-A is best studied withanolide reported to have immunomodulatory and anticancer activity (Ali *et al.*, 1997). Withaferin-A showed marked tumour-inhibitory activity when tested *in vitro* against cells derived from human carcinoma of nasopharynx (KB). It also produced significant retardation of the growth of Ehrlich ascites carcinoma, Sarcoma 180 and E 0771 mammary adenocarcinoma in mice in a dose depended manner where all the mice survived for 100 days without the evidence of growth of the tumor (Devi 1996). Flavonoids have been reported to possess antioxidant and anticancer properties. Flavonoids like luteolin and quercetin were found to inhibit the proliferation of cells of human carcinoma of larynx and sarcoma-180 cell lines (Elangovan *et al.*, 1994).

Phytochemicals as Hepatoprotective Agents

Silybum marianum, *Picrrorhiza kurroa*, *Andrographis paniculata*, *Phyllanthus niruri*, and *Eclipta alba* are proven hepatoprotective medicinal herbs, which have shown genuine utility in liver disorders. These plants are used widely in hepatotprotective preparations where extensive studies have been done (Bisset 1994). Experimental liver damage produced by carbon tetrachloride (CCl_4) has been extensively studied. The profile of damage even after the single administration of this hepatotoxin has been well established (Anand *et al.*, 1992). The critical process underlying CCl_4 hepatotoxicity is the combining effect of both lipid peroxidation and the covalent binding of CCl_4 reactive metabolites to lipids and proteins (Masuda and Nakamura 1990). Cell damage by free radicals has been reported as the predominant mechanism of hepatotoxicity (Gregus and Klaassen 1995). It has been shown that CCl_4 induced lipid peroxidation can be obstructed by natural antioxidants (Wang *et al.*, 2000). The identification of naturally occurring inhibitors of peroxidation resulting in cell damage could therefore lead to important new strategies for disease prevention (Subramanian *et al.*, 1999).

Hepatoprotective activity of silymarin, isolated from *Silybum marianum* has been demonstrated against partial hepatectomy models and toxic models in experimental animals by using acetaminophen, carbon tetrachloride, ethanol, D-galactosamine, and *Amanita phalloides* toxin (Pradhan and Girish 2006). *Andrographis paniculata* Nees. is well known for liver diseases. It contains diterpene lactones (andrographolide, neoandrographolide and kalmeghin). Andrographolide produced dose dependent chloretic effect evidenced by increase in bile flow, bile salt and bile acids in animal models. The cholretic effect of *Andrographis paniculata* was found to be better than silymarin (Shukla *et al.*, 1992).

Saxena *et al.* (1993) demonstrated that the fraction containing apigenin, 4-hydroxybenzoic acid and protocatcheuic acid of ethanolic extract of *Eclipta alba* L., was found to be active against carbon tetrachloride induced hepatotoxicity in rats.

The triterpene saponin, glycyrrhizin present in *Glychyrrhiza glabra* shows potential hepato-protective activity. Experimental hepatitis and cirrhosis studies on rats found that it can promote the regeneration of liver cells and at the same time inhibit fibrosis. Glycyrrhizin can alleviate histological disorder due to inflammation and restore the liver structure and function from the damage due to carbon tetrachloride. Effects of glycyrrhizin have been studied on free radical generation and lipid peroxidation in primary cultured rat hepatocytes. It also helps in lowering the SGPT, reducing the degeneration and necrosis and recovering the glycogen and RNA of liver cells.

Kutkins isolated from *Picrorhiza kurrooa* ex Benth. (Scrophulariaceae) showed significant curative activity *in vitro* in primary cultured rat hepatocytes against toxicity induced by thioacetamide, galactosamine, and carbon tetrachloride in rats (Visen *et al.*, 1998). Ursolic acid isolated from leaves of *Eucalyptus terelicomis* showed hepatoprotective effect which is comparable to silymarin against thiacetamide, galactosamine and carbon tetrachloride in rats (Sarsawat *et al.*, 1996). 3, 4, 5-trihydroxy benzoic acid (gallic acid) isolated from fruit extract of *Terminalia belerica*, is reported as a hepatoprotective principle (Anand *et al.*, 1997). Aqueous seed extract of *Cleome viscosa* and *Pterocarpus marsupium* stem bark extracts showed activity against carbon tetrachloride induced liver damage in Wistar rats (Sengottuvelu *et al.*, 2007, Mankani *et al.*, 2005).

Alkali preparation of the drug *Tephrosia purpurea,* known as sharpunkha in Ayurveda is commonly used in treatment of liver and spleen diseases, offered protective action against carbon tetrachloride and D-galalactosamine poisoning in animal models. Phytochemical analysis of the plant shows the presence of tephrosin, deguelin and quercetin (Murthy and Srinivasan 1993). From all these observations it can be concluded that phytochemicals have the potential to protect the liver from cellular damage and the toxicity by different agents. Some direct correlation has been observed between antioxidant potential and hepatoprotective activity.

Antimicrobial Agents

Plants are rich in a wide variety of secondary metabolites (tannins, terpenoids, alkaloids, and flavonoids) which have antimicrobial properties. They are traditional healers and cure infectious conditions. Increasing drug resistance and production cost of mainstream medicine urge the need to find out new sources of medicine mainly those drugs derived from plants in addition to the traditional antibiotics.

The phenols and polyphenols structurally consist of benzene ring with hydroxyl groups. Catechol and pyrogallol both are hydroxylated phenols of plant origin, shown to be toxic to microorganisms. The site(s) and number of hydroxyl groups on the phenol group are thought to be related to their relative toxicity to microorganisms, with the evidence that increased hydroxylation results in increased toxicity (Ionela and Ion 2007). Tannins are class of poly-phenolic compounds and they are found almost in all plant parts.

Many human physiological activities, such as stimulation of phagocytic cells, host-mediated tumor activity, and a wide range of anti-infective actions, have been assigned to tannins. Their mode of antimicrobial action may be related to their ability to inactivate microbial adhesions, enzymes, cell envelope, transport proteins, etc. Condensed tannins have been determined to bind cell walls of ruminal bacteria, preventing growth and protease activity (Jones *et al.*, 1994). Flavonoids are another class of phenolic compounds known to be synthesized by plants in response to microbial infection, and have been found *in vitro* to be effective antimicrobial substances against a wide array of microorganisms. Their activity is probably due to their ability to complex with extra cellular and soluble proteins and to complex with bacterial cell walls (Ionela and Ion 2007). Essential oils are secondary metabolites that are highly enriched in compounds based on an isoprene structure, called terpenes. Terpenes or terpenoids are active against bacteria, fungi, viruses, and protozoa. Other phytochemical class of compounds like alkaloids and coumarins also shows antimicrobial properties. Several plant extracts and isolated compounds have been tested for their antimicrobial properties (Table 8.1) (Marjorie 1999).

For Metabolic Disorders: Phytochemicals as Anti-diabetic Agents

Diabetes mellitus (DM) is a chronic metabolic disorder characterized by hyperglycemia caused by defective insulin secretion, resistance to insulin action, or a combination of both. It is one of the recent major health problems worldwide. The morbidity and mortality associated with diabetes is anticipated to account for a substantial proportion of health care expenditures. Currently available therapeutic options for non-insulin-dependent diabetes mellitus, such as dietary modification, oral hypoglycemics, and insulin, have limitations of their own. Many kinds of natural products (terpenoids, alkaloids, flavonoids, phenolics) and some others have shown antidiabetic potential, particularly, schulzeines A, B, and C, radicamines A and B, 2,5-imino-l,2,5-trideoxy-L-glucitol, -homofuconojirimycin, myrciacitrin IV and dehydrotrametenolic acid have shown significant antidiabetic activities. Amongst active medicinal herbs *Momordica charantia* L. (Cucurbitaceae), *Pterocarpus marsupium* Roxb. (Leguminosae), and *Trigonella foenum graecum* L. (Leguminosae) have been reported as beneficial for treatment of type 2 diabetes (Mankil *et al.*, 2006).

There are several drug treatments currently available, the need for new herbal agents for treatment of diabetes is necessary due to the increase in diabetic cases and its complexity. Only a few plants have been evaluated by extraction of phytochemicals and their usefulness in experimental diabetic animal models. From plants like *Allium cepa, Allium sativum, Ficus benghalensis, Gymnema sylvestre, Pterocarpus marsupium* etc. active hypoglycemic principles have been isolated and their mechanism of action have studied. Most of them seem to act directly on pancreas (pancreatic effect) and stimulate insulin level in blood. Some have extra pancreatic effect also by acting directly on tissues like liver, muscle etc. and alter favourably the activities of the regulatory enzymes of glycolysis, gluconeogenesis and other pathways. Since the plant products have comparatively lesser side effects, they can act potential hypoglycemic drugs and may provide lead molecules for the development of new and better anti-diabetic drugs.

Swerchirin, xanthone from *Swertia chirayita* (Gentianaceae) shows hypoglycemic activity. Except in rats with severe pancreatic damage, swerchirin showed better

Table 8.1: Plant extracts showing antimicrobial properties

Common Names	*Scientific Names*	*Classes of Compounds*	*Activity Against*
Aloe	*Aloe vera* Tourn. ex Linn.	Complex mixture, mainly flavonoids	*Corynebacterium*, *Salmonella*, *Streptococcus*, *S. aureus*
Bael tree	*Aegle marmelos* Correa ex Koen.	Essential oil; Terpenoid	Fungi
Black pepper	*Piper nigrum* L.	Alkaloid; Piperine	Fungi, *Lactobacillus*, *Micrococcus*, *E. coli*, *E. faecalis*
Clove	*Syzygium aromaticum* (Linn.) Merrill and Perry.	Terpenoid; Eugenol	Bacteria and fungi
Eucalyptus	*Eucalyptus globulus* Labill.	Polyphenols; Tannins	Bacteria, viruses
Garlic	*Allium sativum* Linn.	Sulfated terpenoids; Sulfoxide	Bacteria and fungi
Green tea	*Camellia sinensis* (Linn.) O. Kuntze.	Flavonoids; Catechin	Bacteria and fungi
Lemon balm	*Melissa officinalis* Linn.	Polyphenols; Tannins	Viruses
Orange peel	*Citrus sinensis* (Linn.) Osbeck	Terpenoid	Fungi
Senna	*Cassia angustifolia*	Anthraquinone; Rhein	*S. aureus*

glucose lowering effect compared to tolbutamide, hypoglycemic agent, when studied in normal as well as diabetic rats (Saxena *et al.*1996). Karanjin, a flavone showed significant hypoglycemic activity with a single dose of 0.5 mg/kg in normal rats. However, it was ineffective in alloxan- induced diabetic rats (Mandal *et al.*, 1986).

Summary

The use of medicinal plants, since ancient days in the world and especially in India, contributes significantly to primary health care. India is the home for a number of commercially important medicinal and aromatic crops; its geographical location with salubrious climate makes India virtually the herbarium of the world. Phytochemical components present in herbs or herbal products can protect the tissues from oxidative injury, promote virus elimination, block fibrogenesis, or inhibit tumor growth. Biologically active molecules derived from herbal extracts can serve as suitable primary compounds for effective and targeted drugs. Plants have yielded many important medicines in the past and research groups all over the world look to discover new lead compounds or complex extracts for future drug development. This requires biological testing of plant extracts and isolation of their active components, as well as toxicological, pharmacokinetic and ultimately clinical studies. Very often clinical studies are also conducted. Overall, this is time-consuming and generally very expensive process of uncertain outcome. However, many of the pure natural products and some of the phytotherapeutic preparations used today are derived from plants used in indigenous cultures.

Research till to date proves that phytochemicals, the secondary metabolites, synthesized by the plants for their survival and development are potent remedy for several aliments of human beings. The lesser side effects and efficiency made herbal treatment popular. Plants made much contribution towards the discovery of several synthetic drugs which are in use. Hence the treasure of natural remedy, medicinal plants have to be explored scientifically by chemical and biological assays, which may help to find solutions for several health problems faced by human race. Also it is necessary to protect our natural wealth, plant biodiversity from extinction for future use. Due to global warming and other climatic changes as well as urbanization wiped out several rare plants. Thus it is imperative to take some steps to protect them.

References

Ahmed Iqbal M, Shihab Hasan M, Jamal Uddin S, Ayedur Rahman and Mohammad Mehedi Masud 2007. Antinociceptive and Antioxidant Activities of the Ethanolic Extract of *Excoecaria indica*. *Dhaka Univ. J Pharm Sci* **6:** 51-53.

Ali M and Shuaib M 1997. Withanolides from the stem bark of *Withania somnifera*. *Phytochem* **44:** 1163-1168.

Anand K K, Singh B, Chand D and Bhandan B K 1992. An evaluation of *Lawsonia alba* extract as hepatoprotective agent. *Planta Medica* **58:** 22-25.

Anand K K 1997. 3,4,5-Trihydroxy benzoic acid (gallic acid), the hepatoprotective principle in the fruits of *Terminalia belerica* - bioassay guided activity, *Pharmacol Res* **36:** 315-321.

Anand R, Patnaik G K, Kulshreshtha D K and Dhawan B N 1994. Antiurolithiatic activity of lupeol, the active constituent from *Crateva nurvala*. *Phytotherapy Res* **8:** 417.

Atal C K, Sharma M L and Khariya A 1986. Immunomodulating agents of plant origin. *J Ethnopharmacol* **18:** 133–141.

Barik B R 1976. Premnazole, an isoxazole alkaloid of *Premna integrifolia* L. and *Gmelina arborea* L. with anti-inflammatory activity. *Fitoterapia*. **639:** 395.

Bisset N G 1994. Herbal drugs and phytopharmaceuticals. A handbook for practice on a specific basis. Stuttgart: *Medpharm Scientific* 326-328.

Butlet M S 2004. The role of natural product chemistry in drug discovery. *J Nat Prod.* **67:** 2141-2153.

Cragg G M and Newman D J 2005. Plants as source of anticancer agents. *J Ethnopharmacol.* **100:** 72-79.

Cragg G M, Newman D J and Snader K M 1997. Natural products in drug discovery and development. *J Nat Prod*. **60:** 52-60.

Daniel S Fabricant and Norman R Farnsworth 2001. The Value of Plants Used in Traditional Medicine for Drug Discovery. *Environmental Health Perspectives*. **109:** 69-75.

Devi P U 1996. *Withania somnifera* Dunal (Ashwagandha): potential plant source of a promising drug for cancer chemotherapy and radio sensitization. *Indian J Exp Biol* **34:** 927-932.

Elangovan V, Ramamoorthy N and Balasubramanian S 1994. Studies on the antiproliferative effect of some naturally occurring bioflavonoidal compounds against human carcinoma of larynx and sarcoma-180 cell lines. *Indian J Pharmacol* **26:** 266-9.

Eva Stefanovits-Banyai, Maria H. Tulok, Attila Hegedus, Csilla Renner and Ilona Szollosi Varga 2003. Antioxidant effect of various rosemary (*Rosmarinus officinalis* L.) clones. *Acta Biologica Szegediensis* **47:** 111-113.

Fakeye T O, Pal A, Bawankule D U and Khanuja S P 2008. Immunomodulatory effect of extracts of *Hibiscus sabdariffa* L. (Family Malvaceae) in a mouse model. *Phytother Res* **22**: 664-8.

Gregus Z and Kiwassen C 1995. Mechanism of toxicity. *In*: Klaassen C D (Ed) *The Basic Science of Poisons*, 5th ed. New York, McGraw-Hill. 35-74.

Gryglewski R J, Korbut K and Robak K 1987. On the mechanism of antithrombotic action of flavonoids. *Biochem Pharmacol* **36:** 317-321.

Ionela Daciana Ciocan and Ion I Bara 2007. Plant Products as Antimicrobial Agents. *Analele Stiinþifice ale Universitaþii*, Alexandru Ioan Cuza, Sectiunea Genetica si Biologie Moleculara, TOM **8:** 151-156.

Jones G A, McAllister T A, Muir A D and Cheng K J 1994. *Appl Environ Microbiol* **60:** 1374–1375.

Lee S E, Hwang H J, Jeong H S and Kim J H 2003. Screening of medicinal plant extracts for antioxidant activity. *Life Sci* **73:** 167-79.

Mandal B and Maity C R 1986. Hypoglycemic action of karanjin. *Acta Physiol Pharmacol Bulg* **12:** 42-6.

Mankani K L, Krishna V, Manjunatha B K, Vidya S M, Jagadeesh Singh S D, Manohara, Anees-Ur Raheman and Avinash K R 1995. Evaluation of hepato- protective activity of stem bark of *Pterocarpus marsupium* Roxb. *Ind J Pharmacol* **37:** 165-168.

Mankil Jung, Park Moonsoo, Hyun Chul Lee, Kang Yoon-Ho, Eun Seok Kang and Sang Ki Kim 2006. Antidiabetic agents from medicinal plants. *Curr Med Chem* **13:** 1203-1218.

Marjorie M C 1999. Plant Products as Antimicrobial Agents. *Clinical Microbiol Rev* **12:** 564–582.

Masuda Y and Nakamura Y 1990. Effect of oxygen deficiency and calcium omission on CCl_4 in isolated perfused livers from phenobarbutal pretreated rats. *Biochem Pharm* **40:** 1865-1876.

Mukherjee P K, Saha K, Pal M and Saha B P 1997. Studies on the anti-inflammatory activity of rhizomes of *Nelumbo nucifera* (letter). *Planta Med* **63:** 367-369.

Neogi N C, Rathor R S, Shrestha A D and Banerjee D K 1969. Studies on the anti-inflammatory and antiarthritic activity of achyranthine. *Ind J Pharmacol* **13:** 37-47.

Pallabi D E, Dasgupta S C and Gomes A 1998. Immunopotentiating and Immunoprophylactic Activities of Immue: a Polyherbal Product. *Ind J Pharmacol* **30:** 163-168.

Parmar N S and Ghosh M N 1978. Anti-inflammatory activity of gossypin, bioflavonoid isolated from *Hibiscus vitifolius*. *Ind J Pharm* **10:** 277-293.

Pradhan S C and Girish C 2006. Hepatoprotective herbal drug, silymarin from experimental pharmacology to clinical medicine *Ind J Med Res* **124:** 491-504.

Ramasundaram T, Senniayanallur R A, Paneer S M and Jayaraman B 2007. Antioxidant property of *Emblica officinalis* during experimentally induced restrain stress in rats. *J health Sci* **53:** 496-499.

Ranjith M S, Ranjit Singh A J A, Gokul Shankar S, Vijayalaksmi G S, Deepa K and Sidhu H S 2008. Enhanced Phagocytosis and Antibody Production by *Tinospora cordifolia* - A new dimension in Immunomodulation. *Afr J Biotech* **7:** 081-085.

Rates S M K 2001. Plants as Source of Drugs. *Toxicon* **39:** 603-613.

Rivier L and Bruhn J 1979. Editorial *J Ethnopharmacol*.

Sarsawat B, Visen P K S, Dayal R, Agarwal D P and Patnaik G K 1996. Protective action of Ursolic acid against chemical induced hepatotoxicity in rats. *Ind J Pharmacol* **28:** 232-39

Saxena A K, Singh B and Anand K K 1993. Hepatoprotective effects of *Eclipta alba* on sub cellular levels in rats. *J Ethnopharmacol* **40:** 155-161.

Saxena A M, Murthy P S and Mukherjee S K 1996. Mode of action of three structurally different hypoglycemic agents: a comparative study. *Ind J Exp Biol* **34:** 406-409.

Schultes R E 1978. The kingdom of plants. *Medicines from the Earth*. W.A.R. Thomson (Ed.) McGraw-Hill Book Co. New York 208.

Sengottuvelu S, Duraisamy R and Nandhakumar J and Sivakumar T 2007. Hepatoprotective activity of *Cleome viscosa* against Carbon tetrachloride induced hepatotoxity in rats. *Phcog Mag* **3:** 120-123

Shukla B, Visen P K, Patnaik G K and Dhawan B N 1992. Cholretic effect of Andrographolide in rats and guinea pigs. *Planta Med* **58:** 146-149.

Sree Rama Murthy M and Srinivasan M 1993. Hepatotprotective Effect of *Tephrosia pupurea* in experimental animals. *Indian J Pharmacol* **25:** 34-36.

Subramanian L R, Selvam A L and Mudaliar 1999. Prevention of CCl_4 induced hepatotoxicity by aqueous extract of Turmeric. *Nutr Res* **19:** 429-441.

Sundaram R and Mitra S K 2007. Antioxidant activity of ethyl acetate soluble fraction of *Acacia arabica* bark in rats. *Ind J Pharmacol* **39:** 33-8

Vagelos P R 1991. Are prescription drug prices high? *Sci* **252:**1080–1084.

Visen P K, Saraswat B and Dhawan B N 1998. Curative effect of picroliv on primary cultured rat hepatocytes against different hepatotoxins: an *in vitro* study. *J Pharmacol Toxicol Methods* **40:** 173-179.

Walia H, Kumar S and Arora S 2007. Analysis of antioxidant activity of methanol extract/fractions of *Terminalia chebula* Retz. *J Chinese Clinical Med* **2:** 361-370.

Wang C J, Wang J M, Lin W L, Chu C Y, Chau F P and Tseng T H 2000. Protective effect of *Hibiscus* anthocyanins against tert-butl hydroperoxide-induced hepatic toxicity in rats. *Food Chem Toxicol* **38:** 411-6.

Medicinal Plants: Aspects and Prospects (2014) ***Pages* 126–133**
***Editors:* Mukesh Kumar, Anjali Khare and C.P. Shukla**
***Published by:* BIOTECH BOOKS, NEW DELHI**

Chapter 9

Ethnobotany of Some Medicinal Plants Used by Gujjar Tribe of Sub-Himalayan Tracts, Uttarakhand

Jyotsana Sharma and R.M. Painuli

Department of Botany, H.N.B. Garhwal University, Srinagar Garhwal – 246 174, Uttarakhand

ABSTRACT

The present chapter aims to document the traditional knowledge on different plant species used as folk herbal medicines by Gujjar tribe of Sub- Himalayan tracts, Uttarakhand. The Gujjars are one of the most important migratory tribe of Himalayas, inhabit in close proximity to their environment along with their livestock's. Generally they depend on locally available plant species to fulfil their requirements such as food, fodder, fuel, medicine etc. They have their own traditional system of medicine and still rely on it for the treatment of different ailments affecting humans as well as their livestocks. In this study frequent field trips were made for the collection of plant specimens and information on medicinal aspects was gathered from traditional healers through questionnaire and interviews. The following communication highlights the ethnomedicinal information on 35 plant species belonging to 23 families used as herbal remedies by this tribe.

Keywords: Ethnomedicine, Gujjar tribe, Medicinal plants, Traditional knowledge.

Introduction

The herbal medicines are considered to be of great importance among different communities in many developing countries (Gosh 2003). According to WHO as many as 80 per cent of the world's population depends on traditional medicine and in India 60 per cent of the people in rural areas use herbal medicines (WHO Status Report 2002). In different societies through out the globe medicinal plants makes the base of primary healthcare system and about 85 per cent of traditional medicines are derived from plants (Fransworth 1988). Throughout the world the role of folk herbal medicine has been put forth by different workers on different tribal groups (Lindsa 1978, Hamilton 1995, Ankli *et al.*, 1999, Astaw *et al.*, 1999, Tabuti *et al.*, 2003, Van Wyk *et al.*, 2008). In Indian sub-continent ethnomedicinal activities of plants on various aspects have also been highlighted in recent years (Jain 1991, Painuli 1996, Maheswari 2000, Harsha *et al.*, 2002, Negi *et al.*, 2002, Katewa *et al.*, 2004, Samal *et al.*, 2004, Prusti 2007). The tribal population has a wide knowledge on different aspects of plants as well as their utilization pattern and the management of surrounding environment. They live very close to nature in the areas affording their needs of food, fodder, fuel, medicine etc. Because of lack of modern medical facilities and remoteness the plants are the main source of medicine for them. Still now they believe in their age old traditional knowledge system and medicinal plants are used as herbal remedies for the treatment of different ailments. The traditional knowledge of folk medicine is fast declining in many parts of the world including India because this valuable knowledge is being transferred from one generation to other through oral methods without any written documentation. Therefore, continuous efforts should be made for the conservation of ethnomedicinal plants and to collect the folk information which will provide benefits for future generation.

The Uttarakhand Himalaya is a home of diversity having diverse flora as well as ethnic groups such as Rajis, Jaunsaris, Tharus, Bhoxas, Gujjars and Bhotiyas. Each having different cultures, traditions and various ways of utilization and manipulation of the plants. The state is well known for its cultural heritage and rich medicinal flora. A lot of work on different aspects of medicinal plants in Uttarakhand Himalaya has been done by several workers (Gaur *et al.*, 1984, Kalakoti and Pangety 1988, Purohit 1997, Maikhuri *et al.*, 1998, Singh 2003, Bhatt and Negi 2006, Bhatt and Vashistha 2008). The Sub-Himalayan tract of Uttarakhand Himalaya is bestowed with rich biodiversity. The Gujjar tribe of this tract is a nomadic one and live with the herds of their livestocks. Their economy is totally dependent on selling milk products in the market. Some Gujjar families have their permanent settlement in this zone while others live inside the forest region and migrate from one place to another, as in winter they live in foothill area and in summer they move to higher altitudinal Himalayan zones. They are dependent on their ambient environment to fulfil their needs. Similar to other tribes they also have their own indigenous medicine system and a strong faith on it for the recovery of diseases. It has been observed that there is a wide spread use of medicinal herbs.

A thorough study of literature revealed that various aspects of ethnomedicine used by this tribe of the study area have not been properly carried out. Hence, there is an urgent need to make attempt for the documentation of this precious knowledge for future prospects.

Methodology

The present study is based on extensive field surveys made from July 2007 to September 2008 of the study area for the collection of data on medicinal aspect of various plant species. Information was obtained through a series of interviews with traditional healers or elderly known people who still practiced their indigenous system of medicine. Questionnaire was prepared including vernacular name of plants, plant part used, detailed information about mode of preparation (*i.e.* decoction, paste, powder, infusion), form of usage either fresh or dried, their administration and medicinal use. Data was confirmed by cross checking with different herbal practitioners of study area.

The collected plant materials were identified with the help of standard floras (Babu 1977, Gaur 1999) and identification was confirmed at the herbarium in Botany Department of HNB Garhwal University, Srinagar. Collection and maintenance of plant specimens have been made by following standard methodology (Jain 1989, Rao and Sharma 1990).

Result and Discussion

The present study reveals that the traditional healers of study area have good knowledge on the medicinal property of different plant species that occur in their locality. In the following study 35 plant species belonging to 35 genera and 23 families have been recorded which are arranged in alphabetical order to their botanical name. The vernacular names, habitat, locality and medicinal uses of each species along with their mode of administration are given in Table 9.1.

Usually it was found that for the preparation of herbal medicine, different parts of plant species are used, among them leaf constitute the highest percentage of utilization. Medicines were prepared in the form of powder, decoction, paste, and juice. Mostly the herbal remedies are made from single plant species, or the combination of more than one plant species also occurs. The common mode of administration is oral and generally the plant parts are taken in fresh form, if fresh parts are unavailable then dried parts are also used. Sugar crystals are mixed in the preparation of medicine from *Clerodendrum viscosum* and *Eclipta prostrata* to increase the efficacy of the drug. Interestingly in the treatment of infertility the roots of *Asparagus adscendes* are collected only on Friday night, practitioners believe that if it is taken on any another day then the effectiveness of the drug will be decreased.

It was evident from the observation that the tribal maintain secrecy about the uses of plant wealth and believe that if such knowledge open out for other people then the medicines will lose their power of healing. The knowledge of identifying medicinal plants is confined only few traditional practitioner or elder ones, younger generation takes no interest in their vast knowledge because of modernization and migration from tribal area. The tribal knowledge is useful for future researches in the field of taxonomy, Ethnobotany, pharmacology etc. and also provides a platform for the discovery of new drugs. Hence, strategies should be made to save this wealth of knowledge for the benefits of future generations. It has also been observed that medicinal plants are taken in large quantity for the preparation of medicines by the

Table 9.1: Plants with their traditional uses

S.N.	Botanical Names and Famies	Vernacular Names	Life Forms	Plant Parts and their Uses
1.	*Acorus calamus* L. (Araceae)	Bach	H	About 2-3g powder of tubers is given orally twice a day with hot water in pneumonia.
2.	*Aloe vera* (L.) Burm.f. (Liliaceae)	Chhinar	H	Leaves are roasted in fire and the sap is applied externally on boils.
3.	*Amaranthus spinosus* L. (Amaranthaceae)	Jangli chulai	H	Vegetable of tender leaves is given in stomachache.
4.	*Asparagus adscendes* Buch.-Ham. ex Roxb. (Liliaceae)	Satawar	S	Roots are taken in Friday night, made into paste, tablets (2g each) of this paste is given orally in infertility in women, thrice a day.
5.	*Bombax ceiba* L. (Bombacaceae)	Sembal	T	Powder of calyx and roots of *Asparagus adscendes* is mixed together in equal quantity, 2g of this is given orally in sexual diseases, thrice a day for 8-10 days.
6.	*Boerhavia diffusa* L. (Nyctaginaceae)	Piliya	H	Finally crushed roots are soaked in water for whole night, one glass of this water is given orally for 2 weeks in early morning to cure jaundice.
7.	*Buddleja asiatica* Lour (Buddlejaceae)	Jangli methi	S	Half cup decoction of leaves is given twice a day to get relief from body pain.
8.	*Campylotropis speciosa* (Royle ex Schindler) (Fabaceae)	Batan	S	Half cup decoction of leaves is given orally early in the morning for 3 days.
9.	*Cannabis sativa* L. (Cannabaceae)	Bhang	S	Finally crushed fresh leaves are externally applied on piles.
10.	*Celastrus paniculatus* Willd. (Celastraceae)	Sisiya/Talli	S	Paste of leaves is applied on allergy.
11.	*Centella asiatica* (L.) Urban (Apiaceae)	Bhrami buti	H	Powder of leaves and black pepper *(Piper nigrum)* is given orally with water to increase mental ability, twice a day daily.
12.	*Cissampelos pareira* L. (Menispermaceae)	Katori bel/ Paada	C	Leaf paste is externally applied on boils.Root juice is given orally in dysentery.
13.	*Clerodendrum viscosum* Ventenat (Verbenaceae)	Lojad	S	Root powder with sugar crystals is administered orally with hot water in stomachache.
14.	*Colebrookia oppositifolia* J.E. Smith (Verbenaceae)	Sambhalu	S	Leaf paste is applied externally on wounds. Decoction of leaves is used to bath the lady after delivery to avoid cold.

Contd...

Table 9.1–*Contd...*

S.N.	Botanical Names and Famies	Vernacular Names	Life Forms	Plant Parts and their Uses
15.	*Cuscuta reflexa* Roxb. (Cuscutaceae)	Andar bel/ Akash bel	C	Plant is collected in maximum quantity, roasted in a pan and tied externally on swellings.
16.	*Dendrocalamus strictus* (Roxb.) Nees (Poaceae)	Maanu/Baans	S	Pulp of stem with hot water is administered orally twice a day to get relief from pneumonia.
17.	*Eclipta prostrata* (L.) (Asteraceae)	Sutiya	H	Juice of whole plant with sugar crystals is given in fever. Decoction of leaves is used to bath the lady after delivery.
18.	*Eranthemum pulchellum* Andrews (Acanthaceae)	Kamliya	S	Leaves are fried in *Brassica campestris* oil (sarsoon) and applied externally on cracked feets.
19.	*Gnaphalium hypoleucum* DC. (Asteraceae)	Meliya/Jajoda	H	Leaf paste is applied externally on cuts and wounds.
20.	*Hemigraphis rupestris* Heyne ex T. Anderson (Acanthaceae)	Kethiya	S	Leaf powder is mixed with curd and orally given in piles once a day for one month.
21.	*Holoptelea integrifolia* (Roxb.) Planchon (Ulmaceae)	Papri/Kanju	T	Paste of inner bark is applied on mouth blisters. Tender leaves paste is applied on eczema.
22.	*Inula cappa* Buch.-Ham. ex D.Don (Asteraceae)	Damiya	S	Root powder is mixed with honey and one teaspoonful is given orally in throat infection.
23.	*Ipomoea carnea* Jacquin (Convolvulaceae)	Beshram	S	Leaves are fried in *Brassica campestris* oil (sarsoon) and tied externally on cuts and wounds.
24.	*Lepidagathis incurva* Buch.-Ham. ex D. Don (Acanthaceae)	Phula/Motia	S	Leaf juice is used to treat conjunctivitis.
25.	*Nepeta graciflora* Benth. (Lamiaceae)	Dhamni	H	Leaf paste is applied externally on boils.
26.	*Oxalis corniculata* L. (Oxalidaceae)	Khatti amli	H	Juice of leaves is given orally in jaundice, twice a day for 1-2 week.
27.	*Phyla nodiflora* (L.) Greene (Verbenaceae)	Jakhmi ghaas	H	Paste of leaves is applied on cuts and wounds.
28.	*Pogostemon benghalense* (Burm.f.) Kuntze (Lamiaceae)	Lojad	S	Roots are used as tooth stick to get relief from toothache.
29.	*Plumbago zeylanica* L. (Plumbaginaceae)	Chitra jadi	H	Leaf paste is applied externally in eczema.

Contd...

Table 9.1–*Contd...*

S.N.	*Botanical Names and Famies*	*Vernacular Names*	*Life Forms*	*Plant Parts and their Uses*
30.	*Reinwardtia indica* Dumortier (Linaceae)	Phuliya/Phunli	S	Finely crushed flowers are soaked in water for whole night, this water is given orally in early morning for one week to cure jaundice.
31.	*Ruellia tuberosa* L. (Acanthaceae)	Mashenda	H	Fruits are chewed to get relief from mouth blisters.
32.	*Rungia pectinata* (L.) Nees (Acanthaceae)	Pathri	H	Leaves are soaked with seeds of *Macrotyloma uniflorum* (kulath) for whole night and this water is orally given in stone in kidney.
33.	*Terminalia alata* Heyne ex Roth (Combretaceae)	Asin	T	Leaf powder is given orally with cold water thrice a day in urinary complaints, continuously for 4 days.
34.	*Verbascum chinense* (L.) Santapau (Lamiaceae)	Perkanda	H	Fruits are boiled with milk and orally given to the drug addictive.
35.	*Vetiveria zizainioides* (L.) Nash (Poaceae)	Panni/Khaskhas	S	Root powder is given orally in dysentery.

Note: T: Tree; H: Herb; S: Shrub; C: Climber.

healers, which will cause the degradation of important medicinal flora. Therefore, it is important to create awareness for the conservation of medicinal plants.

Acknowledgement

The authors wish to express their gratitude towards the tribal people and the locals of the area for their active collaboration during field studies. They are also thankful to Prof. R.D. Gaur, Former Head, Department of Botany, HNB Garhwal University, Srinagar Garhwal, for providing necessary help.

References

Ankli A, Sticher O and Heinrich 1999. Medical Ethnobotany of the Yucatec Maya: Healers Consensus as a quantitative criterion, *Eco Bot* **53:** 144-160.

Astaw D, Abebe D and Urga K 1999. Traditional medicine in Ethiopia: Perspective and development efforts, *J Ethiop Med Prac* **1:** 114-117.

Babu C R 1977. *The Herbaceous Flora of Dehradun*, CSIR. Delhi

Bhatt V P and Negi G C S 2006. Ethnomedicinal plant resources of Jaunsari tribe of Garhwal Himalaya, Uttaranchal, *I J Trad Knowledge* **5:** 331-335.

Bhatt V P and Vashistha D P 2008. Indigenous plants in traditional healthcare system in Kedarnath valley of Western Himalaya, *I J Trad Knowledge* **7**: 300-310.

Fransworth N R 1988. Screening Plants for New Medicines, In: Biodiversity, Wilson, E.O. (Ed.) National Acadamey Press, Washington DC, 83-97.

Gaur R D 1999. *Flora of the District Garhwal North-west Himalaya* (*with ethnobotanical notes*). Transmedia, Srinagar Garhwal.

Gaur R D, Semwal J K and Tiwari J K 1984. A survey of high altitude medicinal plants of Garhwal Himalaya, *Bull Med Ethnobot Res* **4:** 102-116.

Gosh A 2003. Herbal folk remedies of Bantura and Medinipur districts, West Bengal (India). *I J Trad Knowledge* **2:** 393-396.

Hamilton A 1995. The people and plants initiative. In: Martin, G.J.(ed.) *Ethnobotany. A Method Manua*, WWF International Chapman and Hall, London.

Harsha V H, Hebbar S V S, Hegde G R and Shripathi V 2002. Ethnomedicinal knowledge of plants used by Kunabi Tribe of Karnataka in India, *Fitoterapia* **73:** 281-287.

Jain S K 1989. *Methods and Approaches in Ethnobotany*, Society of Ethnobotanists, Lucknow.

Jain S K 1991. *Dictionary of Indian Folk Medicine and Ethnobotany*, Deep Pub, New Delhi.

Kalakoti B S and Pangety Y P S 1988. Ethnomedicine of Bhotiya tribe of Kumaun Himalaya, UA, *Bull Med Ethnobo. Res* **9:** 11-20.

Katewa S S, Chaudhary B L, and Jain A 2004. Folk herbal medicines from tribal area of Rajasthan, *I J Ethnopharmacol* **92:** 41-46.

Lindsay R S 1978. Medicinal plants of Marakwet, Kenya. Royal Botanic Garden, Kew, London, UK.

Maheswari J K 2000. *Ethnobotany and Medicinal Plants of Indian Subcontinent*, Scientific Publishers, Jodhpur, India.

Maikhuri R K, Nautiyal S, Rao K S and Saxena K G 1998. Role of medicinal plants in the traditional health care system: a case study from Nanda Devi Biosphere Reserve, *Curr Sci* **75:** 152-157.

Negi C S, Sunil N, Lokesh D, Rao K S and Maikhuri R K 2002. Ethnomedicinal plants: Uses in a small tribe community in a part of Central Himalaya, *I J Hum Eco* **14:** 23-31.

Painuli R M and Maheswari J K 1996. Some interesting Ethnomedicinal Plants used by Sahariya tribe of M.P. *J Econ Taxon Bot* **12:** 179-185.

Prusti A B 2007. Plants used as ethnomedicine by Bondo tribe of Malkangiri district, Orissa, *Ethnobotany* **19:** 105-110.

Purohit A N 1997. Himalayan medicinal plants-focus on Uttarakhand. In: *Himalayan Biodiversity: Action Plan*. Gyanodaya Prakashan, Nainital, pp. 91-110.

Rao R R and Sharma B D 1990. *A Manual of Herbarium Collections*. Botanical Survey of India, Kolkata.

Samal P K, Shah A and Agrawal D K 2004. Indigenous medicinal practices and their linkage in resource conservation and physical well-being of locals in Central Himalayan region of India, *I J Trad Knowledge* **3**: 12-26.

Singh H 2003. Herbal recipes for spermatorrhoea by Bhoxa tribe of Uttaranchal. *Ethnobotany* **15:** 115-117.

Tabuti J R S, Lye K A and Dhillon S S 2003.Traditional herbal drug of Bulamogi, Uganda: plants, use and administration. *J Ethnopharmacol* **88:** 19-44.

Van Wyk B E, Wet H de and Van Heerden F R 2008. An ethnobotanical survey of medicinal plants in the South-Eastern Karoo, South Africa. *South African J Bot* **74:** 696-704.

World Health Organization 2002-2005. WHO. Traditional Medicine Strategy Report, Document WHO/EDM/TRH/2002.1

Medicinal Plants: Aspects and Prospects (2014) *Pages* 134–139
Editors: Mukesh Kumar, Anjali Khare and C.P. Shukla
ISBN: 978-81-7622-309-6
Published by: BIOTECH BOOKS, NEW DELHI

Chapter 10

Influence of Medicinal Plant Products on Household Insects

Kumud Rai[1], Anjali Khare[2] and Bharat Bhushan[3]

[1]Department of Zoology, Nehru Gramin Bharti University, Allahabad – 212 306, U.P.
[2]Department of Botany, Advance Institute of Science and Technology, Dehradun – 248 001
[3]Department of Botany, Sahu Jain P.G. College, Najibabad – 246 763

ABSTRACT

The present study deals with the indigenous pest control practices of home maker of Eastern Uttar Pradesh to maintain the economic injury level, who uses different bio-products to store food grains. Storage of these items is necessary to prevent spoilage, increase usability and also to protect them from household insects pests. Home maker commonly use plant-products against different types of insect pests such as weevils, beetles etc., which affects the life processes of household insect pests commonly found in the grains treated with various plant products. The plant products are more easily available, cheaper and less persistent as compared to existing synthetic chemicals. Besides preventing quantitative and qualitative losses, they don't leave toxic residues in food grains. More than 2100 plant species have been reported to possess insecticidal, antifeedant or repellent properties. The present chapter represents the details of 12 locally distributed plant species with their basic traditional uses. Botanical name, vernacular name, common name, part used, method of use and possible modes of action of each species has been described well.

Keywords: Household pests, Food grain, Plant parts.

Introduction

Cereals play a vital role in agricultural economy of the country. Indigenous people have developed a close and unique connection with land and environment in which they occur. Without having a scientific knowledge the farmers grow cereals all over the globe. Out of them 70 per cent food grains are stored traditionally by the farmer forms at home for their own consumption. About 30 per cent surplus seeds are sold to traders and government agencies in India. 5-10 per cent of the produce losses occur during the storage periods of 3 - 8 months especially by the humidity. The food grains are damaged by *Sitophilus oryzae, Tragoderma granarium, Collosobranchus chinensis* etc. Chemical control of pests in stored grain cause health hazards to human beings and also in non-target organisms. The interest in bio-products has been raised because of the disadvantages associated with the use of chemical pesticides. Home-makers are known for their rich indigenous traditional knowledge for controlling various insect pests which got evolved over generations. Various home-makers emphasize the value of tradition based knowledge. Patro and Pati (1997) have used water extract of turmeric rhizomes (*Curcuma longa* L.) against *Collosobranchus chinensis* as a repellent. The seed powder of custard apple (*Annona squamosa*) is quite effective upon many insect pests found in various stored products and food grains. The present findings support the works of Mishra *et al.* (1992), Dwivedi *et al.* (2005) and Shukla (2010).

The present studies determine the effectiveness of plant products locally used as protectants against the pests by home makers of eastern Uttar Pradesh.

Methodology

In order to collect information on traditional indigenous knowledge related to the plant products used by local people of Eastern Uttar Pradesh, regular personal contacts and group discussions were held with the home makers of different families using plants as pest controls. In this reckoning up, plants have been described following their botanical names followed by families, common names, vernacular names, used parts, methods of use and possible modes of action.

Enumeration

1. *Alium sativum* (L.)

Family: Solanaceae

Common Name: Garlic

Vernacular Name: Lahasun

Parts Used: Cloves

Methods of Use: Put the cloves of garlic in different layers of stored rice and close the container tightly.

Possible Mode of Action: Garlic is insecticidal and antifungal thus controls infestation in rice weevils (*Sitophilus oryzae*)

2. *Annona squamosa* (L.) Merv.

Family: Anonaceae

Common Name: Custard Apple

Vernacular Name: Sarifa

Parts Used: Seeds

Methods of Use: Dried seed powder of custard apple is used to protect the pulses from the infestation of *Collosobranchus maculates* and*Collosobranchus chinensis* Linn.

Possible Mode of Action: This seed powder reduces the egg laying capacity of female pest, population build up and grain damage.

3. *Azadirachta indica* A. Juss (L.)

Family: Meliaceae

Common Name: Neem

Vernacular Name: Neem

Parts Used: Dry leaves

Methods of Use: Make cloth bag of shade dried neem leaves and place them at bottom, middle and top of the container.

Possible Mode of Action: Neem leaves are sterilant, antiparasitic, insect and pest repellent, antifungal, non-toxic and thus prevent spoilage of grains and are commonly used against wheat weevils (*Trogoderma granarium* Everts) and rice weevils (*Sitophilus oryzae* Linn).

4. *Brassica comprestris* (L.)

Family: Brasicasseae

Common Name: Mustard

Vernacular Name: Sarson

Parts Used: Seed Oil

Methods of Use: Seed oil smeared over pulses beetles (*Collosobranchus chinensis* Linn.)

Possible Mode of Action: Seed oil has an antifeedent property or insect repellent property.

5. *Capsicum frutescens* (L.)

Family: Solanaceae

Common Name: Wild chilli

Vernacular Name: Jungle mirchi

Parts Used: Burnt seeds

Methods of Use: Dried burnt seed of wild chilly put with pulse to protect them from all pests.

Possible Mode of Action: The insecticidal property of chilli helps in safe storage.

6. *Cinnamomum aromaticum* Nees

Family: Lauraceae

Common Name: Cinnamon

Vernacular Name: Dal Chinni

Parts Used: Bark

Methods of Use: Mix dry powder of cinnamon bark with food grains (rice and pulses) to keep them protect from the infestation of *Sitophilus oryzae* and *Collosobranchus chinensis*.

Possible Mode of Action: Cinnamon significantly depressed the egg deposition of *S. oryzae* and *C. chinensis*, and reduction of fecundity of *S. oryzae*.

7. *Curcuma domestica* Valeton

Family: Zingiberaceae

Common Name: Turmeric

Vernacular Name: Haldi

Parts Used: Rhizome

Methods of Use: Mix turmeric powder thoroughly with rice grains and fill in air tight containers to protect against rice infestation.

Possible Mode of Action: Turmeric has pesticidal, insecticidal and antifungal properties. It affects as repellent and contact toxicity on a number of stored product beetles (*Collosobranchus chinensis*).

8. *Mentha longifolia* (L.)

Family: Lamiaceae

Common Name: Mint

Vernacular Name: Pudina

Parts Used: Leaves

Methods of Use: Grinded leaves of mint are air dried in shade, ground to a coarse powder and mixed with rice to protect from *S. oryzae*.

Possible Mode of Action: Mint show pesticidal activity against stored grain pest *Sitophilus oryzae*.

9. *Lobelia nicotianifolia* (L.)

Family: Campanulaceae (Bell flower family)

Common Name: Wild Tobacco

Vernacular Name: Jungli Tumabakoo

Methods of Use: Shade dried leaves put with rice to protect from rice weevils (*Sitophilus oryzae*) and rice earhead bug infestation.

Possible Mode of Action: Strong flavor, put repellent and insecticidal properties. Tumbakoo protects the grains from all pests.

10. *Ocimum sanctum* (L.)

Family: Lamiaceae

Common Name: Tulsi

Vernacular Name: Tulsi

Parts Used: Leaves

Methods of Use: Shade dried leaves kept within the dried storage vessel (rice and pulses) protects the pulses from weevils (*Sitophyllus oryzae*) infestation.

Possible Mode of Action: It has insecticidal properties against all pests.

11. *Syzigiuma romaticum* (L.) Merr.

Family: Myrtaceae

Common Name: Clove

Vernacular Name: Laung

Parts Used: Flower

Methods of Use: Put 20-30 cloves at the top layer of rice and pulse container and close the container tightly.

Possible Mode of Action: Strong flavor and bitterness of cloves protects the grain from spoilage and pulse weevils (*Collosobranchus chinensis*) and Rice weevils (*Sitophyllus oryzae*).

12. *Vitex negundo* (L.)

Family: Vitaceae

Common Name: Indian privet

Vernacular Name: Nirgunde

Parts Used: Leaves

Methods of Use: Shade dried leaves powder of *Vitex* is kept inside the container of rice and pulses.

Possible Mode of Action: Vitex show highest mortality against the pulse beetle.

Conclusion

Home makers have been using a large number of biological products to keep the insect pests under economic threshold levels. The plants mentioned in the text have either antifeedent or insect repellent properties. This type of indigenous knowledge can minimize the harmful effects of the chemicals on human body and expenses of the synthetic chemical insecticides. Therefore, the indigenous management of house hold insect pests is quite safe and eco-friendly. The traditional practices followed by home-makers for the storage of their plant products save them from household insect pests. Over the years, these practices have gradually become more popular among the farmers.

References

Dwivedi N D, Shukla S and Chndel B S 2005. Insecticidal prosperities of certain plant extracts as protectants against *Phizopertha dominica*, FABR, *Nat J Life Sci* **2:** 109-111.

Mishra B K Mishra P R and Mahapatre H K 1992. Studies on some plant product mixtures against *Sitophilus oryzae*. L. Infesting wheat seeds. *Ind J Pl Prot* **20:** 178-182.

Patro B and Pati R N 1997. Insecticidal activity of some plant extracts against the pulse beetle *Collosobranchus chinensis* (Linn) informing green gram seeds. *Sci Cult* 63-91.

Shukla C P 2010. Herbal Plant Diversity of Indogangetic Plains. **In:** Kumar A and Das G (eds) Intellectual of Rights: *vis-a-vis* Genome Conservation. Narosa Publ., New Delhi.

Medicinal Plants: Aspects and Prospects (2014) ***Pages*** **140–147**
Editors: **Mukesh Kumar, Anjali Khare and C.P. Shukla**
ISBN: 978-81-7622-309-6
Published by: **BIOTECH BOOKS, NEW DELHI**

Chapter 11

Potential Applications of *Aloe vera* Tourn. ex Linn.: An Increasingly Accepted Medicinal Plant

Anjali Khare

Department of Botany
Advance Institute of Science and Technology,
Dehradun – 248 001, Uttarakhand

ABSTRACT

Aloe vera (=*Aloe barbadensis*) doesn't need any introduction as it is highly popular and widely used in various market products. *Aloe vera* contains high medicinal properties. It is used for the treatment of various ailments and also for cosmetic preparations. Its ointment is available for burns, rashes and cuts. The plant sap is a thick, mucilaginous gel which has a lot of medicinal uses. Oral intake and topical dressing of *Aloe vera* sap helps in curing various problems. Its plants are now being used for the research work related to the cancer and AIDS treatment. In real sense this plant is a boon for the field of medicine.

Keywords: Aloe vera, Medicinal use.

Introduction

Aloe vera, also known as the true or medicinal aloe, is a species of succulent plant that probably originated in the southern half of the Arabian peninsula, Northern Africa, the Canary Islands and Cape Verde. Also called "the elixir of youth" by the

Russians, "the herb of immortality" by the old Egyptians or the "harmonious remedy" by the Chinese, *Aloe vera* is without a doubt the medicinal herb most widely known for its noticeable impacts on health. *Aloe vera* derives its name from the Arabic word 'alloeh', which means bitter. There are nearly 200 varieties of Aloe growing in the dry regions of Africa, Europe, Asia and America but the one that is popularly used in commercial products is *Aloe barbadensis* Miller. It grows in arid climates and is widely distributed in Africa, India and other arid areas. In India, the plant is known as Korphad, Ghrtakumari or Gheekvar and is sometimes used in Ayurvedic healing. The natural range of *Aloe vera* is unclear, as the species has been widely cultivated throughout the world. Naturalised stands of the species occur in the southern half of the Arabian peninsula, through North Africa (Morocco, Mauritania, Egypt) as well as Sudan and neighbouring countries, along with the Canary, Cape Verde and Madeira Islands.

Figure 11.1: Cultivation of *Aloe vera*.

Aloe vera is a stemless or very short-stemmed succulent plant growing to 60–100 cm (24–39 in) tall, spreading by offsets. The leaves are thick and fleshy, green to grey-green, with some varieties showing white flecks on the upper and lower stem surfaces. The margin of the leaf is serrated and has small white teeth (Figure 11.1). The flowers are produced in summer on a spike up to 90 cm (35 in) tall, each flower pendulous, with a yellow tubular corolla 2–3 cm (0.8–1.2 in) long. Like other *Aloe* species, *Aloe vera* forms arbuscular mycorrhiza, a symbiosis that allows the plant better access to mineral nutrients in soil. As *Aloe vera* plants consist mostly of 95 per cent water they are not habitat of cold regions. that's the region they are mostly found in tropical and subtropical regions, mostly in Africa. Hot climate suits the growth of *Aloe vera* plants.

The species are frequently cited as being used in herbal medicines. *Aloe vera* extracts may be useful in the treatment of wound and burn healing, minor skin infections, Sebaceous cyst, diabetes and elevated blood lipids in humans. These positive effects are thought to be due to the presence of compounds such as polysaccharides, mannans, anthraquinones and lectins. The pulp of the plant is highly regarded for its anti-ageing potential. The pulp is used extensively in Siddha medicines for treating constipation, enlargement of spleen, zymotic disease, chengamaari (a type of venereal infection) etc.

Aloe vera plants are known for their medicinal use, various kinds of ailments and ointments are available for burns, rashes and cuts. They are used in various beauty preparations to give the skin a beautiful look. *Aloe vera* plant sap is a thick, mucilaginous gel. This gel has a lot of medicinal uses, that is why it is used in various ailments and ointments. There are more than 200 *Aloe vera* species of genus aloe but here we are mentioning some of the important ones.

1. ***Aloe arborescens*** - *Aloe arborescens* Miller has medicinal properties and is used in healthcare.
2. ***Aloe aristata*** - known as torch plant and lace aloe, generally found in South Africa and surrounding area. Its nectar rich flower attracts birds, wasp and bees easily.
3. ***Aloe dichotoma*** - known as quiver tree or kokerboom, generally found in South Africa and Namibia also. Its bark and branches are used by Bushmen to make quiver for their arrows.
4. ***Aloe ngobitensis***: it's a succulent and member of the aloe genus found in Kenya.
5. ***Aloe variegata*** - known as tiger aloe, is a species of aloe indigenous to South Africa.
6. ***Aloe vera barbados***: Found in northern Africa and has various medicinal uses.
7. ***Aloe wildii***: has a glass-like appearance and is a member of family asphodelaceae. Found in Africa.

Some of the *Aloe vera* species are used for nutritional diets, advantageous for human consumption. Health drinks made from them keeps you cool and provides energy. Commercial beverages and tea are very popular *Aloe vera* products.

Properties of *Aloe vera*

The most often used substance from this herb is the aloe gel, a thick viscid liquid found in the interior of the leaves. The leaves are used in the treatment of burns and the aloine - a bitter milky yellowish liquid is used as a laxative. The herb contains: 20 minerals (Calcium, Magnesium, Zinc, Chromium and Selenium), 12 vitamins (A, B, C, E and folic acid), 20 amino acids from the 22 which are necessary to the human body, over 200 active components including enzymes and polysaccharides. All the active substances enumerated before contribute to the therapeutical value of the herb. The effects that the herb has over the human body are: it toughens up the immune system owing to the 23 peptides contained by the *Aloe vera*, it accelerates and regulates the metabolism, purifies the human body from toxins, bringing about a feeling of calm. Moreover, *Aloe vera* has an antiseptic effect (by destroying the bacteria, viruses and fungi), disinfectant capabilities and can also stimulate the cell-renewing process. *Aloe vera* nourishes and supports the digesting of aliments. It manages to bring the human body to a general balanced state.

Medicinal Uses

Ghratkumari, GheeKanwar, the names are many but they all belong to some miracle plant *Aloe vera*. It s a wonder plant with health benefits so myriad and astounding that hardly any part of human body remains that is not influenced by its healing touch. From being a natural fighter against all sorts of infection, an efficient anti-oxidant to help in all digestion related problems, arthritis, stress, diabetes, cancer, AIDS to being an enhancer of beauty, *Aloe* has been proved by research to be a plant of amazing medicinal properties. The medicinal value of the plant recognized for centuries for its remarkable properties, lies in the gel like pulp obtained on peeling the leaves. Its juice has cooling properties, is anabolic in action, a fighter of pitta and guards against fever, skin diseases, burns, ulcers, boils eruptions etc. Aloe's active principle aloin is responsible for its unique digestive properties. Though it would be too exhaustive to enumerate its health benefits, the areas in which *Aloe* plant extract helps is summarized, in short, as under Antiseptic, Anti-bactericidal - *Aloe vera* produces six anti-septic agents with anti microbial properties and if its juice is taken on daily basis is protective against diseases.

The benefits of *Aloe vera* are manifold. It's gel is used to treat first-degree burns for speedy healing. *Aloe vera* gel is used to shrink warts and lessen the painful effects of shingles. The symptoms of psoriasis can be reduced with *Aloe vera* gel. European folk medicine makes extensive use of *Aloe vera* juice to reduce ulcers and heartburn. It is also used in dental problems such as bleeding gums and denture stomatitis.

Gel from the *Aloe vera* leaf has shown good results in treating facial edema. When used a mouth rinse, it offers benefits for treating lockjaw and cold spores. Recent studies have shown that *Aloe vera* might have a beneficial effect on cardiovascular health. *Aloe vera* juice is made from the nutritious inner gel which contains natural detoxifying powers that ease constipation and cleanse the bowel. It's juice is known to benefit those suffering from IBS (Irritable Bowel Syndrome). *Aloe vera* juice is said to possess soothing properties that help in colitis, peptic ulcers and digestive tract irritations. *Aloe vera* has proved its efficiency from the simplest allergies to the treatment of wounds and skin infections and even to its usage in alleviating more serious afflictions. With the help of this herb a wide variety of internal and external afflictions are controlled, like: asthma, virosis, arthritis, arthrosis, gingivitis, bronchitis, pharyngitis, intestinal inflamations, constipations, obesity, sprains, muscle strains and cutaneous inflamations. The efficiency of the herb has also been proven in the cases of anemia, deficiency illnesses, insomnia and depressions and the B-sisterole from the *Aloe vera* brings about the lowering of the cholesterol level. Also, this herb is used for controlling the side effects of chemotherapy and radiation therapy, diabetes, hepatitis, pancreatitis and multiple sclerosis.

Protector of Human Immune System

The whole leaf extract galvanizes immune system of the cells. The phagocytes increase their scavenging activities, thus cleansing the body and kicking off a whole cascade of protective actions which strengthen immunity.

Improves Digestive System

Research work carried out over the years points conclusively that aloe juice helps in digestive disorders. Constipation, diarrhoea, indigestion, irritable bowel syndrome etc are cured by the flushing action. The deposits of toxins and unwanted substances in our diet which keep accumulating in intestines prevent the absorption of essential nutrients causing nutritional deficiency, lethargy, constipation, lower back ache. *Aloe* juice helps flush out these residues boosting the digestion and giving a greater feeling of well-being.

Aloe vera in Arthritis

Being a stimulant to the immune system, a powerful anti-inflammatory, an analgesic and able to speed up cell growth, it repairs arthritis damaged tissue. While conventional allopathic treatment only relieves pain, *Aloe vera* juice taken internally and applied externally helps in repair process by regenerating cells and detoxifying the affected area. *Aloe vera* fights stress. The stress filled life of today cause bio-chemical and physiological changes in the body, making us susceptible to diseases and disfunction. *Aloe vera* juice is just the thing to get our machinery smoothly and effectively going.

Aloe vera and Cancer

Aloe juice enables the body to heal itself from cancer and the damage done by radio and chemotherapy which destroys healthy immune cells crucial to the recovery.

Aloe vera and Diabetes

It lowers glucose and tri-glyceride levels in diabetic patients. Effects can be seen from the second week of the treatment.

Aloe vera and Hepatitis

Extract of aloe juice has been shown to have beneficial effects on liver and alleviate symptoms considerably in chronic hepatitis patients.

Aloe vera in Heart Disease

Addition of isabgol and *Aloe vera* juice to the diet of patients of angina pectoris, results in marked reduction of serum cholesterol and tri-glycerides and increase in level of HDL.

Aloe vera and AIDS

A daily dose of min. 1200 mg. of active ingredients of *Aloe vera* showed substantial improvement in AIDS symptoms. *Aloe* is as to an AIDS patient as insulin is to a diabetic.

Aloe vera as Wound and Skin Disease Healer

Aloe vera gel is excellent for easing first degree burns, relieves inflammation and accelerates healing. *Aloe vera* gel has anti-fungal, anti-bacterial and anti-viral effects and helps heal minor wounds. It lessens painful effects of shingles, reduces symptoms of psoriasis and eases heartburns and ulcers.

Aloe vera has a long association with herbal medicine, although it is not known when its medical applications were first discovered. Early records of *Aloe vera* use appear in the Ebers Papyrus from 16th century BCE, in both Dioscorides' *De Materia Medica* and Pliny the Elder's *Natural History* written in the mid-first century CE along with the *Juliana Anicia Codex* produced in 512 CE. *Aloe vera* is non-toxic, with no known side effects, provided the aloin has been removed by processing. *Aloe vera* that contains aloin in excess amounts has been associated with various side effects. However, the species is used widely in the traditional herbal medicine of China, Japan, Russia, South Africa, the United States, Jamaica, Latin America and India.

Aloe vera is alleged to be effective in the treatment of wounds, internal intake of its juice has been linked with improved blood glucose levels in diabetics, and with lower blood lipids in hyperlipidaemic patients, but also with acute hepatitis (liver disease). In other diseases, preliminary studies have suggested oral *Aloe vera* gel may reduce symptoms and inflammation in patients with ulcerative colitis. Compounds extracted from *Aloe vera* have been used as an immunostimulant that aids in fighting cancers in cats and dogs; however, this treatment has not been scientifically tested in humans. The injection of *Aloe vera* extracts to treat cancer has resulted in the deaths of several patients.

Topical application of *Aloe vera* may be effective for genital herpes and psoriasis. Its extracts have antibacterial and antifungal activities, which may help in the treatment of minor skin infections, such as boils and benign skin cysts. *Aloe vera* extracts have been shown to inhibit the growth of fungi that cause tinea.

AIDS Cure

Aloe vera is showing a great potential to fight against AIDS. So many researches are going to get the best potential result out of *Aloe vera* plants for cure of AIDS.

Cancer Cure

Aloe vera plants are proving great help to cancer patients that is by the activation of white blood cells which promote growth of non-cancerous cells. Researchers have found the cancer fighting properties of *Aloe vera* and are making it count.

Commodity Uses

A wide array of products with curative and therapeutic effects have been obtained from *Aloe vera*. This herb is one of the main attractions of the pharmaceutical and cosmetic industries and also the most widely used ingredient - starting from vitamins and laxatives to face creams and body care lotions. *Aloe vera* gel contains B-sisterole, powerful anti-inflammatory and anti-cholesterol formulas and lupeol - a strong antiseptic tranquilizer. *Aloe vera* is now widely used on face tissues, where it is promoted as a moisturiser and/or anti-irritant to reduce chafing of the nose of users who suffer hay-fever or cold. It has also been suggested that biofuels could be obtained from *Aloe vera* seeds. It can also be used to retwist dreadlocked hair, a flavourite agent for vegans and those who prefer natural products. *Aloe vera* is also used for soothing the skin, and keeping the skin moist while eliminating the risk of flaky scalp and skin in harsh and dry weather.

Biologically Active Compounds

Aloe vera leaves contain a range of biologically active compounds, the best studied being acetylated mannans, polymannans, anthraquinone C-glycosides, anthrones and anthraquinones and various lectins. *Aloe vera* juice contains some anti-inflammatory fatty acids that alkalize digestive juices and prevent over acidity. Acemannan, found in *Aloe vera* is being studied for its beneficial effects in boosting T-lymphocyte cells that aid the immune system.

Aloe vera Research

Research Proves to be Vital for *Aloe vera*, Exploring New Horizon!

Scientists from the medieval age are researching on *Aloe vera*, in account to take the maximum benefit from the magic plant. Various treatments and cures proved by this plant have compelled them to exhaustively take forward the *Aloe vera* research. Significant contribution to the medicine industry, healing ailments, wounds, skin disorders, stomach problem, cosmetic and beauty problem makes aloe a unique piece. At present the major *Aloe vera* research is being in way for treating AIDS and cancer. Scientists are pretty much sure that *Aloe vera* has the properties of curing these two acute and serious diseases. In addition, various internal stomach diseases including ulcers and bowel diseases are also taken care by *Aloe vera*.

Research on *Aloe vera* has shown that it is also helpful in heart problems with great effect. Diabetic patients benefits from the use of *Aloe vera*. Some of the major researches on this magic herb have been done to find more and more probabilities and scope of these 5 qualities:

(i) Penetration

The ability of *Aloe vera* to reach deepest inside the tissue, a great property of aloe.

(ii) Antiseptic

Great antiseptic properties to cure problems.

(iii) Stimulates Cell Growth

A great property of *Aloe vera*, regeneration of tissues.

(iv) Settles Nerves

Effectively solves problem of body's nervous system.

(v) Cleanness

Detoxifies and normalizes body's metabolism.

Many *Aloe vera* researches have proved that if it is taken internally it has positive effect on immune system, stress, cancer, diabetes mellitus, stomach problem, constipation, AIDS/HIV infection, pain, inflammation, surgical incisions, leukemia and anemia. *Aloe vera* research shows when taken externally it cures problem like Hands, Feet, Sprains and Strains, Sunburn, Scratches, Open Sores, Herpes, Burns, Lines due to aging, Irritations, Ulcerated Skin, Lesions, Eczema etc.

Conclusion

Thus, it can be concluded that the benefits of *Aloe vera* are manifold. There are many more potentials on the research of this miracle plant. As research is an ongoing process and *Aloe vera* research is never ending, we hope to get some more booming news and medicinal ability of *Aloe vera* in future.

References

www.wikipedia.com

http//:www.aloeVeraplant.org

www. google.co.in

Medicinal Plants: Aspects and Prospects (2014) *Pages* 148–165
Editors: Mukesh Kumar, Anjali Khare and C.P. Shukla
ISBN: 978-81-7622-309-6
Published by: BIOTECH BOOKS, NEW DELHI

Chapter 12

Ethno-Taxonomical Studies on the Medicinal Plants of Nanda Devi National Park, Uttarakhand, India

C.S. Rana, J.K. Tiwari and L.R. Dangwal

Department of Botany, H.N.B. Garhwal University, Srinagar – 246 174, Uttarakhand

ABSTRACT

Ethnobotany deals with the applied aspects of plants and has been emerged as an important discipline of traditional botany. With the increasing demands of plant resources in developing world, it has been attracted much attention in recent past. Changing ecosystem functions due to various threats to biological diversity intensify the quest of thorough exploration of plant resources in various applications including medicine. This chapter presents a brief account of ethno-taxonomy of medicinal plants present in alpine and sub alpine regions of core and buffer zone of Nanda Devi National Park (NDNP), district Chamoli, Uttarakhand. Since inhabitants and tribal communities have strong faith and belief in traditional health care system they have been interviewed along with herbal practitioners, priests and shepherds during the entire study. Establishing small scale industry on medicinal plants may be helpful in capacity building of unprivileged inhabitants of this remote region.

Keywords: ***Ethno-taxonomy, Medicinal plants, Nanda Devi National Park, Uttarakhand.***

Introduction

Ethnobotany has attracted much attention not only due to its many economic applications, but also due to its great academic and historical importance. It is obviously a very broad field, including many aspects of botany and many other disciplines (Tiwari 1986). In recent years a number of terms have come into use in varied areas of ethnobotany, relating to all aspects of human utilization and cultural practices (Badoni and Badoni 2001).

Importance of plants in medicine is the most precious gift, nature has provided to human beings. They also provide food, fodder, fuel, timber, resins, and aromatic oil as other essential requirements of the humans (Rana 2007). Society of developing nations always had a deep interest and respect for the plant world. The intrinsic relation between plants and man has been depicted in number of ways (Rawat *et al.*, 2001). Primitive communities and tribes who live in the vicinity of forests, due to being close to the nature, posses a deep practical knowledge on indigenous flora, pertaining to curatives, culture, religious beliefs and myths (Maikhuri *et al.*, 1997, 1998).

The Indian Himalayan Region (IHR) is considered as the repository of biological and cultural diversity and supports about 18,440 species of plant, includes 1748 species of medicinal plants and 675 species of wild edibles (Negi and Gaur 1994). The representative biodiversity rich areas of the IHR have been protected through a Protected Area Network (PAN). At present there are 5 biosphere reserves, 28 national parks, and 98 wildlife sanctuaries in IHR covering 51,899.238 km^2 (Mathur *et al.*, 2000). These protected areas are distributed in tropical, temperate and alpine ecosystems. One of the protected areas Nanda Devi Biosphere Reserve (NDBR) falls in Western Himalayas of Uttarakhand State. It has been attracting worldwide attention from time to time in various disciplines of natural, physical and cultural aspects, not only for its biologically protected sites *viz.* NDNP and Valley of Flowers National Park, but also for the distinct beliefs and traditions in the daily lives of the inhabitants (Samant and Joshi 2000).

The NDNP is the first and foremost highly valued core zone of the NDBR (Adhikari 2003). It falls in greater and Trans-Himalayan region with temperate climatic conditions. It has an area of 624.60 sq. km. and has an average altitude exceeding 4500 m ASL surrounded by high mountain ridges and peaks on all sides except its western sides, which features a deep and virtually inaccessible gorge (Maikhuri *et al.*, 1997). Such geographical but fragile conditions of this remote region represents the best habitat for invaluable medicinal plants.

The taxonomical studies in the Himalayan region started dates back in eighteenth century (Burkill 1965). During the last four decades, botanical explorations in Garhwal Himalaya intensified with the collections of plants from remotest parts of the area. Rau (1975) published *"High Altitude Flowering Plants of West Himalaya"* and Naithani (1984 and 85) published *"Flora of Chamoli"*. Som Deva and Naithani (1987) have published *"The Orchid Flora of Northwest Himalaya"*. Other contemporary contributions are made by Hajra and Jain (1983); Hajra and Balodi (1995) have published *Plant Wealth of Nanda Devi Biosphere Reserve*. Murti (2001) published *Flora of Cold Desert of Western Himalaya* of Garhwal Himalayas, particularly from Nelang, Niti and Mana valleys. More recently Uniyal *et al.* (2007) published *Flowering Plants of Uttarakhand* (A Checklist) and enumerated 4700 species.

Studies on ethnobotanical aspects including medicinal plants have also been conducted in the Himalayan tract, to mention a few include Lal, *et al.*, 1989, Malhotra and Basu 1984, Purohit 1997, Singh *et al.*, 1990 and Uniyal 1989. In Garhwal Himalayas the ethnobotanical studies have gained momentum in the last two and half decades. These studies have brought numerous useful plants into light that were not so well known earlier. Like other regions of the world researchers and ethnobotanists have contributed and generated a lot of significant information about plant utilization in the Garhwal region. An All India Co-ordinated Research Project on Ethnobiology sponsored by the Ministry of Environment and Forests, Government of India (New Delhi) was initiated in 1982 to investigate ethnobotany of UP Himalaya (now Uttarakhand), wherein significant work has been conducted by Bisht and Badoni 1990, Bisht *et al.*, 1988, Gaur 1999, Gaur and Tiwari 1987, Gaur *et al.*, 1984, Nautiyal 1981, Negi and Pant 1992, Negi *et al.*, 1993, 1995. Other significant contributions on Garhwal Himalayan ethnobotany have been made by Ansari and Ghananand 1985, Badoni 1987-1988, 1990, Bhatt and Gaur 1992, Joshi and Pandey 1995 and 1997, Maheshwari and Singh 1984, Mumgain *et al.*, 1990, Negi *et al.*, 1993 and 1995, Pangtey *et al.*, 1989 and Uniyal 1977.

The intensive survey of literature indicates that there is scanty information on the ethnobotany and ethno-taxonomy of medicinal plants of NDBR. Keeping it in mind, ethno-taxonomical explorations of medicinal plants of NDNP were desired to carry out integrated approach of classical taxonomy accompanied with interviews of local inhabitants, *vaidyas*, priests and medicine men. This documentation will bridge up the gap between the folk medicine practitioner, and contemporary natural therapists and researchers of natural product chemistry.

Methodology

Plant Collection and Specimen Preparation

Plants were periodically collected from different parts of Nanda Devi National Park in Nanda Devi Biosphere Reserve, from alpine and sub-alpine zones to cover almost all the months starting from March to November of every year from 2004 to 2007 from an altitude rang of 2000 m to 4500 m above mean sea level (m.asl). Collected plant specimens were pressed, and dried in the blotters or newspapers in the field. Pressed and dried specimens were poisoned with 1 per cent chloride of mercury in ethyl alcohol or denatured spirit, poisoned and dried specimens were mounted on the standard (42x28cm) herbarium sheets adopting the usual herbarium techniques as suggested by Jain and Rao (1977).

Folklore Survey

During the last four years, ethnobotanical surveys and studies were conducted in the tribal inhabitant area of NDNP. The most frequent visited villages were Nitti, Malari, Bampa, Gamsali, Garpac, Jelam, Jumma, Dronagiri, Kaga, Kosa, Phagti, Tolma, Lata, Peng, Suraithoda, Reni, Ringi, Subhain, Karchon, Tugasi, Raigari, etc.

The ethno-medicinal data was gathered from the tribal and local men and women through interviews. In case of the medicinal uses of plants, the data was gathered from the tribal medicine men, 'Vaidyas' and from knowledgeable informants. During the course of this investigation, the data on the names of plants, parts used and their mode of processing and preparation, dosage and mode of various recipes and remedies was also gathered. The medicine men and other informants accompanied the author to collect herbarium specimen from the adjoining forest areas. It has been observed that the experienced inhabitants maintain certain level of secrecy about the medicinal uses of plants. Special efforts were made to develop intimacy with the informants by acquiring their confidence by respecting their customs and rituals, living with them, with the help of village- head and other well known persons of the area. At the time of interview the plant specimens and samples were the center of discussions. Some times fresh plants were collected from the study area and were shown to old knowledgeable informants, who were unable to visit the forest, for accurate information and confirmation of the plant specimens. The discussions regarding various uses of plants in their daily life was noted in field note books. The ethnobotanical data was verified and compared from other informants of the same village and other villages during same trip or in the next trips to get satisfied.

Enumeration

The present study is based on an extensive ethno-taxonomical survey of the plants and their uses especially as curatives in the NDNP. The families have been arranged according to Bentham and Hooker's system of classification (1862-1883), except in few cases where Hutchinson's concept (1973) regarding splitting of families has been followed. Taxonomical categories-genera and species within the family are treated alphabetically and species are described with usual citation-, vernacular-, Hindi-, Sanskrit-, and English names. It is followed by concise taxonomical characteristics, flowering and fruiting periods, distribution status, altitude range, place of collection and herbarium number (CSR-GUH). Finally the ethno-medicinal properties, method of preparation and dosages patterns have been given.

RANUNCULACEAE

Aconitum balfourii Stapf in Ann. Royal Bot.Gard.Calc.**10:** 160. t. 104. 1905; *A. atrox* (Bruhl) Mukerjee in Bull. Bot. Surv. India **3**:101 1961. *A. ferox* var. *atrox* Bruhl in Ann. Royal Bot. Gdn. Calcutta **5**: 110. 1895

Vern. *Rangu-Bhangu, Meetha bish* Sans. *Vatsnabh* Eng. *Indian Aconite.*

Biennial erect herbs with tuberous or fusiform roots. Stem blue purplish – brown. Lower leaves palmately 3-partite, each part deeply 3- lobed. Inflorescence long many flowered raceme, with bluish tomentum. Flowers bluish violet, helmet- shaped. Stamens many. Seeds broadly winged along with raphe.

Fl. and Fr.; Aug. - Nov.

Rare; 3000-4000 m.asl., in open grassy slopes, Chonya near Kuari Pass, CSR-GUH 19378.

The paste of tubers is applied externally on affected body part thrice a day, as an antidote of snake bite and scorpion sting. Root powder is boiled with clarified butter, made into paste and applied externally on affected area twice a day for 2-3 months in the treatment of rheumatic arthritis. Roots purified by heating with cow-dung and boiling with cow-urine and water for 2 to 3 hours separately, dried and made into powder. The powder is roasted with clarified butter and made into pills; one pill is given once a day for a month as tonic in general weakness and for paralytic afflictions.

Aconitum heterophyllum Wall. ex Royle, Illus. Bot. Himal. 56. t. 13. 1834; Hook. f., Fl. Brit. India **1**: 29. 1872

Vern. *Atish, Ateesh.*

Erect biennial herbs with tuberous roots. Leaves heteromorphic; basal leaves petiolate, shining above. Flowers greenish white or pale yellow. Stamens numerous. Seeds obpyramidate with winged angles.

Fl. and Fr.; July. - Sept.

Rare; 3000-4000 m.asl., in grassy moist slopes of alpine meadows on way to Kuari Paas, Dalisera. CSR- GUH 19377.

The aqueous extract of the root 5-10 ml. is given twice a day, early in morning empty stomach and at night after meals for 7 to 28 days in chronic fever, and digestive disorders, in diarrhoea and as cold efficacy. Root mixed with *Ajuga parviflora* leaves and *Podophyllum hexandrum* roots are dried in shade and powdered. The powder is given half teaspoonful twice a day early in the morning and at night after meals up to three months for the treatment of diabetes, leucorrhoea and as carminative.

Delphinium vestitum Royle, Illus. Bot. Himal. 56. 1839; Hook. f., Fl. Brit. India **1**: 26. 1872.

Vern. *Nirbishi.*

Erect perennial herbs with glabrous, slender stem. Leaves sub-orbicular, cordate, 5-7 lobed, hairy dense; basal leaves long stalked upper leaves smaller. Flowers blue or pale- violet spurred. Seeds narrowly winged.

Fl. and Fr.; Aug. - Sept.

Occasional; 3200-3500 m.asl., on open places, in Alpine grassy slopes near forest edges, Kuari Pass, Dalisera. CSR-GUH 19210.

The powdered root mixed with cow-urine made into paste and applied externally on wounds, cuts and blisters. Root paste applied externally thrice a day for 2 to 4 days as antidote in snake bite and Scorpion sting.

Thalictrum foliolosum DC., Syst. Nat. **1**: 175. 1818; Hook. f., and Thomson in Fl. Brit. India **1**: 14. 1872

Vern. *Makdya- ghas, Pilijadi* H. *Mamiri* Sans. *Traymaan, Peetmula* Eng. *Indian meadow Rue.*

Erect, branched, perennial herbs with rhizomatous yellow roots. Leaves pinnately compound; stipules present; leaflets broadly orbicular or ovate- oblong, lobed toothed. Flowers white or greenish- purple in much branched panicles. Stamens many. Achenes sessile, oblong, beaked.

Fl.; Jun.- Aug.; Fr.; Sept.- Oct.

Common; 1500- 2500 m asl., along way side, shady places, Tapovan. CSR-GUH 19337.

Extract of roots is given 1/2 teaspoonful twice a day, early in the morning and at night after meals for 7 to 28 days for the treatment of high fever, headache, as carminative, in piles, and as blood purifier, and in ophthalmic afflictions. The paste of leaves along with cow-urine is applied externally on affected part once a day for a week in the treatment of herpes zoster.

PAEONIACEAE

Paeonia emodi Wall. ex Royle, Ills. Bot. Himal.57. 1734; Hook. f., and Thomson in Fl. Brit. India **1**: 30. 1872.

Vern. *Chandra* H. *Udsalap* Eng. *Himalayan Paeony*.

Perennial, large spreading herbs with thickened root-stock. Stem woody at base. Leaves alternate, ternate, leaflets decurrently, entire; segments oblanceolate, acuminate. Flowers white. Stamens many, orange yellow. Seeds oval, shining black.

Fl.: Mar.-Jun.; Fr. July-Aug.

Common; 1800-3000 m.asl., in shady places, near forest floors, Lamri near Auli. CSR-GUH 19000.

Leaves are dried in shade, washed with hot water thrice, and then used as vegetable twice a day for the treatment of colic, blood dysentery, diabetes and urinary complaints. It is also used as blood purifier, and to improve lactation and treat menstrual problems. Root powder mixed with *Selinum vaginatum* root powder, is given ½ teaspoon twice a day up to 6 months for the treatment of hysteria, convulsion and epilepsy.

BERBERIDACEAE

Berberis aristata DC., Syst. Nat. **2**:8.1821; Hook. f., Thomson in Fl. Brit. India **1**:110. 1872.

Vern. *Kingore, Kirmode* H. *Darul haldi* Sans. *Daruharidra* Eng. *Indian Barbery*.

Deciduous shrubs with pale – brownish to yellowish stem. Leaves obovate or elliptic, sub sessile, cuneate at base acute at apex, entire, spines along margins. Inflorescence drooping raceme 4-6 cm. long, flowers yellow. Berries ovoid to oblong – ovoid, bright red; blue on mature.

Fl. : Mar.- May; Fr. Jun- July

Common; 1800-2700 m.asl., open places, Badagaon, Tapovan. CSR-GUH 19257.

Decoction of root is given 2.5 ml. twice a day for two months in the treatment of urinary complaints, and also as blood purifier. Root juice dropped into eyes thrice a day for a week in ophthalmic infections. Root powder is given 2.5 g twice a day, early in morning and at night after meals for three months in diabetes

PODOPHYLLACEAE

Podophyllum hexandrum Royle, Illus. Bot. Himal.64. 1834; P. *emodi* var. *hexandrum* (Royle) Chat. and Muk. in Rec. Bot. Surv. India. **16(2):** 45. 1953.

Vern. *Shon kakadi*, H. *Ban kakri* Eng, *Indian podophyllum*.

Perennial, succulent, erect herbs with rhizomatous rootstocks. Leaves long stalked, deeply cut into 3-5 toothed lobes; brown colour spotted. Flowers pinkish white, cup-shaped; solitary. Fruit a large red berry; seeds many; brown; embedded in pulp.

Fl. and Fr.; Apr. – Oct.

Rare; 3000-4000 m. asl., on alpine meadows near forests edges, along shady ravines, near boulders, on way to Kuari Pass. CSR-GUH 19282.

Rhizome powder is given 2.5 g. twice a day, early in the morning prior meals and at night after meals for 14 to 28 days in bile complaints, as carminative, in leucorrhoea and chronic fever. Root powder with *Ajuga parviflora* given 2.5 g. twice a day for three months in the treatment of diabetes.

PAPAVERACEAE

Meconopsis aculeata Royle, Illus. Bot. Himal. 67, t. 45. 1839; Hook. f., Fl. Brit. India. **1**: 118. 1872.

Vern. *Laxmana* Eng. *Blue poppy*.

Biennial, prickly herbs with rhizomatous brown, yellow rootstock. Leaves pinnatifid, sparsely bristly- haired; lobes usually rounded, toothed and widely spaced. Flowers solitary, fewer, bluish; capsules prickly and bristly, short.

Fl. and Fr.; Jun.- Oct.

Rare; 3000- 4500 m. asl., on hill slopes amidst boulders, in alpine meadows, Kuari Paas, Dalisera. CSR-GUH 19323.

Root is purified, powdered and roasted with clarified butter and taken 1g once a day for the treatment of unconsciousness, and as tonic in general weakness, and nervous irritation.

Meconopsis robusta Hook. f. and Thomsom. Fl. Ind. 253. 1855; Hook. f., Fl. Brit. India. **1**: 118. 1872.

Vern. *Laxmana* Eng. *Yellow poppy*.

Biennial, prickly herbs with yellowish rhizomatous roots. Leaves pinnatifid, sparsely bristly- haired; lobes usually rounded, toothed and widely spaced. Flowers solitary, yellow shows; capsules prickly and bristly, short.

Fl. and Fr.; Jun. - Oct.

Rare; 3000- 4500 m. asl., on hill slopes amidst boulders, in alpine meadows, Kuari Paas, Dalisera. CSR-GUH 19396.

Root is purified, powdered and roasted with clarified butter and made into pills, one pill is given once a day or night after menstruation in sterility.

GERANIACEAE

Geranium wallichianum D. Don ex Sweet, Geran. 1: t. 90. 1820; Edgew. and Hook. f., in Fl. Brit. India 1: 430. 1874.

Vern. *Ratanjot, Laljari* Eng. *Robert Geranium*.

Perennial, trailing, pubescent herbs with woody rootstock. Leaves orbicular, palmately lobed; toothed, tip acute, hairy. Flowers light- purple, solitary. Stamens 10; anther blue- black. Capsules long including beak. Seed minute.

Fl. Jul. – Sept.; Fr.: Sept. - Oct.

Common; 2500-3000 m. asl., in open moist localities near forest edges, Unile near Auli. CSR-GUH 19519.

Root sap filtered and dropped into eyes and ears thrice a day for 7 to 21 days for the treatment of earache and eye ailments and to improve vision. The root paste is given twice a day for a week in the stomach disorders.

RUTACEAE

Skimmia anquetilia Taylor and Airy Shaw in Curtis, Bot. Mag, n.s, 182. 4: t. 789. 1979. *S. laureola* (DC.) Siebold and Zuccarini ex Walpers, Repert. 5: 405. 1842; Hook. f., in Fl. Brit. India 1: 499. 1875.

Vern. *Nairpatti* H. *Patrang*.

Evergreen, aromatic glabrous shrubs. Leaves oblong lanceolate, thick, acute, entire margins; petioles long, alternate, shining above, yellowish beneath. Flowers yellowish- white. Stamens 5 yellow. Drupes ovoid, red, long.

Fl.: Apr. - Sept.; Fr.: Oct. - Nov.

Common; 1500-3000 m. asl., in moist localities, near water channels, Lamri near Auli. CSR-GUH 18992.

Leaf paste mixed with cow's urine and the paste applied twice a day for 14 to 28 days for the treatment of psoriasis and leucoderma. Leaves chewed for cooling effect.

FABACEAE

Astragalus condoleanus Royle ex Benth. in Royle Illus. Bot. Himal. 199. 1835; Baker in Hook. f., Fl. Brit. India 2: 132. 1876.

Vern. *Rudravanti* Sans. *Rudravanti* Eng. *Milk – vetch*.

Sub erect or prostrate pubescent herbs with thick roots. Leaflets oblong or sub orbicular obtuse, sparsely silky. Flowers sub sessile yellow, on compact head like racemes. Vexillum emarginated; wings feathery; keel beaked. Pods shortly stalked 8-12 seeds.

Fl. Apr. – Jun; Fr.; Aug.-Sept.

Rare, 3000-4000 m. asl., open hill slopes in sandy soils, on way to Malari, Dalisera. CSR-GUH 19436.

Root powder mixed with *Bombax ceiba* root powder, roasted with clarified butter and made into pills, one pill is given twice a day, for 30 to 90 days for the treatment of abdominal complaints, leucorrhoea, as aphrodisiac, to improve fertility in men, as tonic in general weakness, to regulate menstruation and in diabetes.

Potentilla fulgens Wall. ex Hook. in Bot. Mog. 53: t. 2700.1826; Hook. f., in Fl. Brit. India **2**: 349. 1878.

Vern. *Bajardantii.*

Perennial, pubescent diffused herbs with rhizomatous pale- brown roots. Leaves imparipinnate, leaflets several, alternate, ovate or obovate, toothed, upper surface silky- green hairy; lower white tomentom. Flowers yellow crowded in terminal corymbs. Achens many, glabrous.

Fr.: Jul. - Sept.; Fr.: Sept. – Oct.

Common; 2000- 3000 m. asl., open, exposed, moist places, forest edges, Dalisera. CSR-GUH 19458.

The paste of root is filled in cavity of teeth in toothache, and for strong tooth. Decoction of roots applied on affected area twice a day in burns as antiseptic. Leaf and root paste applied externally on wounds twice day for a long time.

ROSACEAE

Prunus cerasoides D.Don, Prodr, Fl. Nep. 239.1825; Hook. f., in Fl. Brit. India **2**: 314. 1878.

Vern. *Payan* H. *Padam* Sans. *Padmaka* Eng. *Himalayan Wild- Cherry*.

Deciduous trees, with reddish- brown bark. Leaves elliptic or ovate- lanceolate, apex acuminate, glabrous, shining above, stipule long, subulate. Flowers pinkish white in umbellate fascicles. Fruits red or yellow ovoid, shining glabrous; 1-seeded, red.

Fl.: Oct. - Dec.; Fr.: Feb. - Mar.

Common; 2000-2500 m. asl., on road sides, near boulders, rocky hills, Panya, CSR-GUH 19434.

The decoction of bark and leaves is given a teaspoonful twice a day with cow's milk for 14 to 28 days in internal injury, in sprain and in chest pain. Seed oil applied externally in arthritis. Bark paste is applied on dislocated joints and fractured bones as plaster for a month.

SAXIFRAGACEAE

Bergenia stracheyi (Hook. f. et. Thoms) Engl. in Bot. Zeit. **26**:842.1868. *Saxifraga stracheyi* Hook. f., and Thomas. In J. Linn. Soc. **2**:61.1857; Hook. f., Fl. Brit. India. **2**:398.1878.

Vern. *Silpadi.*

Perennial herbs with very thick rootstocks; stem very short, fleshy. Leaves, oblong ovate, glabrous, above shining beneath, margin serrate, hairy, toothed. Flowers pinkish- white in corymb panicles.

Fl. and Fr.; May – Jul.

Rare; 3000-4500 m. asl., on moist shady rocks, wet boulders, Dalisera. CSR-GUH 19329.

Root dipped into mustard oil for a night, extraction of root applied on hairs twice a day for a month, as a hair tonic. The decoction of root is given ½ teaspoonfuls twice a day, early in the morning and at night after meals for 30 to 45 days for the treatment of cough and cold, abdominal ailments, liver complaints, asthma, piles. The dried roots and leaves are used for preparing tea or powder of above parts is given thrice a day, for a month to the patient of kidney stone, and for abdominal and urinary complaints.

CUCURBITACEAE

Trichosanthes tricuspidata Lour., Fl. Cochinch. 589. 1790; Naithani, Fl. Chamoli **1**: 246. 1984; Hook. f., Fl. Brit. India **2**: 606. 1879.

Vern. *Alaru, Aladu*, H. *Indrian* S. *Sweet pushpi*.

Perennial, subdeciduos glabrous climbers. Leaves broadly ovate, 3-7 lobed, apex acuminate, base cordate, margins denticulate, glabrous above, sparsely hairy beneath. Flowers creamy – white in axillary stalks. Fruits smooth, globose or ellipsoid, scarlet red on ripening. Seeds many, embedded in blackish, green pulp.

Fl.: Aug.- Sept. Fr. Sept.- Dec.

Uncommon; 1800-2200 m. asl., waste places, climbing on bushes, rocks, Chormi. CSR-GUH 19452.

Dried seeds and roots powder is given approximately ½ teaspoonfuls twice a day, for a week in the treatment of pneumonia. Extract of root is given ½ teaspoonful twice a day early in the morning and at night after meals for 30 to 90 days in bronchitis, and cardiac complaints. Seed paste along with mustard oil is applied externally as an anti-allergic.

APIACEAE

Angelica glauca Edgew. in Trans. Linn. Soc. **20**: 53. 1846; C.B. Clarke in Hook. f., Fl. Brit. India **2**: 706. 1879.

Vern. *Choru* H.*Chora, Chorak* San. *Gandrayan*.

Perennial glabrous herbs, with thick, aromatic rootstocks. Stem stout fistular. Leaves pinnatiparitite leaflets 3-5, lanceolate, irregularly serrate, toothed, leaf base sheathing. Flowers white in racemes. Stamens 5. Fruits oblong, flattened.

Fl.: Jul. - Aug.; Fr.: Sept.-Oct.

Rare; 3000-3500 m. asl., moist places, forest edges, on way to Dalisera. CSR-GUH 19369.

The decoction of root is given ½ teaspoonfuls with milk, twice a day, for 2-3 weeks to treat constipation, dyspepsia and bronchitis. Extract of rootstock is given a tablespoonful thrice a day, for 3 to 5 days, to the treatment of toxic effect in livestock.

VALERIANACEAE

Nardostachys grandiflora DC., Prodr. **4**: 624. 1830; Hook. f., Fl. Brit. India **3**: 211.1881.

Vern. *Maasi, Jatamasi.*

Erect, perennial herbs with aromatic fibers rootstock.Leaves radical and cauline, petiolate, glabrous pubescent, sessile, elliptic, oblong, lanceolate. Flowers pink- purple in terminal clusters, stamen 4. Fruits long; seeds obovate, compressed.

Fl. and Fr.; Aug. - Oct.

Rare; 3200-4000 m. asl., among rocks and boulders, grassy slopes, Gorson. CSR-GUH 19132.

Roots are dried in shade, and powdered with *Betula utilis* bark, and roasted with clarified butter, applied in arthritis twice a day. Dried root and leaf powder is given ½ teaspoonful twice a day in morning and at night for 30 to 90 days in cardiac complaints, piles, hysteria, epilepsy, dysmenorrhoea, bronchitis and respiratory complaints, also used as blood purifier, and as aphrodisiac.

ASTERACEAE

Jurinea dolomaiea Boiss. Fl. Or. Suppl. 311.1888; Hook. f., Fl. Brit. India **3**: 378. 1881.

Vern. *Guggle dhoop.*

Perennial stem less glabrous herbs, with rhizomatous thick rootstock. Leaves oblong blunt, pinnate, lobed, toothed, often with purple mid-vein, wooly, white beneath. Flowers – heads purple short- stalked; pappus brown; Achens ashy green.

Fl. and Fr.: July – Oct.

Uncommon; 3500m. asl., in open slopes of alpine meadows, Dalisera. CSR-GUH 19400.

Extract of root is given ½ teaspoonful twice a day early in morning and at night after meals for 14 to 28 days for the treatment of leucorrhoea, menstruational disorders, and as an analgesic. Root powder is given ½ teaspoonful twice a day for 3 month in the treatment of mental disorders, gout, urinary complaints, and in loss of appetite.

Saussurea lappa (Dcne.) Sch.-Bip. in Lannaea **19**: 331.1846; Hook. f., Fl. Brit. India **3**: 376. 1881.

Vern. *Kuth*

Erect, perennial, pubescent herbs with thick rootstock. Basal leaves pinnately lobed; winged leaf-stalk; upper leaves smaller, cordate, ovate. Stem clasping, glabrous. Flower heads purple in dense rounded terminal cluster.

Fl. and Fr.: Aug.-Sep.

Rare; 3500 m. asl., cultivated Dunagiri and Ganeshpur. CSR-GUH 19278.

Root dried in shade powdered mixed with honey and given half teaspoonful twice a day early in the morning and at night after meals for a month in the treatment of asthma, bronchitis, and as a tonic. Root paste applied externally thrice a day for 2-3 weeks in eczema, ring worm.

Saussurea gossypiphora D.Don, Prodr. Fl. Nep. 168. 1825; Hook. f., Fl. Brit. India **3**: 376. 1881.

Vern. *Hiyun-Kaun.*

Perennial or biennial, erect, wooly herbs, with stout hollow stem. Leaves linea-lanceolate, coarsely toothed; obovate, lanceolate, embedded in dense wooly hairs. Flower heads white, odorous, cylindrical.

Fl. and Fr.: Sept.-Oct.

Rare; 4000-4500 m. asl., near boulders, rocks, Panger Chula. CSR-GUH 19100.

Root dried, powdered, roasted and mixed with clarified butter made into paste. The paste is applied externally on affected part twice a day for 3-4 weeks in the treatment of piles. Decoction of root and flower is given ½ teaspoonful twice a day, as well as applied externally on chest for a month in the treatment of cardiac complaints and chest pain. Extract of leaves is given a teaspoonful thrice a day for a week in the treatment of snake bite and scorpion sting.

Tanacetum dolichophyllum (Kitamura) Kitamura in Hara *et al.*, Enum. Fl. Pl. Nep. **3**: 45. 1982. *Chrysanthemum dolichaphyllum* Kitamura in Acta Phyt. Geobot. **23**: 73. 1968. *T. longifoleum* Wallich ex DC., Prodr.6: 130. 1838 pp. (non Thnub.); Hook. f., in Fl. Brit. India **3**: 320. 1881.

Vern. *Guggl dhoop* Eng. *Long Leaved Tansy.*

Perennial herbs with slender stem, rootstock thick; Leaves linear lanceolate, middle leaves smaller, sessile. Capitula many, yellow; bract hairy.

Fl. and Fr.: Jul. - Sept.

Occasional; 3000-4500 m, asl., on open grassy slopes of alpine meadows, Dalisera. CSR-GUH 19359

Root dried in shade, powdered, and boiled with water. The decoction is applied or massaged thrice a day for a month in the treatment of chest pain and paralytic affliction, in arthritis, Root and leaf paste applied on cuts and wounds.

Tanacetum nubigenum DC. Podr. **6**: 130, 1830. *Chrysanthemum nubigenum* (DC.) Hand Maz. Ssymb. Sin.7: 1113.1936. Hook, f., Fl. Brit. India **3**: 318. 1881.

Vern. *Doana, Doan.*

Erect, perennial, aromatic herbs, with tomentose wooly stem; rootstock woody.Basal leaves petiolate, lamina oblong-ovate, subsessile whitish on both sides. Capitula yellow, capsule narrowly oblong, ribbed.

Fl. and Fr.; Jul -Sep

Rare; 3500- 4500 m, asl., on rocky slopes, sandy soils, on way to Dhakwani. CSR-GUH 19109.

Roots are dried in shade, powdered, and mixed with a drop of Ginger juice and Safron. The paste is given approximately ½ teaspoonful twice a day early in the morning and at night after meals for three month as a tonic in general weakness and to increase immunity. Extract of leaves is given approximately ½ teaspoonful twice a day for a week in chronic fever.

BORAGINACEAE

Arnebia benthamii Wall. ex G. Don I.M. Johnston, J. Arn. Arb. 25. 56. 1954; Hook. f., Fl. Brit. India **4**:177. 1883.

Vern. *Balchhari, Laljadi.*

Erect perennial herbs with thick red rhizomatous roots. Stem solitary, haired. Leaves long lanceolate, apex acute, glabrous both side, margin entire. Flowers crowded in terminal dense spikes, pale-pink or rosy. Seed minute black brown.

Fl. and Fr.; June - Nov.

Rare; 3000 – 4500 m. asl., in alpine slopes, near rocks and boulder, on way to Kuari Pass, Dalisera. CSR-GUH 19270, 19425.

Decoction of root is applied on burnt parts thrice a day for a week for early healing as antiseptic. Extract of root with mustard oil is applied on hairs as hair tonic. The root powder with *Dactylorhiza hatagirea* and *Polygonatum verticillatum* tubers powder is given ½ teaspoonful twice a day for a month in the treatment of leucorrhoea, piles, diabetes, mental disorders, as a tonic in general weakness.

SCROPHULARIACEAE

Picrorrhiza scrophulariflora Pennell in Monogr. Acad. Nat. Sci. Philad. **5:** 65. t. 613. 1943. *P. Kurrooa* aeceet. *non* Royle *sensu* Hook. f., Fl. Brit. India **4:** 290. 1884.

Vern. *Katuki.*

Perennial herbs with creeping thin rootstock. Stem stout. Leaves sub radical, spathulate, ovate, lanceolate, margins serrate, toothed shining above, glabrous beneath. Flowers sky blue in terminal racemes, seeds brownish.

Fl. and Fr.; June – July.

Rare; 3000 – 4500 m. asl., on hill slopes, near rocks, boulders, on way to Dalisera, Chitarkhana. CSR-GUH 19364.

The extract of root with *Carum carvi* seed powder is given approximately half teaspoonful twice a day, early in the morning and, at night after meals for 45-60 days in the treatment of chronic fever, as liver stimulant, in bronchial asthma, as stomachic. In some cases powder taken with hot water and cold water for hot and cooling effect.

LAMIACEAE

Thymus linearis Benth. in Wall. Pl. As. Rar. **1:** 31. 1830. *T. serpyllum auct. non* L. *sensu* Hook. f., Fl. Brit. India **4**:649. 1885.

Vern. *Marchya ghass* H. *Ban-Ajwani* Eng. *Wild Thyme.*

Perennial, prostrate tufted, aromatic herb with woody rootstock. Stem slender branched. Leaves sessile, gland – dotted, oblong- ovate – lancealate. Flowers crowded in small terminal spikes; pink – purple.

Fl. and Fr.; Jun.-Oct.

Common: 2700-3000 m. asl., on open hill slopes, grassy localities Gorson. CSR-GUH 19384.

Powdered leaf along with honey is applied externally on affected area thrice a day for a week in the treatment of eczema and psoriasis. Roasted plant is given twice a day for a month to treat bronchial asthma, and abdominal disease. Whole plant is powdered and given approximately half teaspoonful twice a day for 30-45 days in the treatment of weak vision, and regulating menstruation.

POLYGONACEAE

Rheum australe D. Don., Prodr. Fl. Nep. 75. 1825; Hook. f., Fl. Brit. India **5**: 56. 1886.

Vern. *Dolu.*

Robust, perennial, glabrous herbs, with thick yellow rhizomatous rootstock. Stem tall, hollow leaves large, orbicular, ovate, entire, apex obtuse, pubescent beneath. Flowers purple or pink in axillary and terminal panicles Fruits ovoid, oblong.

Fl. and Fr.; June – Aug.

Rare; 3000 – 4000 m. asl., on open grassy slopes, rock boulders, in alpine meadows, on way to Dalisera. CSR- GUH 19188.

Extract of root along with turmeric and milk is taken thrice a day for a week in internal injury. Root decoction along with clarified butter is applied externally on affected area twice a day for 45-60 days in joint pain. Root paste is applied on cuts and wounds. Root powder mixed with *Bombex ceiba, Picrorhiza kurooa, Dactylorhiza hatagirea* root powder is given half teaspoonful twice a day for a month in the treatment of leucorrhoea, abdominal complaints, as a tonic in general weakness, sex tonic and in mental disorders.

Rheum moorcroftianum Royle, Ill. Bot. Himal. 315, 318. 1836; Hook. f., Fl. Brit. India **5**: 56. 1886.

Vern. *Archu.*

Robust, perennial, stem less herbs with stout, woody rootstock. Leaves large, orbicular, ovate, pubescent beneath, glabrous above, veined, margins red. Flowers pinkish in terminal spikes. Nutlets winged.

Fl. and Fr.; Jun. – Sept.

Uncommon; 3500 – 4500 m. asl., on hill slopes of alpine meadows, Dalisera, CSR- GUH 19531.

Root powder roasted with clarified butter and made into pills, one pill is given twice a day for 30-45 days in the treatment of chronic bronchitis, asthma, abdominal disease, and as tonic in general weakness.

BETULACEAE

Betula utilis D. Don, Prodr. Fl. Nep. 58. 1825; Hook. f., in Fl. Brit. India **5**: 599. 1888.

Vern. *Bhoj* H. *Bhojpatra* Sans. *Bhurj* Eng. *Himalayan Silver Birch.*

Large deciduous trees, with white or brown bark peeling off in papery layers. Leaves ovate or ovate-lanceolate, base cordate, acute, serrate, lateral nerves 7-9 pairs. Flowers in spikes; female catkins solitary fruits winged nut. Bark peeling off in papery layers

Fl. and Fr.; May – Oct.

Uncommon; 3500 – 4000 m. asl., edges of alpine meadows,mixed with fir forest, on way to Dalisera. CSR GUH 19538.

Bark powder mixed with cow urine made into paste, applied externally on affected part twice a day for a week in skin ailments. The bark peeling is used as plaster for dislocated joint. Roasted bark with clarified butter is applied externally in joint pains.

Discussions

The present work deals with detailed ethno-taxonomical studies on the medicinal plants of the NDNP Chamoli, Uttarakhand. With the objectives of medicinal uses in mind, an intensive ethno-taxonomical study on the medicinal plants used by the inhabitants of NDNP has been undertaken. The study has yielded some promising results, like unknown or less known uses of plants, etc. It is hoped that the detailed observation on the traditional uses of plants in medicine by the tribal communities and primitive societies of this region will result in strengthening our knowledge on the Indian System of Medicine (ISM). A total 29 important threatened medicinal plants, followed by 24 Genera and 29 species have been mentioned in present investigation. *Aconitum balfourii* in arthritis, *A. heterophyllum* in chronic fever, *Astragalus candoleanus* in abdominal complaints, *Arnebia benthamii* in mental disorders, *Berberis aristata* in diabetes, *Podophyllum hexandrum* in leucorrhoea, *Potentilla fulgens* in toothache, *Thalictrum foliolosum* as carminative, *Trichosanthes tricuspidata* in diabetes, *Paeonia emodi* in diabetes and blood dysentery, *Rheum australe* in internal injury, *Saussurea costus* in asthma and *Saussurea gossipiphora* in cardiac complaints etc. have potential to be used as the emergency drugs. This investigation elucidate several indigenous practices of medicinal plants in the adjoining the area of NDNP and would be useful to students, naturalists, researchers, environmentalists, phytochemists and practitioners of indigenous system of medicine. It highlights new and noteworthy information on plants used in medicines which can be utilized to uplift the economy of the inhabitants of NDNP. Therefore, we strongly recommend the establishment of medicinal plant based industries in the region.

Acknowledgements

The authors are grateful to the inhabitants of NDBR for providing the information and hospitality and to Prof. R.D. Gaur, Emeritus Fellow, UGC for critical evaluation of manuscript.

References

Adhikari B S 2003. Ecological Attributes of Vegetation in Nanda Devi National Park. *Biodiversity Monitoring Expedition* 15-38.

Ansari A A and Ghana Nand 1985. Some medicinal plants of Pauri, Garhwal *Himalayan Chem Pharm Bull* **2:** 42-44.

Badoni A K 1987-1988. Ethnobotany of hill tribes of Uttarkashi: plants used in rituals and psychomedicinal practices. *J Himal Studies and Regional Develop* **11-12:** 103-115.

Badoni, A.K. 1989-90. Remarks on the high altitudinal medicinal plants of Garhwal Himalaya. *J Himal Studies and Regional Develop* **13-14**: 37-45.

Badoni, A K 1990. An ethnobotanical study of Pinsari community: A preliminary survey *Bull of Bot Surv India* **32:** 103-115.

Badoni A K and Badoni K 2001. Ethnobotanical heritage. In: Kandari O P and Gusain O P (eds.). *Garhwal Himalaya: Nature, Culture and Society* Transmedia, Media House, Srinagar, Garhwal 127-147.

Bhatt K C and Gaur R D 1992. A contribution to ethnobotany of Rajis in Pithoragarh district. *Acta Bot Indica* **20:** 76-83.

Bisht M K, Bhatt K C and Gaur R D 1988. Folk medicine of Arakot valley in district Uttarkashi: An ethnobotanical study. ***In:*** P. Kaushik (ed.). *Indigenous Medicinal Plants* Today and Tomorrow's Printers and Publishers, New Delhi 157-166.

Bisht M K and Badoni A K 1990. Araceae in the folk life of the tribal populace in Garhwal Himalaya. *J Econ Bot* **1:** 21-24.

Burkill I H 1965. *Chapters on the History of Botany in India*. Manager of Publication, Govt of India, Calcutta.

Gaur R D 1999. *Flora of the District Garhwal: North West Himalaya (with Ethnobotanical Notes)*. Transmedia, Srinagar (Garhwal).

Gaur R D and Tiwari J K 1987. Some little known medicinal plants of Garhwal Himalaya: An ethnobotanical study. ***In:*** Leeuwenberg A G M (ed.) *Medicinal and Poisonous Plants of the Tropics* Netherlands 139-142.

Gaur R D, Semwal J K and Tiwari J K 1984. A survey of high altitude medicinal plants of Garhwal Himalaya. *Bull Medic Ethnobot Res* **3:** 102-116.

Hajra P K and Balodi B S 1995. *Plant Wealth of Nanda Devi Biosphere Reserve.* BSI. Howrah.

Jain S K and Rao R R 1997. *A Hand Book of Field and Herbarium Methods*. Todays and Tomorrow's Printers and Publishers, New Delhi.

Joshi P and Pande P C 1997. Ethnobotany of Bhotia tribe of Kumaun Himalaya ***In:*** Sharma S K and Dhauondiyal N C (eds.) *Studies on Kumaun Himalaya* Indus Publishing Co. New Delhi 63-89.

Joshi P and Pande P C 1995. Ethnobotany of the Rajis of Kumaun Himalaya: a preliminary study **In:** Joshi M P (ed.) *Rishi-II. Uttaranchal Himalaya* Shree Almora Book Depot Almora 241-263.

Kritikar K R and Basu B D 1984. *Indian Medicinal Plants*. Bishen Singh Mahendra Pal Singh, Dehra Dun.

Lal B, Dhasmana H, Nigam R K 1989. A contribution to the medicinal plant – lore of Garhwal region. *Himal Res Develop* **8:** 28-30.

Mathur V B, Kathyat J S and Rath D P 2000. *Wildlife and Protected Areas*. Wildlife Institute of India, Dehradun, *Envis Bull:* **3:** 1.

Maheshwari J K and Singh J P 1984. Contribution to the ethnobotany of Boxa tribe of Bijnor and Pauri Garhwal district U P *J Econ Tax Bot* **5:** 253-259.

Maikhuri R K, Nautiyal S, Rao K S and Saxena K G 1997. Medicinal plants cultivation and biosphere reserve management: A case study from the Nanda Devi Biosphere Reserve, West Himalaya. *Curr Sci* **74:** 157-163.

Maikhuri R K, Nauityal S, Rao K S and Saxena K G 1998. Role of medicinal plants in the traditional health care system: a case study from Nanda Devi Biosphere Reserve. *Curr Sci* **75:** 152-157.

Malhotra C L and Balodi B 1984. Wild medicinal plants in the use of Johari tribals. *J Econ Tax Bot* **5:** 881-883.

Mumgain S K and Rao R R 1990. Some medicinal plants of Pauri Garhwal. *J Econ Tax Bot* **14:** 633-640.

Murti S K 2001. *Flora of Cold Deserts of Western Himalaya*. BSI, Calcutta.

Naithani B D 1984-1985. *Flora of Chamoli* 2 Vols. BSI, Howarah.

Nautiyal S 1981. Some medicinal plants of Garhwal hills: A traditional use. *J Scientific Res Plants and Med* **2:** 12-18.

Negi K S and Pant K C 1992. Less known wild species of *Allium* Linn (Amaryllidaceae) form mountainous region of India. *Econ Bot* **46:** 112-116.

Negi K S, Tiwari J K and Gaur R D 1985. Ethnobotanical importance of some trees in Garhwal Himalayas. *J Ind Bot Soc* **61:** 32.

Negi K S, Tiwari J K, Gaur R D and Pant K C 1993. Notes on ethnobotany of five districts of Garhwal Himalaya U P India. *Ethnobot* **5:** 73-81.

Pangtey Y P S, Samant S S and Rawat G S 1989. Ethnobotanical notes on Bhotia tribes of Kumaun Himalaya. *Ind J Fores* **12:** 191-196.

Purohit A N 1997. Himalayan medicinal plants– Focus on Uttrakhand ***In:*** *Himalayan Biodiversity: Action Plan*. Gyanodaya Prakashan Nainital 91-110.

Rana C S 2007. Ethnobotanical Studies on the Medicinal Plants of Nanda Devi Biosphere Reserve, Uttaranchal. D. Phill. Thesis HNB Garhwal University, Srinagar Garhwal, Uttarakhand.

Rau M A 1975. *High Altitude Flowering Plants of West Himalaya*. Howrah.

Rawat D S, Bhandari B S and Gaur R D 2001. *Vegetation Wealth of Garhwal Himalaya* **In:** Kandari O P and Gusain O P (eds.) *Nature, Culture and Society* Transmedia, Srinagar Garhwal. 69-92.

Samant S S and Joshi H C 2003. Floristic Diversity, Community Pattern and Changes of Vegetation in Nanda Devi National Park. *Biodiversity Monitoring Expedition*. Uttaranchal Forest Department 39-44.

Singh H, Saklani A and Lal B 1990. Ethnobotanical observations on some Gymnosperms of Garhwal Himalaya, Uttar Pradesh, India. *Econ Bot* **44:** 349-354.

Tiwari J K 1986. Ethnobotanical study of the Medicinal Plants of Garhwal. Unpublished D. Phil. thesis Garhwal University, Srinagar Garhwal.

Uniyal B P, Sharma J R, Chaudhary U and Singh D K 2007. *Flowering Plants of Uttarakhand* (A Checklist) Bishan Singh Mahendra Pal Singh Dehradun.

Uniyal M R 1989. *Medicinal Flora of Garhwal Himalaya*. Sri Baidyanath Ayurved Bhawan Pvt Ltd, Nagar.

Medicinal Plants: Aspects and Prospects (2014) Pages 166–190
Editors: Mukesh Kumar, Anjali Khare and C.P. Shukla
ISBN: 978-81-7622-309-6
Published by: BIOTECH BOOKS, NEW DELHI

Chapter 13

Economics of Plant Diversity of India: Medicinal Profile

Deepak Kumar[1], Deepika[1] and Bharat Bhushan[2]

[1]Department of Botany, R.S.M. Degree College, Rampur Ghogher – 244 602, Moradabad
[2]Department of Botany, Sahu Jain College, Najibabad – 246 763, Bijnor

ABSTRACT

The use of medicinal plants is increasing worldwide. According to World Health Organization (WHO), approximately 80 per cent of world population currently uses herbal medicines directly as tea, decocts or extracts with easily accessible liquids such as water, milk or alcohol. Although modern synthetic drugs are mostly used in developed countries, the use of herbal drugs in Western world is well accepted and a continuously high demand for plant material and received greatest interest in the USA and Europe over the past 30 years and account for over 50 per cent of the over the counter market. The present chapter includes enumeration and details of 67 commonly occurring and widely used medicinal plants.

Keywords: Ethnobotany, Medicinal Plants.

Introduction

The distribution of plants in India, though a very general one, has as far as possible, been verified from herbarium materials and authentic literature. The information on drugs and their properties have been studied very well from authentic publications and only those uses and economic importance of medicinal plants have been described which have been recognized in British Pharmaceutical Codex and/or United States Dispensatory, or whose properties have been shown experimentally on animals or in clinical tests. Information related to medicinal plants and traditional medicines can be found in documents and databases aimed at readers in the range of disciplines including Botany, Ecology, Chemistry, Medicine, Veterinary Science etc. However there are a few publications reporting current work or reviewing and analyzing recent advances. Access to relevant information by the public, decision makers and local communities is still very limited. The use of medicinal plants is increasing worldwide. According to World Health Organization (WHO), approximately 80 per cent of world population currently uses herbal medicines directly as tea, decocts or extracts with easily accessible liquids such as water, milk or alcohol (Farnsworth 1990). Although modern synthetic drugs are mostly used in developed countries, the use of herbal drugs in western world is well accepted and a continuously high demand for plant material and received greatest interest in the USA and Europe over the past 30 years and account for over 50 per cent of the over the counter market. In an attempt to over-come these problems for both drugs, intensive research is being carried out worldwide including combinatorial biosynthesis or improved bio-processing in bioreactors for both artemisinin and paclitaxel. The application of biotechnological techniques to medicinal plants has received considerable interest, especially when the final product is defined, purified and naturalized. The manipulation of medicinal plants is well-known and accepted both by scientists and consumers, if the pathways and product yields can be optimized to create precursors for semi-syntheses, food components, pesticide residence and cellular storage conditions as shown for *Mentha piperata* with enhanced resistance against fungal attack and abiotic stress (Julsing *et al.*, 2007).

A Summary of Selected Concerned Literature

Research on herbal medicinal plants and associated issues are very limited in India, Rashid *et al.* (1987) worked out into the nature and functional dynamics of crude drug market in different parts of India. They observed that different *Ayurvedic*, Unani and other pharmaceutical industries of the country commonly use 142 different crude drugs. Chowdhury *et al.* (1996) documented 42 folk formularies, which had long been used traditionally against dysentery and diarrhoea in different parts of India and Bangladesh. Alam *et al.* (1996) also documented 143 folk formularies against 53 common diseases. Ara *et al.* (1990) studied into this phenomenon in some details.

Major Medicinal Plants in the Study

1. Indian Acalypha (*Acalypha indica* L.)

Family: Euphorbiaceae

Indian names: Khokali, kuppi, muktabarsi, venchhi kanto, tuppakire, kuppamani, khokhali, indramoris etc.

An annual herb, up to about 75 cm high, leaves 3-8 cm long, ovate, thin usually 3 nerved, margins of leaves toothed leaf stalks longer than leaves. Flowers in axillary erect spikes, female flower supported by conspicuous wedge shaped bracts, male flowers minute, fruits small, hairy, concealed in the bracts. This plant is found almost throughout India in the plains, it grows as a weed in gardens and agricultural fields or as a plant of waste places and on road sides. *Acalypha* drug is obtained by this plant. This drug is useful in bronchitis, asthma, pneumonia and rheumatism. Its roots and leaves have laxative characters. A poultice of fresh leaves is useful on ulcers.

2. Brahmi [*Bacopa monnieri* (L.) Penn syn. *Herbestis monniers* (L.)]

Family: Scrophulariaceae

Indian names: Safed chamni, brahmi, nirubrahmi, barna, ghola, sambra ne alchlu.

This herb spreads on ground, its stems and small leaves are succulent *i.e.* fleshy. Flowers arise in the axils of the leaves and are borne on short pedicels. One of the five sepals is larger than others. The corolla is bluish white in colour. This herb is distributed in moist or wet places, such as on borders of water channels, wells, irrigated fields etc., in all parts of India. The drug brahmi is constituted by whole parts of plant. This is used as a tonic for nerves, mental diseases, constipation and as a diuretic. This herb contains an alkaloid bramhine, which is a cardiac tonic *i.e.* provides strength and tone to the heart.

3. Indian Barberry (*Berberis aristata* DC.)

Family: Berberidaceae

Indian names: Daruhaldi, daru haridra, rosvat, mara darisina, maramanjal.

A large thorny shrub, yellow, branches whitish or pale grey. Leaves characteristics, fascicled in axils of branched or simple spines, coriaceous usually sharp toothed, vein very fine. Flowers yellow in short branches. Fruit ovoid and seeds few. This plant occurs in the Himalayas between 2000 and 3000 m. It is also grown in the Nilgiri hills in south India. The chief constituent of the drug is the alkaloid berberine. It is used in curing fevers and as mild laxative and tonic.

4. Punarnava (*Boerhavia diffusa* L.)

Family: Nyctaginaceae

Indian names: Bishkopra, punarnava, motostudo, gonajali, ataki, patthar chatta.

A much branched herb, generally spreading on the ground or partly ascending. Leaves two on each node, one smaller than the other, base of the leaves cordate, lower surface of leaves whitish, upper green. Flowers very small, reddish, in short clusters on long axillary stalks. Fruits with five ridges, glandular. This plant is distributed in all parts of India. Punarnava drug is obtained by whole plant particularly the roots, leaves and seeds. This drug includes punarnavine alkaloid and used as a diuretic *i.e.* to promote urination in dropsy and in jaundice and gonorrhoea. The diuretic activity of the drug has been experimentally confirmed on animals.

5. Butea [*Butea monosperma* (Lamk)]

Family: Fabaceae

Indian names: Dhak, palas, kesoodo, muttugo, palasu, palus papra.

This plant is a well known tree of India. It is medium sized tree with compound leaves, each leaf comprising three leaflets, flowers appear in about February to March in small but dense clusters generally on leafless branches. The fruit is a flat pod having a single seed. The red coloured gum called Bengal Kino or Butea gum are obtained from this tree. The gum contains tannins and is valuable for treatment of diarrhoea. The seeds are used as anthelmintic in treatment of roundworms and tapeworms. This tree is highly valued as a host tree for the lac insect. The flowers yield a yellow dye.

6. Cassia (*Cassia fistula* L.)

Family: Caesalpiniaceae

Indian names: Amaltas, kilvali, kirala, sinar, sonaru, bandarlati, garmalo, kakke, kritamalam, suvamaka, konnai, rela.

It is a small or medium size tree with compound leaves and large, shining, dark green leaflets. Flowers bright yellow, in very large, hanging bunches. Fruits 50-60 cm long, black or shining dark brown, almost cylindric. The tree is a conspicuous sight in flowers as well as in fruits and can be spotted in a forest even from long distance. This tree is found throughout India up to an altitude of about 1,500 m, and is more common is moist or evergreen forests. Though medicinal properties have been attributed to nearly all parts of the tree, the fruits are most important and are included in the indian Pharmaceutical Codex. The bark of this tree is known as 'Sumari' and is rich in tannin. The timber of the tree is strong and tough and is suitable for house and bridge post and agricultural implements.

7. Catharanthus [*Vinca rosea* (L.)]

Family: Apocynaceae

Indian names: Sadabahar gul feringi, sadabahar, kasikanigalu, ainskati, pillaganneru

An erect, herb up to about 1 m high, leaves ovate, opposite, flowers in axillary clusters of 2 or 3, petals white or rose-pink, in one variety petals white with pink or reddish tinge at base, fruits many-seeded, follicles. The plant is a native of Madagascar, but has become naturalised throughout the tropics of both hemispheres. The roots of the plant constitute the drug. They were known to posses toxic and stomachic properties, but recently they have gained repute as a source of valuable alkaloids resembling those from *Rauwolfia* species. Roots of *Catharanthus* have more ajmalicine and serpentine than even *Rauwolfia serpentina*, they also possess reserpine. The alkaloids process hypotensive, sedative and tranquillizing properties. The plant is a handsome garden plant and flowers throughout the year, this gives the plant the name 'Sadabahar'. The plant is quite hardy and is popular for ornamental planting.

8. Centella [*Centella asiatica* (L.)]

Family: Apiaceae

Indian names: Brahnnanduki, thankhuria, barmi, vondelega, brahmi, brahmi buti, mandukparni, vallarai elai, bokkudu.

The plant trails on he ground and its creeping stems bear roots on their nodes. Leaves small, 2-4 diameters, some times more rounded or broad kidney- shaped, their margins toothed. Flowers minute, pinkish red, 3-6 in cluster. Fruits small, like a grain of barley, 7-9 ridged. This plant grows in moist places throughout India; it is commonly seen in marshy banks of rivers, streams and ponds and in irrigated lawns, fields, etc. The drug comprises fresh and dried leaves and stems of this plant. Roots and seeds are also used in medicine. The leaves or the entire plant parts are boiled in water and this decoction is given in treatment of leprosy. The plant is also considered useful in certain kinds of tuberculosis and as tonic for brain.

9. Centratherum (*Vernonia anthelmintica* Willd.)

Family: Asteraceae

Indian names: Banjira, somraj, kalijiri, kadujirage, kalen jiri, somaraji, kattuchiragam, kala zira

An erect tall herb, stems and leaves covered with minute hairs. Leaves 6-10 cm long, their margins toothed, base tapering into a petiole. Flower heads 1.5-2.5 cm diameter, in small clusters, each head with 30-40 minute purplish flowers. Fruits which are scientifically called achenes, 4.5-6.0 mm long, cylindric, hairy with 10 narrow ridges, the tuft of hairs on top of achenes, reddish. The drug comprises fresh (as fresh as possible) dried seeds of the plant. As the scientific name of the plant indicates, it is a valuable medicine as anthelmintic, that is, to destroy worms. It is useful in thread worm infections.

10. Cinchona (*Cinchona* sp.)

Family: Rubiaceae

Indian names: The commonest local name is cunain, it is based on quinine, the main active principle in the bark of *Cinchona* trees.

The genus *Cinchona* is not indigenous to India. The dried bark of the above mentioned species constitutes the drug *Cinchona*. *Cinchona* bark yields several active principles, of which quinine is most important, it is well known for its effective use in malarial fevers. Higher doses of quinine preparation can cause temporary (or even permanent) deafness, blindness, giddiness and nausea. The *Cinchona* bark from which quinine has been extracted, serves as a tanning material.

11. Cinnamon (*Cinnamomum verum* Presl. Syn. *C. zeylanicum* Blume)

Family: Lauraceae

Indian names: Dalchini, taj, lavangpatti, darushila.

The scientific name *zeylanicum* refers to Sri Lanka, where the tree is found growing naturally. It is an evergreen tree about 6-8 m high. leaves large, ovate, thick, leathery, flowers minute, in large hairy clusters. This tree occurs in South India up to an altitude of about 1,500 m, is more common at lower altitudes, even below 200 m. It is also cultivated in certain other parts of India. The branches of the trees are lopped and their bark removed. The dried inner bark constitutes the drug *Cinnamon*. The drug is used in diarrhoea, nausea, and vomiting. It is commonly used as a condiment. It is used as a stomachic and carminative, it cures gastric debility and flatulence, and also has the property of destroying certain germs and fungi. The oil obtained from leaves is used as a flavouring agent and preservative for sweets, soaps, etc. and for local application on certain rheumatic pains.

12. Colchicum (*Colchicum luteum* Baker)

Family: Liliaceae

Indian names: Hirantutia, hiranyatooth, irkim, hiranyatuth.

An annual herb, corms brownish in colour, almost conical in shape, with one side flat, the other roundish. Leaves very narrow but broader towards the tip, increasing in size as the plant approaches fruiting stage, 15-30 cm long, 0.8-1.5 cm broad. The plant is found in north-western Himalayas from about 700 to 2,800 m altitude, usually in outskirts of forest or in open grassy place. The fresh corms of the plant collected before its flowering constitute the drug *Colchicum corm*. and the ripe dried seeds, the drug *Colchicum seed.* The corms contain the active principle Colchicine. Clinical experiments with Colchicum in small doses over a long period have shown success in about sixty per cent of patients. The use of the drug can, however, cause severe irritation in intestines and to counteract this. The plant can be grown at higher altitude (1500-3000 m) in the Himalayas. Colchicine has been largely used in scientific research in plant breeding, it induces polypolidy, that is, multiplication of basic chromosome numbers.

13. Coptis (*Coptis teeta* Wall)

Family: Ranunculaceae

Indian names: Mamira, tita.

A small stem less herb, root-stock woody, golden yellow, densely fibrous, very bitter. Leaves compound, glabrous, leaflets 5.0-7.5 cm long, ovate, lanceolate, pinnatifid. Flowers 1-3, small, regular, white, fruit many-seeded follicles, seeds black. It is mainly found in temperate regions of Mishmee hills in Arunachal Pradesh in India. The dried rhizomes commonly known as Tita roots constitute the drug. It is a tonic, stomachic and efficaceous in debility, also used in intermittent fevers.

14. Cymbopogon (*Cymbopogon* species)

Family: Poaceae

Several species of the genus *Cymbopogon* grow in India, some are cultivated. Though medicinal properties have been attributed to some of these species, none of them is of much importance. *Cymbopogon citratus* (DC.) Stapf is grown in several parts of India. It yields oil called West Indian lemon-grass oil. An infusion of its leaves is sometimes taken as a substitute for tea, as a refreshing beverage.

15. Datura (*Datura stramonium* L.)

Family: Solanaceae

Indian names: Dhatura: shet dhatura, dholo dhaturo.

A bushy plant up to about 1m high, leaves large, ovate, toothed. This plant occurs in temperate himalayas up to 2,500 m and in hilly regions of central and southern India. The drug consists of dried leaves, flowering tops and seeds of the plant. The chief active principle in the leaves is hyoscymine, the drug is, therefore, useful in the same manner as belladonna or hyoscyamus. The drug is useful in bronchitis or asthma and controls salivation in mouth, it is antispasmodic and narcotic.

16. Digitalis (*Digitalis purpurea* L.)

Family: Scrophulariaceae

Indian names: Tilpushpi (Tilpushpee).

A biennial or perennial herb up to about 1-2 m high, lower basal leaves long-stalked, hairy, ovate, 15-30 cm long, the upper leaves almost without stalks, becoming smaller in size upwards. This plant is not indigenous to India, it is cultivated chiefly in hilly regions of northern India, such as in Kashmir, it was grown in Darjeeling and in Nilgiri hills also but these cultivations are now abandoned. The dried leaves of the plant constitute the drug. The leaves must be dried at about 60° C temperatures and as soon as possible after collection. The main use of this drug is in heart diseases. The drug promotes and stimulates the activity of all muscle tissues. Digitalis is used in some ointments for local application on wounds and burns.

17. Dioscorea (*Dioscorea* sp.)

Family: Dioscoreaceae

This is a large genus of twining herbs. In some species the stems twine to the right. Leaves are simple or compound, mostly the nerves are prominent. The genus has recently gained much repute as a source of steroidal sapogenins, like diosgenin. These are promising starting material for synthesis of cortisone. Yams are a common article of food particularly among the hill tribes and the poor. *D. alata* L is an important taxa.

18. Embelia (*Embelia tsjeriam-* cottam)

Family: Myrsinaceae

Indian names: Baibirang, dhadki jhanti, ambati, nununiya.

A shrub or small tree, young branches covered with brownish hairs, mature branches glabrous. Leaves up to 12 cm long, narrowed at both ends, dotted all over with glands, margins simpler or toothed, leaves sometimes reddish on lower surface, hairy on nerves. This plant occurs chiefly in Eastern India and the Deccan penissula. The dried fruit of the plant constitutes the drug. The main use of the drug is in treatment of tapeworms. Given in suitable doses, the drug kills the tapeworms, they are then expelled with the help of a purgative *Embelia* itself has mild laxative property. The drug also showed some antibacterial and anti-tubercular activity.

19. *Emblica* [*Phyllanthus officinalis* L. (Gaertn.)]

Family: Euphorbiaceae

Indian names: Aonla, chukna amlok, nellikayi, amla, aora, nellikai.

A small or middle-sized deciduous tree. Leaves small, 10-13mm long, 2-3 mm broad, very closely set in pinnate fashion, branchlets look rather feathery in general appearance. Male and female flowers born on same tree, flowers pale green usually in small dense clusters below the leaves, male flowers small, numerous, on short stalks, female flowers also small, fewer. Fruit 1.5-2.5 cm diameter, fleshy, roundish, rather indistinctly marked into 6 lobes. The plant occurs throughout tropical and sub-tropical India, The fresh or dried fruits of this tree constitute the drug. The fruits are on of the three constituents of the well-known Indian preparation *Triphala* (the other two constituents being *Bahera- Terminalia bellirica* and Harrad, *T. chebula*). *Triphala* is used as laxative and in treatment of enlarged liver, stomach complaints, pain in eyes, etc. Emblica fruits are a good liver tonic, raw fruits are cooling and mild laxative. Even pickled fruits are prescribed in Indian medicine. Flowers roots and bark of the tree are also medicinal, seeds are reported to cure asthma and stomach disorders. Fruits are largely used in inks and dyes and for hair shampoos, fruits along with twig bark are employed for tanning and dyeing. Leaves are used as manure in cardamom plantations, they also help in improvement of alkali soils.

20. Ephedra (*Ephedra gerardiana* Wallich)

Family: Gnetaceae

Indian names: Asmania, phok, busdhur

A small shrub up to about 1m high. Stems woody below much branched, branches almost whorled, erect or initially a little spreading and then ascending upwards. Leaves reduced to minute 2 toothed sheaths. Male flowers few together in ovate spikes. Female flower 1-2, in short spikes. Fruits 7-10 mm long, ovoid, red, surrounded by succulent bracts. This plant occurs in the drier regions of higher Himalayan ranges, at about 2,000-4,000m. It is also cultivated. The dried stems of *Ephedra gerardiana*, collected in autumn, constitute the drug *Ephedra*. The main use of this drug, which yields ephedrine, is in treatment of asthma, particularly bronchial asthma. *Ephedra* proves a very effective cardiac stimulant. It has also been used in hay-fever, rashes, etc, of allergic origin. As the drug has some effect on the urinary bladders also, it is used to control night wetting. Overdoses of ephedrine can cause nausea, sweating, certain skin diseases, etc

21. Euphorbia (*Euphorbia hirta* L.)

Family: Euphorbiaceae

Indian names: Lal dudhi, barakeru, dudheli, accegida, moti dudhi, nanabala.

An annual herb ascending or erect up to 50cm high, stems round, covered with yellowish hairs. Leaves in opposite pairs, small, up to about 4 cm long, dark green on upper side, pale on lower side, margins faintly toothed. The plant occurs in waste place throughout the warmer regions of India, and in a variety of soil and moisture conditions. The entire plant, collected in flowering and fruiting stage and dried, constitutes the drug. *Euphorbia* causes relaxation of bronchioles, and has a depressant action on heart and respiration. It is useful in removing worms in children, in bowel complaints, asthma and cough. The milky juice of the plant is applied on warts. The antibacterial and antitubercular activity of the plant has been shown experimentally. The leaves of *Euphorbia hirta* are eaten as vegetable.

22. Asafoetida (*Ferula narthex* Boiss.)

Family: Apiacceae

Indian names: Hing, sap, yang, perungayam, inguva.

A tall perennial herb with robust carrot-shaped roots. Leaves of two kinds, lower simple, 30-60cm long, ovate, upper much divided into numerous segments. This plant grown in Kashmir. *Asafoetida* is the resinous and scented substance obtained by incision in the living rhizomes or roots of this plant and its allied species. It is a stimulant for respiratory and nervous system and is very effective in pneumonia and bronchitis in children. Experiments have shown that *Asafoetida* has some sedative properties and its possible use in heart disease has been suggested. *Asafoetida* is largely used as a spice, it is an essential ingredient in most pulses and vegetable curry preparation, pickles, etc.

23. Wintergreen (*Gaultheria fragrantissima* Wallich)

Family: Ericaceae

Indian names: Gandhpura-ka-tel, jirhap, hemant-harit,

An evergreen shrub about 3 m high. Stems much-branched, bark orange-brown. Leaves up to 13 cm long, rather broad, leathery, dotted with glands, margins toothed. Flowers small, greenish white, in short axillary bunches. Fruit roundish, enclosed in blue sepals. This plant occurs in the hilly regions of northern, eastern and southern India, between 1,500 and 2,500m altitude. The wintergreen oil is the volatile oil obtained from fresh plants of this species. This oil is used in treatment of various forms of rheumatism. It is applied locally, it is also added to certain ointments and liniments to counteract their irritating effects. It is useful in destroying hookworms and also has stimulant and carminative properties. The oils is used as a flavouring agent in soft drinks, lemon drops, toothpastes, etc. The oil is also used in many preparations for killing or repelling mosquitoes, flies and other insects. The plant is grown as an ornament in hill stations of India.

24. Indian Gentian (*Gentian kurroo* Royle)

Family: Gentianaceae

Indian names: Kamalphul, kutki, nilkanth, karu, pakhanbhed, nilkanthi

A small perennial herb with tufted decumbent stems, 10-30 cm high, rootstock thick. Lower basal leaves 7-13 cm long, tufted, upper leaves up to 2.5 cm long, narrow. Flowers blue, with white spots, 4.5 cm long, 2-25 cm diameter, arising singly or in small bunches of 2-4 flowers. Fruit about 2 cm long. This plant occurs in north-western Himalayas, between 1,500 and 3,500 m altitude, in Kashmir and adjoining hills. The dried rhizomes of the plant constitute the drug. *Gentian* is used for improving appetite and stimulating gastric secretion. It is an ingredient of many tonic and stomachic preparations. It has pleasant flavour and is agreeable to taste. In large doses it causes purging.

25. Indian Sarsaprilla (*Hemidesmus indicus* (L.) Schult.)

Family: Periplocaceae

Indian names: Anantmul, salsa, sogadeberu, anantavel, nannari.

A perennial twiner of creeper, rootstock woody, fragrant, stems slender, hairless, leaves vary greatly in shape and size, they are 5-10 cm long, but their breadth varies from 5 to about 4 cm, dark green, often with whitish blotches, pale or whitish hairy on lower surface. Flower very small, greenish, in small, compact cluster. Seeds small, black, with a tuft of white hairs at top, all parts of the plant have white milky juice. The plant occurs almost throughout India. The dried roots of the plants constitute the drug. The drug is useful in fever, skin diseases, loss of appetite, syphilis, leucorhoea and other urinary complaints. The drug is largely used as a blood purifier and in rheumatism. The leaves of the narrow-leaved from of the plant are chewed and are refreshing.

26. Kurchi [*Holarrhena antidysenterica* (Roth) A.DC.]

Family: Apocynaceae

Indian names: Indrajau, kurchi, kada, dudhari kewar, indrabam, pala kodsa.

A tall shrub or small tree, sometimes up to 10 m tall, Leaves 10-30 cm long, ovate, thin, nerves on the leaves conspicuous. Leaf stalks very small. Flowers white, fragrant, 1-1.5 cm diameter, in large terminal bunches. Fruits slender, cylindric, 20-45 cm long, 6-8 mm thick, dark grey with white specks all over. The plant occurs throughout India up to an altitude of about 1,200 m. It was found to be very common on forest fringes in Madhya Pradesh and Maharashtra. The dried bark of the plant constitutes the drug *Kurchi.* The chief use of this drug is in amoebic dysentery. Either an extract of the bark is used singly or several other preparations in combination with chemical compounds are used. The alkaloid Conessine present in this bark has been found to retard growth of tubercular bacilli. Seeds also possess alkaloids which are effective in dysentery. Certain medicinal properties have been attributed to leaves also.

27. Chalmogra [*Hydnocarpus kurzii* (King) Warb.]

Family: Flacourtiaceae

Indian names: Chalmogra, lamtani, maravetti,

A tree about 15 m high, sometimes more, with a tall trunk and a crown of drooping branches. Leaves about 20cm long, pointed at tip, leathery. Flowers small, yellow, in small axillary clusters. Fruit 6-7 cm diameter, round, brown, seeds many. The tree is found in evergreen forests of Assam and Tripura, at certain spots it is very abundant. The oil obtained from the fresh ripe seeds of the tree is medicinal. The oil is used in treatment of leprosy. Formerly the oil was given orally, now certain preparations based on the oil are injected. The bark of the tree contains certain tannins and is considered useful in fever. The seed-cake of *Hydnocarpus kurzii* is used as manure. Some animals eat the fruits of this tree; it has been recommended that the flesh of such animals should not be eaten. Even fish stupefied or killed by Chalmogra fruits are considered unsuitable for human consumption.

28. Talmakhana [*Hygrophila auriculata* (Schum.) Heine]

Family: Acanthaceae

Indian names: Kuliakanta, ekhro, vayelchulli, kokilaksha, nirugobbi

A stout erect herb, 60-150 cm high, stems unbranched, erect, straight, four-angled, hairy, more so below the swollen nodes. Leaves whorled, 6 at a node, each having a sharp, yellow spine in its axil. Flowers 8 in a whorl, purple-blue, about 3 cm long, 2-lipped, lower lip characteristically folded. The plant occurs in marshes and other moist spots throughout India, often filling the shallow, partially dried ditches and roadside water channels. The drug consists of the entire plant including the roots. It is useful in dropsy, jaundice, rheumatism and diseases of urinogenital system, seeds are considered useful also in venereal diseases. The diuretic activity of the drug is believed to be due to combination of both inorganic and organic contents of the plants. Seeds and roots, singly also, have diuretic property. Leaves are useful in cough and urethral discharges.

29. Kaladana [*Ipomoea nil* (L.) Roth, syn. *I. hederacea* auct. non jacq.]

Family: Convolvulaceae

Indian names: Nilkulmi, kaladana, gouribija, kanikhondo jiriki.

An annual twining plant, stems sparsely covered with hairs. Leaves 5-12 cm diameter, ovate, shallowly or deeply 3-lobed. Flowers 4-5 cm long, funnel-shaped, blue, slightly orange coloured below, few growing together in small stalked clusters. Fruit 8 mm diameter, almost roundish or ovoid. Seeds, small, glabrous. This plant occurs throughout India up to an altitude of about 1800 m, it is truly wild in India, at times it is cultivated. The dried seeds of the plant constitute the drug. Kaladana is used as a purgative. Overdoses of the drug cause irritation. Fresh fruits of the plant are eaten as vegetable. The plant is often grown for its beautiful flowers.

30. Henna (*Lawsonia inermis* L. syn. *L. alba* L.)

Family: Lythraceae

Indian names: Mehendi, mendi, mohuz, hinah, goranti

A middle-sized or large, much-branched shrub, sometimes tree-like, branches 4-angled, usually ending in a sharp point. Leaves opposite, 2-3 cm long, often acute and sharp-pointed. Flowers small, white or pinkish, fragrant, in terminal large bunches. Fruit small. The plant occurs in several parts of India, chiefly in the drier parts of the peninsula, but is usually cultivated in hedges. It is also cultivated for commerce in Punjab, Gujarat, Madhya Pradesh and Rajasthan. The leaves of the plants have certain medicinal properties. They are astringent and are used as a prophylactic against skin diseases. The leaves have also been shown to have some action against tubercular and other bacteria, and in typhoid and haemorrhagia. The bark and seeds of the plant are also reported to be used in Ayurvedic and Unani medicines. The chief use of the Henna plant is as pleasant orange dye for colouring palms, nails, hair, beard, even tails and limbs of animals. The oil obtained from its flowers is used in perfumery.

31. Lobelia (*Lobelia nicotianifolia* Heyne ex Roth)

Family: Lobeliaceae

Indian names: Narasal, bantamaku, naali, devanala, adavipogaku.

A large herb, stem stout, hollow, 3-5 m high, occasionally branched in the upper part. Leaves very large, up to 45 cm long in lower part of stem, smaller upwards, margins not entire, the main nerve of the leaf is whitish. Flowers large, white, at very large terminal bunches. Fruit 8 mm, roundish. Seeds many, small, yellowish, brown. The plants occur in the hilly regions of the Deccan peninsula, also in adjoining plains. The drug consists of aerial part of the plant collected in October to November and dried in shade. The properties of this drug are somewhat similar to nicotine. Lobeline, the alkaloid contained in this drug, stimulates respiration and is used for reviving respiration in cases of failures of respiration caused by excessive narcotics, anesthesia's, gases, etc. Aerial parts of *L. nicotianaefolia* exude a white juice, this juice causes blisters on skin. The dried plants seem to cause irritation of mucous membranes even from some distance.

32. Madhuca (*Madhuca indica* Gmel. syn. *Bassia latifolia* Roxb.)

Family: Sapotaceae

Indian names: Mahua, ippe, mohuka, ippa.

A large deciduous tree with rather shorter bole, but larger crown. Leaves 12-25 cm long, thick leathery, pointed at tip, nerves quite prominent. Flowers small, fleshy, pale or dull white, in clusters near the ends of branches, stalks of flowers bent downwards, covered with brownish hair. Flowers strongly musk-scented, falling at dawn. Fruit 2.5-5 cm long, fleshy, greenish, seeds 1-4, brown, shining, 2.5-3.5 cm long. The tree occurs in all the plants and lower hills of India up to 1,200 m, it is very common in submontane regions of the Himalayas, and is, at certain places, a chief constituent of the forest vegetation. The bark, leaves, flowers and seeds of the tree constitute the drug. Leaves are astringent, their ash, mixed with butter (ghee), is applied on burns and scalds. Flowers are used in cough and bronchitis, they are cooling and nutritive. Seeds promote formation and flow of milk. Oil from seeds is good on skin diseases and as laxative. *Madhuca* flowers are eaten raw or cooked, their excessive use is harmful. Flowers are largely employed for making alcohol, vinegar, syrups, jams, etc. Mahua oil is largely used in manufacture of soaps, in cooking, etc. Mahua cake is used as manure, sometimes it is fed to cattle.

33. Kamela [*Mallotus philippensis* (Lamk.) Muell. Arg.]

Family: Euphorbiaceae

Indian names: Kamela, gangai, kapila, shendri, kapila podi, kunkuma.

A small or middle-sized evergreen tree, sometimes only a much-branched large shrub, young parts densely covered with minute red hairs. Leaves alternate, borne on long stalks, variable in shape, 7-20 cm long, their lower side dotted with reddish glands, prominently veined. Flowers minute, male and female on separate plants. Female flowers in erect 5-8 cm long spikes, male flowers yellow, in 8-15 cm long, drooping bunches. The tree is found throughout tropical regions of India from an altitude of 1,500 m in the Himalayas, southwards up to Kerala. The red glandular and hairy substance separated from the fruits forms the drug. Kamela is chiefly used for destroying tapeworms. The Kamela powder is taken with milk or curd, etc. If one dose of Kamela does not expel the worm, the dose is repeated. Sometimes a dose of castor oil is necessary to expel the dead worm. The hairs on the fruits have been tried (on animals) as oral contraceptives, they have been found to reduce fertility in female rats and guinea-pigs. The tree is a well-known source of Kamela dye and used in colouring textiles. The oil extracted from seeds is being used in paints and varnishes, and in several other ways. Due to excellent drying properties, the oil is much valued in painting work and is in great demand. The wood of the tree is used for minor domestic articles and fuel.

34. Mint (*Mentha* sp.)

Family: Lamiaceae

Indian names: Most of the Indian names of *Mentha* species are based on the word pudina.

The plants belonging to the genus *Mentha* are aromatic herbs. Several species grow wild, some are cultivated. The chief constituents for which these plants are valued are Menthol and Peppermint oil. *Mentha arvensis* L. (English-Field Mint, Corn Mint, Indian name-Pudina) is an erect branched herb up to about 60 cm high. Leaves up to 5 cm long, leaf-stalk small or none, margins toothed Flowers small, lilac, in small bunches, borne on axils of leaves. This herb occurs in Kashmir at an altitude of about 1,500-3,00 m, it is also cultivated. Infusion of its leaves is used in rheumatic pains and indigestion. *Mentha longifolia* (L.) Huds. (Syn. *M. sylvestris* L., English-Horsemint, Hindi-Jangli Podina, Punjabi-Koshu) is a larger herb and occurs in Kashmir and northern parts of Punjab and Uttar Pradesh. This Plant is considered antiseptic, stimulant, and useful in digestive complaints and fever. The most important plant of this group is *Mentha piperita* L. (English-Peppermint, Punjabi-Vilayati Pudina). This plant is cultivated in several parts of India, *e.g.* in Kashmir, Uttar Pradesh, Mysore, Madras, etc. The dried leaves and flowering tops of the plant constitute the drug Peppermint. The drug is used in treatment of flatulence, vomiting, diarrhoea and nausea. Bruised leaves are applied in headache and other pains.

35. Jatamansi (*Nardostachys grandiflora* DC. syn. *N. jatamansi* DC.)

Family: Valerianaceae

Indian names: Jatamansi, bhutijatt.

A perennial herb up to about 60 cm tall, rhizomes woody, long, covered with fibres from petioles of withered leaves. Lower basal leaves up to about 20 cm long, narrowed into the petiole, upper leaves much smaller, almost ovate. Flowers small, several in small bunches. Fruit small, covered with minute hair. The plant occurs in the alpine Himalayas between 3,000 m and 4,500 m altitude, eastwards to Bhutan. The dried rhizomes and roots of this plant are medicinal. *Jatamansi* has tonic, antispasmodic and stimulant properties, it is therefore, useful in treatment of certain types of fits, convulsions and palpitation of the heart. It is used also as laxative and for improving urination, menstruation and digestion. It is used as a substitute for the drug Valerian (*Valeriana officianalis* L.)

36. Tulsi (*Ocimum sanctum* L.)

Familly: Lamiaceae

Indian names: Tulsi, vishnu tulsi, trittavu, krishna thulasi, thulasi.

This is well- known sacred plant of the Indians. It is a much branched erect herb, up to about 75cm high, hairy all over. Leaves opposite, about 5 cm long, margins entire or toothed, hairy on upper as well as lower surface, dotted with minute glands, aromatic. This plant is grown in houses, gardens, and temples all over India and is often found as escape. The leaves and seeds of the plant are medicinal. The oil obtained from leaves has the property of destroying bacteria and insects. The juice or infusion of the leaves is useful in bronchitis, catarrh, digestive complaints, it is applied locally on ringworm and other skin diseases. The plant is rather an essential article in the worship of Hindu gods and goddesses.

37. Turpeth [*Operculina turpethum* (L.)]

Family: Convolvulaceae

Indian names: Nisoth, dudiya kalmi, nishotara, nisot, sivathi, tellategada.

A large twining herb with milky juice, root long, branched, fleshy, stems winged. Leaves 4-10 by 1.5-7.0 cm, ovate, cordate. Flowers white, 4-5 cm long, funnel-shaped, in few-flowered bunches. Sepals about 2 cm long, but when plant is in fruit, they become much larger and brittle and enclose the fruits. The plant occurs almost throughout India up to an altitude of about 1000 m, it is sometimes grown in gardens for its beautiful flowers. The dried roots of the plant constitute the drug. The drug contains *Turpethin*. The Turpeth has almost same properties as the true Jalap, obtained from *Exogonium purga* and is a suitable substitute for it. It is used as a purgative. Powdered roots are suitable for this. The samples of Turpeth commonly available in markets have stems and twigs of the plant mixed with the roots.

38. Harmal (*Peganum harmala* L.)

Family: Rutaceae

Indian names: Harmal, isband, harmar, simagoranti.

A Shrub about 30-40 cm high, leaves 5-8 cm long, divided into numerous narrow segments. Flowers 2-3 cm diameter, white, single in axils of leaves. Fruits capsular, globose, 5-8 mm diameter, deeply lobed. Seeds 2.5-4 mm long, brownish, of various shapes and with reticulated seed coat. The plant is found almost throughout northern and north-western India, and in drier regions of the Deccan. The seeds contain several alkaloids and are useful in asthma, hysteria, rheumatism, gallstones, colic pains, fever, jaundice and complaints of difficult and painful menstruation. They are also used as narcotic, anthelmentic and emetic. The Alkaloids Harmaline, Yageine and Harmine, obtained from this drug, are all psychomimetics, *i.e.*, act as hallucinogens.

Experiments have confirmed bactericidal action of the drug, but tests in hospitals did not show any useful effect of the drug in malaria. The seeds yield a red dye. Powdered roots mixed in mustard oil kill lice in hair. The plant, kept in a room repels mosquitoes.

39. Pergularia [*Pergularia daemia* (Forsk.) Chiov]

Family: Asclepiadaceae

Indian names: Utarmi, chagulbati, utami, karail, yugaphala, uttamani.

A twining herb, stem hairy and with milky juice. Leaves 4-6 cm long, sometimes more, ovate or round, hairy on lower surface, deeply cordate at base. Flowers pale or white, small, in short clusters. Fruits reflexed in pairs, 5-8 cm long, 1.3 cm broad, covered with spinous outgrowths. The plant is found almost throughout India up to 1000 m. The entire plant is used in medicine. The juice of the leaves is given in catarrhal affection and infantile diarrhoea. It forms a constituent of a purgative preparation given in rheumatism and amenorrhoea. It is used as a uterine tonic, expectorant and emetic. Its action as a uterine tonic has been confirmed experimentally, it is useful is curing gynecological conditions.

40. Picrorhiza (*Picrorhiza kurrooa* Royle-ex Benth.)

Family: Scrophulariaceae

Indian Names: Kuru, kutki, kataki, kadu, karu, kutaki, katuka, katuku-romi

It is a small herb with spathula-shaped, 5-10 cm long leaves. Rhizomes of the plants are 15-25 cm long and woody. Flowers small, in cylindric spikes, spikes borne on almost leafless erect stems. Flowers are of two kinds, some are with 8 mm long filaments others with 2 cm long filaments. The plant occurs only in highly mountains, *i.e.* at about 3000-4000 m altitude in the Himalayas from Kashmir to Sikkim. The drug comprises the dried rhizomes of the plant. It is bitter tonic and is useful as antiperiodic and promotes secretion of bile, it is believed to have same properties as Gentian. The antibiotic activity of the drug *Picrorhiza* has been shown experimentally. The Himalayan mountains between the altitudes of 3000 and 4000 m are considered to be suitable for its cultivation.

41. Pine (*Pinus roxiburghii* Sargent syn. *P. longifolia* Roxb.)

Family: Coniferae

Indian names: Chir, tellia, suralagach chil,

It is a large tree with typical needle leaves, leaves 25-30 cm long, in clusters of three. Male cones small, female fruiting cones 10-20 cm long, conical. The tree is found in the lower Himalayas and other hills of India, it is commonly planted for ornament. The drug oil of turpentine is obtained by purification from turpentine, an oleoresin obtained from *P. roxburghii* and certain other species of this genus. The oil has local irritant action and most of its medicinal uses are due to that property. In controlled small doses, it acts as a stimulant expectorant and it is useful in chronic bronchitis. It cures flatulent colic. It has limited use also in typhoid, minor hemorrhages. The timber of the tree is largely used for various purposes.

42. Long Pepper (*Piper longum* L.)

Family: Piperaceae

Indian names: Pipal, pipali, piplu, pipli, magadhi.

A small aromatic plant trailing on ground, also climbing. Lower leaves 6-10 cm long, broadly ovate, deeply cordate with big lobes at base, upper leaves oblong-ovate, cordate, all dark green and shining above, pale on lower surface, stipules conspicuous, 1.3 cm long, but soon falling. Spikes of flowers solitary, bracts of male spikes narrow, of female circular. Long Pepper consists of the dried fruits of the plant. It is used as a tonic, and in making irritating snuffs. It is also used in liniments for rheumatic pains and paralysis. A decoction made from the dried immature fruits is useful in chronic bronchitis. Ripe fruits are aromatic, stomachic and carminative. The antibiotic activity of the leaves and fruits has been shown experimentally. Another very important species of the genus is *Piper nigrum* L. or Black Pepper, Mirch, Kali Mirch, Dakhnimirch. Most of the vernacular names of the plant describe the shape or colour of the fruits. The Sanskrit name indicates the use of the plant. *P. nigrum* is a robust woody climber with large, broad ovate or round leaves. Flowers borne in long spikes, fruits small, globose, 6-7 mm diameter, yellow, turning red when ripe.

43. Ishabgul (*Plantago ovata* Forsk.)

Family: Plantaginaceae

Indian names: The local names of this plant in different regions and languages of India are only minor variations of the Sanskrit word *Ishapgola* or *Isapgol*.

An almost stem less small herb, covered with dense or soft hairy growth. Leaves 8-25 cm long, very narrow. Flowers minute, in oval or cylindric spikes, 1.5-4 cm long. Fruit 8 mm long, its upper half separates like a lid. Seeds boat-shaped. The plant grows naturally in only restricted areas in north-western India, but is largely cultivated elsewhere. The seeds of this plant constitute the drug. The useful properties of the seeds are due chiefly to the large amount of mucilage and albuminous matter present in them. Ishabgul is very useful in several kinds of chronic dysentery, such as of amoebic and bacillary origin and chronic diarrhoea. It is useful as a soothing agent for mucous membranes, and is useful in constipation. Seeds should be soked in water before use, so that they soon get disintegrated in the alimentary canal, else the whole seeds can cause irritation or even mechanical obstruction in intestines. *Ishabgul husk* (Hindi– *Ishabgul-ki-bhusi*) is the dry seed coat of *Plantago ovata*, obtained by crushing the seeds and separating the husk by winnowing. The embryo oil of seeds having 50 per cent linoleic acid prevents arteriosclerosis. Oil is more active than safflower oil and was found to reduce the serum cholesterol level in rabbits.

44. Indian Podophyllum (*Podophyllum hexandrum* Royle syn. *P. emodi* Wallich ex Hook F. and Th.)

Family: Berberidaceae

Indian names: Papri, venivel, padwal, gulkakri.

A succulent erect herb with a creeping rootstock. The flower bearing branch is erect, leafy at top, leaves two, 15-25 cm diameter, deeply divided in 3-5 lobes, toothed purple spotted. Flowers 2.5-5 cm diameter, white or pinkish, cup shaped, usually solitary. Fruit ovoid, 2.5-5 cm, scarlet. This plant occurs at high altitudes in the inner ranges of Himalayas, usually at about 3000-4000 m. In Kashmir, it comes down to about 1800 m. The dried rhizomes of the plant are used in medicine. The rhizomes contain a resin Podophyllin. It is purgative, its action is slow but severe, and large doses can cause acute irritation and griping. The drug is usually administered in mixture with Belladonna or aloes, etc. *Podophyllum* is reported to be useful in many skin disease and tumorous growths, its use in curing cancerous tissues is now in experimental stages.

45. Psoralea (*Psoralea corylifolia* L.)

Family: Fabaceae

Indian names: Babchi, latakasturi, bavancigida, kala gija.

It is an erect herb, with densely gland-dotted branches. Leaves round, dotted with black glands on both surfaces. Flowers small, bluish purple, 10-30 in a bunch, arising in axils of leaves. Fruits black, roundish or oblong, closely pitted, seed one, smooth. This plant is found throughout India as a weed in waste places, it is also cultivated in some places. The seeds of the plant consitute the drug. The seeds contain an essential oil which is very effective on certain bacteria causing skin diseases. The drug is, therefore, useful in leucoderma and leprosy as an external application in the form of ointment, as well as for taking internally. The oleoresinous extracts of the seeds are useful for local application on leucoderma of nonsyphilitic origin. Due to its use in leprosy, the drug has been called in our indigenous system as Kushtha nashini. Roots of the plant are reported to be useful in caries of teeth, and the leaves in diarrhoea.

46. Indian Kino (*Pterocarpus marsupium* Roxb.)

Family: Fabaceae

Indian names: Bijasal, piasal, bibla, honne, honi, Bijaysar.

A large handsome tree. Leaves compound, having 5-7 leaflets, leaflets 8-13 cm long, oblong or elliptic. Flowers about 1.5 cm long, yellow, scented, in very large, dense bunches. Fruit 2-5 cm long, roundish, winged, with one seed. The tree is common in central and peninsular India, chiefly in mixed deciduous forests. It ascents to about 1100 m in Gujarat, Madhya Pradesh, and sub-Himalayan tracts, etc. The tree in flower is a beautiful sight. Leaves, flowers and gum of the tree are medicinal. The gum called Kino exudes from incisions in bark. It is astringent and useful in diarrhoea. Its action in milder than Catechu (Katha), it is also used for toothache. The gum is considered to be useful in fevers and urinary discharges. Bruised leaves are applied on boils, sores and other skin disease. In clinical experiment, it was shown that only 7 per cent of the diabetic patients treated with this drug showed improvement. The timber of the tree is valued for high-class furniture construction work, etc.

47. Rauvolfia [*Rauvolfia serpentina* (L.) Benthum]

Family: Apocynaceae

Indian names: Chandra, sarpgandha, chandrika, dhanbarua,

An erect glabrous shrub, 30-75 cm high, leaves whorled, 8-20 cm long, gradually tapering into a short petiole. Flowers about 1.5 cm long, petals white or pinkish, peduncle deep red, in small clusters. Fruits small, round, dark purple or blackish when ripe. The plant is found in almost all parts of India up to an altitude of about 1,000 m, it is more common in submontane regions of Himalayas and in lower ranges of Eastern and Western Ghats, also collected from several places in plains of Bengal, Bihar, Uttar Pradesh, Maharashtra, etc. The plant is now being cultivated at several places. The drug consists of the dried roots with bark intact, preferably collected in autumn and from plants of about 3-4 years age. It is believed that this plant has been known in Indian medicine for about 4,000 years. A mention of the plant if found in Charak's work. The roots contain several alkaloids. The chief use of the drug is as a sedative and hypnotic and for reducing blood pressure. The drug is now largely used in insanity and high blood pressure. The roots of the plant are useful also in diseases of bowels and in fever.

48. Rhubarb (*Rheum emodi* Wallich ex. Meissn)

Family: Polygonaceae

Indian names: Revandchini, kokima, revat-chini.

A tall herb with very stout stem and roots. Lower leaves very large, about 60 cm diameter, round, on very stout, 30-40 cm long petioles. Flowers minute, dark purple, in a very large bunch. Fruit 1.3 cm long, purple. The plant occurs in the Himalayas at about 3,000-4,000 m altitude, in Kashmir, Himachal Pradesh, Punjab, Uttar Pradesh and also Nepal. The dried rhizomes of *Rheum emodi* constitute the drug. The rhizomes of *R. webbianum* Royle (Archu) are also accepted. The rhizomes should be collected from 6-7 year old plants just before the flowering season, they should not be decorticated. Rhubrab is used as purgative. The drug also has tannins and therefore after purgation it creates an astringent effect, which causes constipation. Rhubarb is not suitable in cases of chronic constipation, but only in mild ones. Due to astringent action, Rhubarb is also given in certain types of diarrhoea.

49. Castor-Oil Seed (*Ricinus communis* L.)

Family: Euphorbiaceae

Indian names: Arandi, bherenda, diveli, haralu, erand.

It is a tall shrub, sometimes becoming tree-like. Leaves very large, broad, roundish in outline but partly divided into 7 (sometimes 9) lobes, margins toothed. Flowers large, in big terminal bunches. Fruit is prickly capsule, rather marked into six parts. The plant is largely cultivated on borders of agricultural fields, gardens, etc., and also found wild in fields, gardens, near habitations and in waste places. The seeds of the plant are used in medicine. The seeds are poisonous and even 2 or 3 seeds can be fatal. The oil obtained from the seeds, called Castor oil, is used as a purgative. The use of Castor oil for facilitating child-birth is doubtful, rather it should be sparingly used even as purgative for pregnant women and during menses. A gel prepared from Castor oil is useful in dermatosis and is a good protective in occupational eczema and dermatitis.

50. Sandalwood (*Santalum album* L.)

Family: Santalaceae

Indian names: Chandan, sukhad, gandham chakka.

A middle-sized evergreen tree, branches almost drooping, bark dark, rough, with vertical cracks mature wood scented. Leaves 4-7 cm long, opposite, shining on upper surface. Flowers small, dull, purplish, in small bunches. Fruits foundish, 6 mm diameter, purple black succulent. The tree grows wild in the Deccan peninsula, particularly in the southern region. The oil obtained from the heartwood of this tree is medicinal. The oil is used in treatment of dysuria, *i.e.* to promote and facilitate urination, cystitis (inflammation of bladder), gonorrhoea and cough. The drug is useful in tuberculosis of gall bladder. Oil from the seed is used on skin diseases.

51. Asoka [*Saraca asoca* (Roxb.) De wilde syn. *S. indica* auct. non L.]

Family: Caesalpiniaceae

Indian names: Ashoka, aasopalav, asogam.

A small tree, leaves compound, evergreen, forming a dense crown, leaflets 7-25 cm long, slightly leathery. Flowers bright orange-coloured due to coloured bracts, in small dense bunches. Fruits 15-25 cm long, flat, seeds many. The tree occurs in central and eastern Himalayas, eastern Indian and in the south, it is often cultivated for its showy flowers. The dried bark of the tree is of medicinal value. It is used as an astringent in treatment of excessive menstruation as a uterine sedative. It can be used as a substitute for Ergot in treatment of uterine hemorrhages. Flowers pounded and mixed with water are useful in hemorrhagic dysentery. Seeds are reported to be useful in urinary discharges.

52. Sida (*Sida cordifolia* L.)

Family: Malvaceae

Indian names: Kungi, bala, hiretutti, chikana, jayanti,

A small much-branched shrub, minute star-shaped hair present all over the plant, leaves 2-5 cm, ovate or roundish, thick, margins toothed, petioles shorter than leaves. Flowers, yellow small, one or a few together, fruits 6-8 mm diameter, divided into 7-10 parts, each strongly reticulated, and with two awns (spiny projections) on tip. The plant occurs throughout India as a common weed, usually in waste places and open scrub forests. The entire plant is used as medicine. It is used as a general tonic and for improving sexual strength.

53. Belleric Myrobalan [*Terminalia bellirica* (Gaertn.) Roxb.]

Family: Combretaceae

Indian names: Bahera, bhovian, telaphala, akkam, tadi,

A large tree, often with buttresses. Leaves large, 10-25 cm long, clustered near ends of branches. Flowers small, pale green, bad-smelling, in simple spikes. Fruit 2-3 long, ovoid, brownish, densely covered with hairs. This tree occurs almost throughout India up to about 1,000 m, exception the dry regions of western India, it is more common in mixed deciduous forests. The dried fruits of the tree constitute the drug. The fruits are useful in stomach disorders such as indigestion, diarrhoea. The Bahera fruit is one of the three constituents of the famous Indian preparation Triphala, the other two being amloki (or Aonla) Harra. The timber of this tree keeps well under water and is used for boats etc., also for miscellaneous agricultural tools.

54. Tinospora [*Tinospora cordifolia* (Willd.) Hook. f. and Th.]

Family: Menispermaceae

Indian names: Gilo, gulancha, gulvel, galo, amritaballi, gulvel,

A large climber with succulent stems, throwing aerial roots like the common Banyan (*Ficus*) tree. Stems and branches specked with white glands. Leaves 5-10 cm. ovate or roundish, 7-9 nerved, petioles slightly shorter than leaves. Flowers minute, male and female separate, male flowers grouped in axils of bracts, female solitary. The plant occurs throughout tropical regions of India. The stems of the plant collected in hot season and dried, with bark intact, constitute the drug. The drug is believed to be useful as a tonic and antiperiodic, if is also considered aphrodisiac. The starch obtained from the roots and stems of the plant is useful in diarrhoea and dysentery, it is also a nutrient.

55. Bishkhapra (*Trianthema portulacastrum* L. syn. T. monogyna L.)

Family: Aizoaceae

Indian names: Lalsabuni, sabuni, satodi, bishkapra,

A succulent spreading herb, stems much-branched, angular. Leaves broader towards tip. Flowers very small, almost concealed in the base of leafstalk. Fruits small, like flowers, concealed in base of leafstalk. Seeds several, black, kidney shaped, covered with minute outgrowth. The plant is found throughout India. The leaves of the plant are medicinal. Leaves contain an alkaloid Punarnavine. It is diuretic, *i.e.* promotes urination and is, therefore, useful in dropsy. It is beneficial in swellings of body caused by disorders of liver or kidney, it is particularly helpful in early stages of the diseases.

56. Gokhru (*Tribulus terrestris* L.)

Family: Zygophyllaceae

Indian names: Chhoto gokhru, neringil, nerunji, palleru.

A prostrate spreading herb, densely covered with minute hair. Leaves in opposite pairs, 5-8 cm long, compound, leaflets 4-7 pairs, 8-12 mm long. Flowers pale yellow, 1-15 cm diameter, growing solitary opposite to the leaves or in axils of leaves. The plant occurs throughout Indian almost up to 3,000 m altitude. The fruits of the plant constitute the drug. The fruits are useful in urinary complaints and sexual weakness. It is cooling. An infusion of the fruits is useful in gout and diseases of kidney, it promotes urination. Clinical tests have confirmed efficacy of the drug in promoting urination.

57. Ashvagandha (*Withania somnifera* Dunal)

Family: Solanaceae

Indian names: Ashvagandha, asan, amukkiram,

A small or middle-sized under shrub, up to 1.5 m high, stem and branches covered with minute star-shaped hair. Leaves up to 10 cm long, ovate, hairy-like branches. Flowers pale green, small about 1 cm long, few flowers borne together in short axillary clusters. Fruit 6 mm diameter, globose, smooth red, enclosed in the inflated and membranous calyx. The plant occurs in drier regions of India, it is also cultivated. The drug consists of the dried roots of the plant. Ashvagandha is useful in consumption, sexual and general weakness and rheumatism. It is diuretic, *i.e.* it promotes urination, acts as a narcotic, and removes functional obstructions of body. The root powder is applied locally on ulcers and inflammations. The antibiotic and antibacterial activity of the roots as well as leaves recently been shown experimentally.

58. Aconite (*Aconitum* sp.)

Family: Ranunculaceae

Indian names: Aconitum chasmanthum, mohri

The trade of common name *Aconite* refers to the scientific name of this group of plants. The drug Aconite comprises tuberous roots of these plants, it has been well-known in medicine as it possesses certain very toxic and poisonous substances. The British Pharmacopoeia recognises the drug obtained from *A. napellus* L. This speies is not found in India, but Indian species contain similar substances or are suitable substitutes.

59. Vasaka (*Adhatoda zeylanica* Medik syn. *A. vasica*)

Family: Acanthaceae

Indian names: Adusa, bansa bahaka, vasaka, alduso, adusoge.

A tall, much-branched, dense, evergreen shrub, with large lance-shaped leaves. Flowers in dense, short spikes, stalks of the spikes shorter than leaves. Leaf-like structures, called bracts, present on the spikes, these are conspicously veined, corolla (the whorl of petals) of the flowers is white with few purplish markings. Fruits capsular, 4-seeded. This plant occurs throughout the plains and submontane regions of India, it is common near habitations. The drug Vasaka comprises the fresh or dried leaves of the plant. Leaves contain an alkaloid vasicine, and an essential oil. The expectorant activity is due to stimulation of bronchial glands. Larger doses can, however, cause irritation and vomiting. Recent experiments have confirmed the usefulness of Vasaka. The leaves of this plant are also utilised as green manure and for yielding a yellow dye. Due to the presence of certain alkaloids, the leaves are not easily attacked by fungi and insects and are, therefore, used in packing or storing fruits.

60. Bel [*Aegle marmelos* (L.) Correa]

Family: Rutaceae

Indian names: Bel, bili, bilpatre, vilvam, vilva, maredu.

A medium-size deciduous tree bearing strong axillary thorns. Leaves with 3 ro 5 leaflets. Flowers greenish-white, sweet-scented, about 2.5 cm across, in small bunches. Fruit 8-20 cm diameter, globose, green, finally greyish, rind woody, pulp orange-coloured, sweet, aromatic. This tree occurs in the submontane regions and plains almost throughout India. It is also planted. The drug called Bel (*or Belae fructus*) comprises fresh, ripe or half-ripe fruits of the tree. The Bel fruit is valuable chiefly for its mucilage and pectin, it is very useful in chronic diarrhoea and dysentery, particularly for patients having diarrhoea alternating with spells of constipation. The wood is suitable for making charcoal for producer-gas plants. The gummy substance, in which the seeds remain embedded, is used as an adhesive, and in varnishes and cementing mixtures.

61. Greater Galangal [*Alpinia galanga* (L.) Willd.]

Family: Zingiberaceae

Indian names: Kulinjan, dumpurasme, kulanjan

A herbaceous plant up 2 m high, leaves long, narrow, green above, pale beneath, whitish on margins, median nerve very strong. Flowers about 3 cm long, greenish white, in compound dense bunches, lip of corolla white, streaked with red. The plant occurs naturally in the eastern Himalayas and the Western Ghats, it is also cultivated at several places. *Greater Galangal* is the dried rhizome of this plant. It is useful in rheumatism, respiratory complaints, specially of children, and in bronchial catarrh. The drug is also considered useful in stomach complaints and as a tonic, deodorant and disinfectant. It is a stimulant aromatic like ginger.

62. Chhatim [*Alstonia scholaris* (L.) R. Br.]

Family: Apocynaceae

Indian names: Chhatin, chhattim, pala, chatinan, palaigh.

A large evergreen tree reaching up to about 25 m high having bitter milky juice, bark rough, dark grey, branches whored, base of the tree often fluited or buttressed. Leaves leathery, 10-20 cm long, 4-7 in a whorl. Flowers small, greenish white, spice-scented in many-flowered clusters. This tree is found in the more moist regions throughout the country. This drug is considered very efficacious in chronic diarrhoea and dysentery. It is useful in malarial fever and brings down the temperature gradually and without causing perspiration or exhaustion, which usually follow other medicines for malaria. The drug is reported to cause paralyzing effect on the motor nerves and consequent fall in blood pressure. Some experiments have contradicted antibiotic and physiological activity of the drug.

63. Kalmegh [*Andrographis paniculata* (Burm. f.) Wall. ex Nees]

Family: Acanthaceae

Indian names: Kirayat, alui, nelabevu, nelavepu, nila vembu.

An erect branched annual herb, branches sharply for angled. Leaves lens shaped. Flowers small, in large, spreading and sparse bunches. The plant occurs throughout India, chiefly in the plains. The drug *Kalmegh* consists of all parts of the plant above ground, *i.e.* excluding only the roots. *Kalmegh* is a bitter tonic, it is useful in curing fever, worms, dysentery, general weakness and excessive gas formation in stomach. It is also reported to be useful for children suffering from liver and digestion complaints.

64. Indian Birthwort (*Aristolochia indica* L.)

Family: Aristolochiaceae

Indian names: Isharmul, toppalu, sapasan, peru hanunthu.

A twining shrub. Stems woody below, slender and soft above. Main rootstock woody. Leaves variable in size and shape from linear, that is, very narrow, to obovate, that is broader towards tip. Flowers are small. It is found throughout India in lower hilly regions and in plains, it is more common in southern and eastern India. *Aristolochia* consists of the dried stems and roots of the plant. This drug is useful only in very controlled doses, in small doses it promotes digestion and regulates menstruation.

65. Wormseed (*Artemisia maritima* L.)

Family: Asteraceae

Indian names: Kirmani ajvain, murni, gadadhar.

A stout, much-branched, perennial, aromatic shrub, about 1 m high. Leaves 2-5 cm long, whitish in colour, divided into numerous fine, linear segments, upper leaves simple, undivided linear. Flower heads small, in short spikes. The plant occurs in northern India from Kashmir to Kumaon at about 2,000 and 3,000 m altitude. The drug consists of the dried immature leaves and flower heads of the plant. It has been recommended that the drug should be collected in early summer or late spring, by which time the flower heads have not fully matured. The medicinal properties of this plant are based chiefly on Santonin, contained in the young leaves and flower heads, the proportion of Santonin is reported to be maximum just before the flowers begin to open, it diminishes soon after. The chief use of the drug is for expulsion of worms from the stomach.

66. Belladonna (*Atropa acuminata* Royle *ex* Lindley)

Family: Solanaceae

Indian names: Angurshafa, sagangur.

An erect, branched perennial herb, 60-90 cm high. Leaves brownish green, 7-15 cm long, narrowed at both ends. Flowers in axils of leaves, solitary or in pairs, about 2.5 cm long, bell-shaped, yellowish brown. It is found in Kashmir at altitudes of 2,000 to 3,500 m. It is also cultivated. The drug *Belladonna* consists of the dried leaves and other aerial parts of the plant, roots are also included. Total alkaloid content depends upon stage of development of plant, low during flowering and very high when bearing green berries. The drug obtained from the leaves and other aerial parts brings about a decrease in secretion of sweat and salivary and gastric glands, it acts as strong anti-spasmodic in intestinal problems. The drug obtained from the root of the plant has similar properties as the one obtained from leaves and twigs. Roots, are however, believed to have certain poisonous substances and are employed chiefly in preparation prescribed for external applications of rheumatism, neuralgia, inflammations, etc.

67. Neem (*Azadirachta indica* A. Juss)

Family: Meliaceae

Indian names: Neem, Vembu, Vepa.

Neem is a very well-known tree of India. The tree has pinnate leaves, *i.e.* its leaves are divided into numerous smaller segments called leaflets, each leaflet looking like an ordinary leaf. Flowers small, white, in short axillary bunches. Fruits 1.2-1.8 cm long, green or yellow, seed one in each fruit. The tree occurs naturally in Deccan Peninsula, but is cultivated all over India near habitations and on road sides, etc. The drug consists of dried stem bark, leaves, and root bark. The bark is a bitter tonic, astringent and antiperiodic, *i.e.*, it is useful in fever, it breaks the periodic sequence of fever (like malaria) and is useful in skin diseases. The leaves are bitter and are largely applied on skin diseases and boils, a decoction of leaves is also internally. The antibiotic activity of leaves and roots of the tree and their utility in skin diseases have been confirmed experimentally.

Conclusion

The age-old practice of *Vaidya*s is currently threatened by most of the problems including limited availability of the required plants, and herbs, rapid destruction of natural forests, lack of formal arrangement or institution to train and nurture this knowledge, lack of organised propagation nurseries, inadequate institutional and external support and patronisation (especially from the government), low quality and poor stock of raw materials in the open market, and unwillingness among the youngsters to learn and adopt the practice. Despite the rather dismal present state of affairs, this deeply rooted social practice, which has significant value as a community service, still holds great potential and remains too important to be ignored. Drawing on the respondents comments and our observation during the fieldwork, the following ideas and clues on possible improvement may be considered:

1. Experimental propagation nurseries may be established under government and non-government initiatives to ensure sustained supply of seedlings.
2. Organised motivational and awareness raising campaign regarding medicinal plants and their benefits (*e.g.* free from negative side effects, low cost) may be carried out at the community level, especially amongst the younger population, by involving the community leaders and local community based organisation (*e.g.* schools and religious institutions) and NGOs.
3. With the active participation of the local people, the existing medicinal plants should be systematically documented and recorded, the document may also be made available in major local languages in a simple and user-friendly manner.

4. Local base and community relations– two of the major benefits of some of the local NGOs and community based organisations may also be exploited for initiating a network or platform to bring the *Baidyas* together.
5. The local press, media and folk cultural practices (*e.g.* folk theatres) may be utilised as community-based extension and dissemination media to highlight the importance of conserving this traditional practice and heritage.
6. The mainstream research institutions in the country, especially the forest and agricultural research institutes and universities may be encouraged to provide the much-needed research support for proper documentation and dissemination of the knowledge on medicinal plants and associated folk and herbal treatment methods.

References

Alam M K, Chowdhury J U and Hasan M A 1996. Some folk formularies from Bangladesh. *Bangladesh J Life Sci* **81:** 49-69.

Ara R, Mohiuddin M, Alam M K and Rashid M H 1990. Bangladesher oushnadhio Gach Gachrar full of faler Dinpanji (in Bengali) NTFP series-3 Bangladesh Forest Resrarch Institute, Chittagong.

Chowdhury J U, Alam M K and Hasan M A 1996. Some traditional folk formularies against dysentery and diarrhoea in Bangladesh *J Econ Tax Bot* **12:** 20-23.

Faransworth N R 1990. The role of ethnopharmacology in drug development. Ciba foundation symposium **154:** 2-11.

Julsing M K, Quax W J and Kayser O 2007. The engineering of medicinal plants: prospects and limitations of medicinal plant biotechnology. Wiley-VCH Verlag GMbH and Co. KGaA. Weinheing.

Rashid M H, Alam M J, Ara R and Merry S R 1987. Crude drug market of survey, Bangladesh. Bangladesh FRI, Chittagong.

Medicinal Plants: Aspects and Prospects (2014) ***Pages* 191–217**
***Editors:* Mukesh Kumar, Anjali Khare and C.P. Shukla**
***Published by:* BIOTECH BOOKS, NEW DELHI**

Chapter 14

Notes on Some Common Medicinal Plants of Sikkim Himalayas and Pharmacological Support for their Traditional Uses

Kumar Avinash Bharati[1], *Manish Kumar Singh*[2] *and Rajesh Kumar*[3]

[1] Raw Materials Herbarium and Museum Delhi (RHMD), NISCAIR, New Delhi – 110 012
[2]Arid Forest Research Institute, Jodhpur – 342 005, Rajasthan
[3]Department of Botany, Bareilly College, Bareilly – 243 005, U.P.

ABSTRACT

The aim of the present communication is to find out the use of ethnomedicinal plants in alternative medicines and their scientific validation through literature review. A total of 32 medicinal plants have been selected by random sampling method in the study area. The selected medicinal plants have been studied for pharmacological properties and their uses in different alternative systems of medicines in India *viz.* Ayurveda, Unani, Homeopathy, Siddha and Tibetan. The pharmacological properties and their earlier studies have shown that the percentage of natural products in modern drug is considerable, with estimates varying from 35–50 per cent. The data from indigenous systems of Sikkim have 59.37 per cent similarity with Ayurveda, 37.50 per cent similarity with Unani, 18.75 per cent with Homeopathy, 15.62 per cent with Tibetan and 12.50 per cent with Siddha. Almost 72 per cent of ethnomedicinal plant species show scientific validations for their ethnic uses. The present study indicates that considerable number of ethnomedicinal plant species are used in two or more alternative medicinal systems for the treatment of same or similar ailment, suggesting potential pharmacological opportunities in the future.

Keywords: ***Ethnomedicine, Pharmacology, Ayurveda, Unani, Homeopathy, Siddha, Tibetan, Sikkim.***

Introduction

Ethnic medicines are informal systems that include folk belief, skills, techniques and tactics relating to the health care and are passed from generation to generation mainly through oral traditions (Gotage and Ramdas 2008). In contrast, the principals and practices of alternative medicines are formulated and catalogued in Ayurveda, Siddha, Unani, Homeopathy, Chinese and Tibetan records. With the exception, of Chinese, all of these medicinal systems are used in India, but Ayurveda, Homeopathy and Unani, are the most common and provide health care for more than 60 per cent of the population.

A comparative study was carried out to find out the correlation between ethnic uses of plants Sikkim Himalaya with various alternative medicinal practices including, Ayurveda, Siddha, Unani, Homeopathy and Tibetan. The new challenges in the form of drug resistant malaria and TB due to mutant microorganisms are major challenge for pharmacologists, it compels us to find out new drugs (Lewis 2003). Traditional knowledge always plays major role in medicinal chemistry from starting point, morphine from *Papaver sominiferum*, atropine from *Atropa belladonna*, ephedrine from *Ephedra sinica* etc.(Vogel 1999) to present day, artimisinin from *Artimisia annua* (Klayman 1985).

Traditional medicine using herbal drugs exists in every part of the world (Vogel 1991). However, few parts of the word have preserved the treasure of ancient medicinal tradition due to remote locations, poor infrastructure and poverty. Sikkim Himalayas is an area known for its wealth of diverse medicinal plants. The state of Sikkim is situated on the flanks of Eastern Himalayas between 27° 10' – 28° 5' N latitude and 88° 30' – 89° E longitude. The ethnic composition is unique. Apart from the three major ethnics–Bhutia, Lepchas, and Nepalese, a conglomerate of over 20 ethnic tribes and numerous sub-tribes inhabit the region (Rai and Sharma 1994).

Our goal was to find out the efficacy, scientific validation and the similarities between Sikkim ethnomedicine and other medicinal systems, as well as to demonstrate the value of integrating ethnobotanical and pharmacological studies. It will provide scientific support for the ethnic use of plants (if ethnic uses are contradictory to pharmacological studies then such use should be discouraged), relationship between various alternative systems of medicines and raw material for new drug.

Methodology

The plants selected for this study are frequently used in ethnomedicines by indigenous people. A random survey was conducted to find out commonly used plants and their uses by interview of local people through open conversation rather than closed questionnaire, but always bore in mind the information we wanted to obtain. The present survey involved 35 interviewees (60 per cent male and 40 per cent female). A total of 35 plants were selected on the basis of their frequency of use. A comparative study was done with various alternative medicines and pharmacological literatures.

Results

Please see Table 14.1.

Discussion

The ethno-medicinal system of Sikkim Himalaya shows moderate to very little affinities sharing at least one common use with other alternative medicinal practices in India, as for example, 59.37 per cent with Ayurveda, 37.50 per cent with Unani, 18.75 per cent with Homeopathy, 15.62 per cent with Tibetan and 12.50 per cent with Siddha. The present finding indicates that it is an independent medicinal system. Further, 71.87 per cent of species shows pharmacological support for their ethnic uses and in rest of the case pharmacological work is not available. It is very healthy result to support the ethno-medical system of Sikkim Himalaya for benefit of rural mankind. When compared the data with other traditional and alternative medicinal systems of India with scientific validations, it is found that *Swertia chirata* has at least one common use in all the systems that is Ayurveda, Homeopathy, Siddha, Tibetan and Unani. The *Terminalia chebula* is another important species; it has at least one common use in Siddha, Tibetan, Homeopathy and Unani. There are a number of species showing affinities with Ayurveda and Unani systems, like *Asparagus racemosus, Centella asiatica, Embelia ribes, Holarrhena antidys enterica, Mallotus philippinensis, Terminalia bellirica* and *Zingiber officinale*. The species sharing affinities with two or more systems of medicines with pharmacological validation may play as a source of new drugs for modern medicine. Further, medicinal properties of such species should be propagated through extension education in remote areas after proper evaluation and standardization. Ethnic drugs have potential to play very crucial role in primary health care in rural areas of third world countries. New emerging diseases and the development of resistance by microorganisms to current drugs will require novel compounds to control these inevitable events. A broad interdisciplinary effort involving experts in a number of fields embracing plant taxonomy, ethnobotany, pharmacognosy, biochemistry, analytical chemistry, pharmacology, pharmaceutics, and medicine is required to achieve the goal. This interdisciplinary effort will continue to be important into next millennium, until ultimately disease as we know it no longer exists (Lewis 2003).

The present study demonstrates the value of integrating ethnobotanical and pharmacological studies. Even we do not develop new drugs but able to demonstrate the medicinal value of some of the plants and make more effective alternative medicines. It will provide new resources to find out new modern drugs. Suppose, ethnic use of a plants have similar healthcare use in different alternative medicinal systems and it has also been confirmed by pharmacology like, *Asparagus racemosus, Cannabis sativa,*

Table 14.1: Some medicinal plants of Sikkim that are used in alternative medicines in India. Taxa are listed alphabetically by botanical names, families in parenthesis, local names (ethnic groups in parenthesis). Ethnomedicinal uses in alternative systems of medicines in India and pharmacological activities. A= Ayurveda, B= Bhutia, H= Homeopathy, L=Lepcha, N= Nepali, S= Siddha, T= Tibetan, U= Unani.

Ethnomedicine at Sikkim	*Use in Alternative Medicine in India*	*Pharmacological Activity*
1. *Acorus calamus* Linn. (Acoraceae) Bojo, Bojho (N), Vacha, Vaca (A), shu-dag nag-po (T), Waj-e-Turki (U).		
The rhizome used in epilepsy and other mental ailments, intermittent fever, chronic diarrhoea, asthma, cough, sore throat, colic pain and brain tonic (Hussain and Hore, 2007); skin disease, fever, cough (Pradhan and Badola, 2008).	A: (Rhizome) Voice clearance, CNS depressant (Anonymous, 2000); vatakapha disorders, pain and for the purification of stool and urine. It is used in constipation, abdominal pain, epilepsy and insanity and it also increase the memory, strength and intelligence of child (Panda, 2007). H: (rhizome) Gastric and respiratory diseases, dyspepsia, vomiting, spasmodic complaints (Nadkarani, 1976). T: Cures sores, aches and pains (Kunzang, 2001); boils on skin and discharge of pus and fluid, this disease known as *Shu-ba* (Lhawang *et al.*, 1995). U: (rhizome) Desicant/Siccative, Inspissant to Semen, Demulscent, Thermogenic, Diuretic (Ahmed *et al.*, 2005).	Sedative and analgesic effect, depression in blood pressure and respiration, hypotensive, hypothermic, CNS depressant, anticonvulsant, antimicrobial, carcinogenic, anthelmintic, insecticidal, sedative-tranquillizing (Anonymous, 2000).
2. *Alstonia scholaris* (L.) R. Br. (Apocynaceae) Chation (N)		
Bark in rheumatism, malaria, and skin disease. Root juice is used with milk in Leprosy. Latex used as vermifuge (Pandey, 1991).	A: (stem bark) Anti-pyretic, anthelmintic, astringent, cardiotonic, depurative, digestive, febrifige, galactagogue, stomachic, thermogenic, tonic (Anonymous, 2000); Catarrhal and malarial fever, Chronic diarrhoea and dysentery (Nadkarni, 1976). H: (stem bark) Debility, Diarrhoea, Dysentery, Lactation, Leucorrhoea, Hyperemesis gravidrum (Anonymous, 2007).	Anti-ascariasis, antidysentric, anti-pyretic, antibacterial, astringent, antimalarial, CNS depressant (picrinine), antimalarial (Anonymous, 2000).
3. *Amomum subulatum* Roxb. (Zingiberaceae) Alaichi (N), Sthoolaila (A), ma-ko-la (T), Qaqlahkibar, QaqlahZakar (U).		
Seed oil allay irritability of the stomach, decoction of fruit used as a gargle in	A: (Seeds) Halitosis, Skin disease, wounds, Ulcers, Cough, Fever and Gonorrhoea (Anonymous, 2000).	Antioxidant, hypoglycaemic, antimicrobial, antifungal (Gupta and Tandon, 2004); Ethanolic

Contd...

Table 14.1–*Contd...*

Ethnomedicine at Sikkim	*Use in Alternative Medicine in India*	*Pharmacological Activity*
affections of the teeth and gums, in the combination with the seeds of melons it is used as a diuretic in the cases of gravel of kidneys (Biswas and Chopra, 1982).	T: Mental disease, fainting due to ill health (Kunzang, 2001). U: (Dried ripen fruits and Seeds) Tonic for heart and liver, astringent to bowels, hypnotic, and appetiser and cause belching, decoction of seed is used as a gargle in affection of gum and teeth. Seed in conjugation with quinine as an antidote in either snake or scorpion venom (Anonymous, 1987).	extract (50 per cent) of the rhizome and roots showed hypoglycaemic activity (Anonymous, 2000).
4. *Artemisia vulgaris* Linn. (Asteraceae) Tetaypati, Teil (N), Damanaka, Topadhana (A), mkhan-dkar (T), Biranjasif,Shuwela (U).		
Leaf juice use to stop nose bleeding, asthma and disease pod brain (Hussain and Hore, 2007). Leaves use externally as an antiseptic and orally as anthelminthic (Pandey, 1991).	A: (whole plant) Skin diseases, irritable bowel syndrome, bleeding, various toxic condition and to maintain the body humors (Panda, 2007). H: (whole plant) Congestion of Brain (Hydrocephalus), Coloured light produces dizziness in eye, Pain and blurring of vision, profuse menses in female, profuse sweat and smells like garlic (Boericke, 2007). T: Bleeding from nose (Kunzang, 2001). U: (Whole plant) Inflammation, Amenorrhoea, Retention of Urine, fever, inflammation of visceral organs (Anonymous, 1992).	Anti-worm, anti-estrogenic, and progestational and anti-progestational effects (Khare, 2004); pollen is allergic (Caballero, 1997), anti-inflammatory (Tigno *et al.*, 2000); analgesic (Pires *et al.*, 2009), Hepatoprotective (Gilani *et al.*, 2005); Antispasmodic and bronchodilator (Khan and Gilani, 2009).
5. *Asparagus racemosus* Willd. (Aspragaceae) Satamuli (N), Shatavari (A), Sawa-gaaya (T), Satawar (U).		
Root in stomach trouble (Chhetri, 2007); as tonic and aphrodisiac (Pandey, 1991); anti-dysentery, diuretic, root powder taken with milk to cure piles (Biswas, 1956).	A: (root) Galactagouge, Nervine tonic, Vigour, weight gain (Anonymous, 2000); It increase the semen, milk, memory and used in abdominal discomfort, dysentery, inflammation and all vata pitta condition (Panda, 2007). T: Gout (Kunzang, 2001).U: (root) Galactogouge, Spermatogenic, diabetes (Ahmed *et al.*, 2005).	Galactagogue, nematicidal, anticancer, antidysentric, antiabortifacient (Shatavarin I), antioxytoxic (Shatavarin II), antiviral, diuretic, antiamoebic, hypoglycaemic, hypotensive, anticoagulant, enzymatic (Anonymous, 2000); antidiabetic, anabolic, antyimicrobial, antiallergic, anthelmintic (Gupta and Tandon, 2004).

Contd...

Table 14.1–*Contd...*

Ethnomedicine at Sikkim	*Use in Alternative Medicine in India*	*Pharmacological Activity*
6. *Bauhinia variegata* Linn. (Saxifragaceae) Koeralo, Takki (N), Kancanara (A)		
The bark is a wound healer also used in cough, diarrhoea; root decoction is used in indigestion and in goitre (Pandey, 1991), dysepepsia, flatulence and dried buds used for cough, bleeding piles, haematuria (Biswas, 1956).	A: (stem bark, flower) Psychotic syndrome, Lymph tissue disorder, tumour like growth (Anonymous, 2008). S: flatulence, skin disease (Anonymous, 2008).	Antitumor, anti-inflammatory, anti-ulcer, antimicrobial (Anonymous, 2000).
7. *Bergenia ciliata* (Har.) Stenb. (Saxifragaceae) Pakhanbhed, Pakhin Bet (N), Pashanabheda (A), Pakhan Bed (U)		
The root is given in ulcer, tuberculosis, cough and spleen enlargement, cut and burn (Hussain and Hore, 2007); toothache, bronchitis (Chhetri, 2007).	A: (root) Tonic infever, diarrhoea and cough, cuts and burns, ophthalmia, dissolving, kidney stone. Leaf juice is used for earache (Anonymous, 2000). U: (root) Anti-inflammatory, Diabetes (Ahmed *et al.*, 2005).	Spasmogenic, antiprotozoal, anticancer, antilithic, cardiotoxic, CNS depressant, anti-inflammatory, diuretic (in mild doses), antidiuretic (higher doses), prevention of stress induced erosions (bergenin), lowering of gastric output (Anonymous, 2000).
8. *Cannabis sativa* Linn. (Cannabaceae) Ganja, Bhang (N), Vijaya (A), Kanca (S), Qinnab (U).		
Leaves are used as sedative (Chhetri, 2007); dried inflorescence is powdered into paste with warm water and taken orally to cure severe stomache caused due to indigestion (Dash *et al.*, 2003); leaves used in body ache (Pradhan and Badola, 2008).	A: (Whole plant) Narcotic, Hypotonic, diarrhoea in children (Anonymous, 2000). H: (Male and female flowering tops) Ascites, Asthma, Cataract, Corneal opacity, Cystitis, Glaucoma, Gonorrhoea, Hysteria, Impotence, Lecucorrhoea, Nephritis, Nose bleeds, Phimosis, Pneumonia, Priapism, Sexual disorders, Urethral discharge (Anonymous, 2007). S: (leaf and tops) Cough with bouts, Hunger, Pain in the nerve and nerve supplying areas, One sided headache (Uni lateral it may be right or left), Dysfunctional uterine bleeding - Bleeding disorder in Aged women (40 - 50), Vomiting and Diarrhoea (Anonymous, 2008). U: (fruit, leaf) Insomnia, indigestion, spermatorrhoea (Anonymous, 2006); Headcheese, Migrain, insomnis, fever, Orchitis, spermatorrhoea, premature ejaculation, acute pain (Anonymous, 2006).	CNS depressant, analgestic, antiepileptic, Nematicidal, abortifacient, sedative, anticonvulsant, antibacterial, antifungal, antitumor, diuretic, anti-emetic, anti-inflammatory, antipyretic, hypothermic, antiestrogenic, euphoric, anti spasmodiac (Anonymous, 2000).

Contd...

Table 14.1–*Contd...*

Ethnomedicine at Sikkim	*Use in Alternative Medicine in India*	*Pharmacological Activity*
9. *Celastrus paniculatus* Willd. (Celastraceae) Ruglin (L), Jyotishmati (A), Malkangni (U).		
The seeds are used in rheumatism, paralysis and leprosy. Leaf juice is given as an anti-dote in overdoses of opium. Seeds made into a paste with cow's urine are applied to cure scabies, oil taken internally in beri-beri, seeds are used in chronic lumbago (Biswas, 1956).	A: (Bark, leaf, seed) The bark is brain tonic, abortifacient. Leaf juice used in dysentery, antidote for opium poisoning. Seed are useful in abdominal disorders, leprosy, pruritus, leucoderma, skin disease, paralysis, cerebral disorders, leprosy, pruritus, leucoderma, skin disease, paralysis, cerebral disorders, arthritis, asthma, cardiac debility, inflammation. Nephropathy Tonic infever, diarrhoea and cough, cuts and burns, ophthalmia, dissolving, kidney stone. Leaf juice is used for earache. Seed oil is used in fever, sharper memeory, beri-beri, wound, eczema (Anonymous, 2000). U: (leaf) Digestive, carminative, expectorant, aphrodisiac, Brain tonic, Stomachic and Intestinal tonic, Blood puifier, Laxatives, Thermogenics, Stimulant (Ahmed *et al.*, 2005).	Antihistaminic, sedative, anticonvulsant, antiprotozoal, antiviral, antipyretic, antiulcerogenic, anti-emetic, antibacterial, schizontocidal, emmenagogue, hypotensive, stimulant, central muscle relaxant, hypolipidaemic, anti-atherosclerotic, spasmolytic, transquillizer, anti-inflammatory, antifertility (Anonymous, 2000).
10. *Centella asiatica* (L.) Urban (Apiaceae) Ghod Tapre, BhramJhar (N), Mandookaparni (A), Brahmi (U).		
The whole plant is used as brain tonic (Pandey, 1991).	A: (whole plant) Nervine tonic, Memory enhancer (Anonymous, 2000).H: (whole fresh plant) *Acne rosacea*, Constipation, Elephantiasis arabum. Favus, Gangrene after amputation, Ichthyosia, Lupus, Uterus, Follicular inflammation, Vagina pruritis (Anonymous, 2007). U: (whole plant) Nervine tonic, teeth and gum tonic, Brain tonic (Ahmed *et al.*, 2005).	Anti-inflammatory, anti protozoal, spasmolytic, alternative, astringent, antifertility, sedative, CNS depressant, antitubercular, antileprotic, hepatoprotective, anti spasmodic, hypotensive (Anonymous, 2000).
11. *Cissampelos pareira* Linn. (Menispermaceae) Tamshaprip (L), BatulPati (N), Patha (A)		
The root is used in diarrhoea, dysentery, indigestion and urinary disorder. Paste of leaves applied externally on wound (Pandey, 1991); remedy against wasp, bees and scorpion string (Biswas, 1956).	A: (root, leaf) Fever, analgesic, anti-inflammatory (Sharma, 1969). The root used in fever, dysepsis, diarrhoea, dysentery, blood disorders, oedema, leprosy, asthma, lactation disorders. Leves used in eye problems, skin disorders Tonic infever, diarrhoea and	Hypoglycaemic, a potant neuromuscular blockingent, muscle relaxant, antibacterial, CNS depressant, curariform like activity, antileukemic, antifertility, fungitoxic, antitumour, activity aganist human carcinoma cells of nasopharynx in cell

Contd...

Table 14.1–*Contd...*

Ethnomedicine at Sikkim	*Use in Alternative Medicine in India*	*Pharmacological Activity*
	cough, cuts and burns, ophthalmia, dissolving, kidney stone. Leaf juice is used for earache (Anonymous, 2000).	culture (Anonymous, 2000).
12. *Costus speciosus* Sm. (Costaceae) Bet Lauree (N), Kebuka (A)		
Leaves are used in fever (Krishna and Singh, 1987); rhizome in urinary tract infection (Chhetri, 2007; Pradhan and Badola, 2008); rhizome is used in chest pain (Biswas, 1956).	A: (rhizome, root) cough, bronchitis, fever, rheumatism, urinary disorders, loss of appetite, loose motion and skin diseases (Panda, 2007).	Antifertility, estrogenic, ecbolic, abortifacient, anti-inflammatory, Cardio tonic, anti arthritic, oxitocic, antimicrobial, spasmolytic (Anonymous, 2000).
13. *Datura metel* Linn. (Solanaceae) Kalo Dhaturo (N)		
Four to five seeds are taken orally for seven days to cure mad dog bite (Dash *et al.*, 2003).	A: (whole plant, leaf, flower, seed) The plant is used in asthma, cough, fever, inflammation, oedema, insanity, duodenal ulcer, renal colic, calculi. The root used in bites of rabid dogs. The leaf poultice in lumbago, sciatica, neuralgia, painful swellings (Deshpande, 2006). H: (seed) Convulsions, Delirium, Epilepsy, Eye affections of Mania, Timidity (Anonymous, 2007).	Anthelmintic, anticancer, antispasmogenic, blood pressure depressant, strong nematicidal, anticholinergic, antiviral, analgesic (Anonymous, 2000).
14. *Embelia ribes* Burm. f. (Myrsinaceae) Buibidans, Pierlahara (N), SangrikAsumbu (L), Vidanga (N), Vidanga (A), Vaivitankam (S), byi-tan-ga (T), Baobarang (U).		
The fruits are used as anti-worm (Pandey, 1991).	A: (fruit) Worm infestation (Anonymous, 2000). S: (fruit) Acid peptic disease,Toxic substances, Worm infestations, Toxic substances, Worm infestations, Due to gas obstruction and which creates pain in the related region, Vayvu, Anaemia (Anonymous, 2008). T: Swelling of abdomen due to indigestion and strengthens the digestion (Kunzang, 2001); acne (Lhawang *et al.*, 1995).U: (Fruit, leaves and root) Kill and expel intestinal worms (Anonymous, 1997).	Nematicidal estrogenic, hypoglycaemic, anthelmintic, anti-inlammatory, hypotensive, antipyretic, diuretic, hepatoprotective, antiandrogenic, immunostimulant (Anonymous, 2000).

Contd...

Table 14.1–*Contd...*

Ethnomedicine at Sikkim	*Use in Alternative Medicine in India*	*Pharmacological Activity*
15. *Hedychium spicatum* Sm. (Zingiberaceae) Pankhaphool (N), Shati (A), Sgaskya (T)		
The rhizome used in abdominal disorders and as stimulant (Pandey, 1991); stomache, tonic, cuminative (Biswas, 1956).	A: (rhizome) Respiratory problems, Cough and cold, diarrhoea, breathlessness, piles, ulcers, promote growth of hair, liver disorder, hiccough, fever, rheumatoid arthritis, inflammation, pain, skin disease (Anonymous, 2000). T: Swollen stomach, indigestion, vomiting of phlegm, pain immediate after eating, perspiration (Bzhi and Sngog-pog, 1981).	Hypotensive, hypoglycaemic, anti-inflammatory, vascodilatory, anti-spasmodic, tranquillizer, CNS depressant, hypothermic, spasmolytic, spasmolytic, analgesic, antimicrobial (Anonymous, 2000).
16. *Holarrhena antidysenterica* (L.) Wall. (Apocynaceae) Indrajow, Kurchi, Aulay Khirrn (N), Kutaja, Indrayava (A), dug-mo-nyun, dug-nyun (T), InderjaoTalkh (U).		
The stem bark is used in diarrhoea and dysentery (Chhetri, 2007); amoebic dysentery (Rai and Sharma, 1994).	A: (Bark and seeds) dysentery and diarrhoea, bark rubbed over body in dropsy (Anonymous, 2000); Bleeding (Anonymous, 2008).H: (stem bark) Acute and chronic dysentery, colicky pain, Tenesmus, Passing of blood and mucus with stools (Anonymous, 2007).T: Fever (Kunzang, 2001); vomiting, thrist, dryness and a bitter taste of mouth, vomiting of bile, cramp known as inflammatory *glang-thabs* (Bzhi and Sngog-pog, 1981). U: (stem bark) Dysentery, diarrhoea, anti-worm (Anonymous, 1987).	Antitubercular, hypotensive, antiprotozal, hypoglycaemic, antispasmodic, antigiardiastic, antifungal, antiamoebicidal, antidiarrhoeal, antiamoebicidal, antidiarrhoeal, anticancer, anti spinochetal (Anonymous, 2000).
17. *Lycopodium clavatum* Linn. (Lycopodiaceae) Nagebeli (N)		
The roots used in indigestion (Chhetri, 2007); diuretic, demulcent, anti-septic and pulmonary disorder, chronic kidney diseases, stop haemorrhage after child birth (Biswas, 1956).	H: (Spores, Fresh plant) Albuminuria, Aneurysm, Angina pectoris, Aphasia, Asthma, Impotency, Metrorrhagia, Nymphomania, Otorrhoea, Parkinson's disease, Peritonitis, Prostatitis, Renal colic, Rheumatism, Taste abnormal, Typhoid, Water brash, Warts, Hernia (Anonymous, 2007).	Anti-inflammatory (Orhan *et al.*, 2007), an acetylcholinesterase inhibitor (Orhan *et al.*, 2007), anti-cancerous (Mandal *et al.*, 2010).

Contd...

Table 14.1–*Contd...*

Ethnomedicine at Sikkim	*Use in Alternative Medicine in India*	*Pharmacological Activity*
18. *Mallotus philippinensis* Muell.-Arg. (Euphorbiaceae) Numboongkor, Purva, Tukla (L)Sinduri (N), Kampillaka (A), Qinbeel, Kambila (U)		
The ripen fruits are used as vermifuge (Pandey, 1991).	A: (Glandular hair) Against worm and parasite, tumour problem (Anonymous, 2000); worm constipation, infestation and abdominal diseases (Panda, 2007). H: (Fruit, Red powder on seeds) Anthelmintic (Anonymous, 2007). U: (Glandular hair) Remedy of guinea worm, cure wound (Anonymous, 1992).	Antifilarial, antifertility, anthelmintic, antibacterial, hypoglycaemic, anticancer, antiapasmodic, haemostatic, anti-inflammatory, wound healing, cardiac depressant, antimicrobial (Anonymous, 2000).
19. *Mentha viridis* Linn. (Lamiaceae) Mentha, Babri (N), Pippermint (A)		
The leaves given in fever and bronchitis. Decoction used as lotion for aphthae. The oil is distilled from fresh flowering spearment. Oil is used in the rheumatism (Biswas, 1956).	A: (whole plant) Sickness, Flatulence (Nadkarni, 1976). H: (whole plant) Scanty urine with frequent desire (Boericke, 2007).	Antioxidant (Mkaddem *et al.*, 2009; Arumugam, 2006), antimicrobial (Mkaddem *et al.*, 2009).
20. *Oroxylum indicum* Vent. (Bignoniaceae) Totilla, Tatelo, Shivnak (N), Rip (L), Syenaka (A)		
The bark and seeds are used in fever and pneumonia (Pradhan and Badola, 2008).	A: Urinary bladder problems and used in stones, diarrhoea and anorexia (Panda, 2007).	Diuretic, spasmogenic, anti-inflammatory, antifungal (Deshpande, 2006). Acridic, astringent, anodyne, anti-inflammatory, aphrodisiac, appetizing, anthelmintic, constipating, digestive, diuretic, expectorant, felrifuge, refrigerant, stomachis (Anonymous, 2000).
21. *Picrorhiza kurroa* Royle ex Benth. (Scrophulariaceae) Kutki (N), Katuka (A), Katuku Rokini (S), kat-bee (T), Kutki (U).		
The rhizome and root are used in fever, cough and asthma (Pradhan and Badola, 2008).	A: (rhizome) Hepatic disorder (Anonymous, 2000); tonic, cathartic, stomachic, given in fever, dyspepsia, as strong purgative and also applied in scorpion and other insect bites (Panda, 2007).S: (rhizome) Fever, All type of Lung diseases, Eczema, gastro intestinal disorder in	Antipyretic, anti-inflammatory, antiviral, hepatoprotctive, smooth muscle relaxant, anti spasmodic diuretic, antibacterial, antiasthamatic, antihepatotoxin (Anonymous, 2000).

Contd...

Table 14.1–*Contd...*

Ethnomedicine at Sikkim	Use in Alternative Medicine in India	Pharmacological Activity
	infants, A group of Ulcers over the skin Surface (Anonymous, 2008). T: (rhizome) Suppress burning sensation due to acidity (Kunzang, 2001). U: (root) Antipyretic, makes skin pores clean, carminative, cause sneezing, stomachic and intestinal tonic, Analgesic, laxative (Ahmed *et al.,* 2005).	
22. *Plantago major* Linn. (Plantaginaceae) Nasha Jhar (N), Lisan-ul-Hamal, Kaseer-ul-Azla (U)		
The whole plant is used in Pneumonia (Pandey, 1991).	H: (Whole fresh plant) Bed wetting, Mastitis, Ciliary neuraligia, Erysipelas, Dysentery, Toothache, Wounds (Anonymous, 2007). U: (seed) Diarrhoea, dysentery, epitaxis, menorrhagia (Anonymous, 1992).	Antiviral (Chiang *et al.,* 2002, Chiang *et al.,* 2003); antitumor (Ozaslan, 2007); immunoenhancing (Gomez-Flores *et al.,* 2000); Hepatoprotective and anti-inflammatory activities (Turel *et al.,* 2009); antidiarrhoeal (Atta and Mouneir, 2005); Analgesic (Guillén, 1997).
23. *Rauvolfia serpentina* Benth. ex Kurz. (Apocynaceae) Sarpgandha (N), Sarpagandha (A), Asrol (U)		
The root is used in fever (Chhetri, 2007); antidote to the bites of poisonous reptiles and stings of insects, root decoction is helpful during child birth, root is remedy in painful affections of bowels, insomnia (Biswas, 1956).	A: (root) Decrease Blood pressure, nervine tonic (Anonymous, 2000).H: (Roots) Addison's disease, Angina pectoris, Basedow's disease, Coitis, Dystonia, Hypotension, Parkinson's disease, Thyroid disorders, Vasomotor complaints (Anonymous, 2007). U: (root, leaf) Depressant to heart, decoction facilitate child birth, dysentery, painful affection of the bowles, insomnia, leaf Juice in the treatment of opacities of cornea (Mehr-e-Alam Khan, 2004).	Anticholinergic, hypotensive, anticontractile, sedative, relaxant hyperthermic, antidiuretic, sympathomimetic, hypnotic, vasodilator, antiemetic, antiarrhythmic, nematicidal (Anonymous, 2000).
24. *Rubia cordifolia* Linn. (Rubiaceae) *Lepcha*- Vhyem, Vhyeni (L), Soth (B), Manjito (N), Manjistitha (A), btsod (T), Majeeth (U).		
The root is used in jaundice (Chhetri, 2007); urinary tract infection, skin disease (Pradhan and Badola, 2008); irregular monthly courses (Biswas, 1956).	A: (stem) Hormonal therapy in women, blood purifier, Skin disorder (Anonymous, 2000); Kaphapitta disorders. It has analgesic and inflammatory properties. It is used in the diseases of the uterus, pains in the joint,	Antioxidant, antibacterial, anticancer, anti-inflammatory, antiviral, haemostatic, anti-lipid peroxidative activity, hypoglycaemic (Anonymous, 2000).

Contd...

Table 14.1–*Contd...*

Ethnomedicine at Sikkim	*Use in Alternative Medicine in India*	*Pharmacological Activity*
	rheumatic conditions, leucorrhoea, blood disorder, etc. Also used as febrifuge and consider as best drug in gout (Panda, 2007). T: Fever, dysentery (Kunzang, 2001); wart (Lhawang *et al.,* 1995). U: (Dried root) Amenorrhoea, diuretic, deobstruent (Anonymous, 1997).	
25. *Sinopodophyllum hexandrum* (Royle) T.S. Ying syn. *Podophyllum hexandrum* Royle (Berberidaceae) Papari, Panchpatey (N), Banakarkatee (A), ol-mo-se (T)		
The rhizome and root is used in diarrhoea, skin disease and as tonic (Hussain and Hore, 2007).	A: (root) blood purifier, purgative and alterative. It is considered as a cardiac tonic in small doses. It also finds use as a stimulant in peristalsis, allergy and skin inflammations (Panda, 2007). T: Gynaecological disorder, blood disorder, skin disease (Kletteret. al., 1995).	Antioxidant (Arora *et al.,* 2005); anticancerous (Giri and Narasu, 2000)
26. *Swertia chirata* C.B. Clarke (Gentianaceae) Chirowto (N), Kiratatikta (A), tig-ta, rgya-tig (T), Chiraita (U).		
The leaves and stem are used in liver disorder, cough, constipation, fever, skin disease, worms and as tonic (Hussain and Hore, 2007).	A: (whole plant except root) Fever, tonic, astringent, stomachic, improves eye sight, pain in the joints, scabies (Anonymous, 2000).H: (Whole plant excluding roots) Dullness of mind, Headache, Halitosis, Pain in throat, Fever, Liver and Spleen enlarged, Burning while urinating (Anonymous, 2007). T: Fever (Kunzang, 2001). U: (Whole plant except root) Anthelmentic, antipyretic, laxative, galactogue (Anonymous, 1992).	Antispasmodic, anti-inflammatory, antimalarial, hepatoprotective, antiulcerogenic, CNS depressant, laxative, stomachic, antidiarrhoeal, hydrocholeratic, cardiostimulant, antileishmanial, anthelmintic, anticarcinogenic (Anonymous, 2000).
27. *Symplocos racemosa* Roxb. (Symplocaceae) Palyok (L), Kaidai, Khoidai, Chumlane (N), Lodh Pathani (U).		
The bark is used in bowl complains, dysentery, dropsy and ulcers, used in stopping haemorrhage from teeth or prolonged bleeding of women, cures wound in vagina and prevents chance of abortion in right months (Biswas, 1956).	U: (Stem bark) Cicatrizant, Inspissant to Semen, Analgesic, Astringents and Habisat (Ahmed *et al.,* 2005).	Antimicrobial, antidiarrhoeal, spasmogenic, heart depressant, blood pressure, depressant (Despande, 2006)

Table 14.1–*Contd...*

Ethnomedicine at Sikkim	*Use in Alternative Medicine in India*	*Pharmacological Activity*
28. *Taxus wallichiana* Zucc. (Taxaceae) Cheongbu (L), Dhengresalla, Chhareysalla (N), Talispatra (A).		
A tincture of young shoots is used in headache, giddiness, diarrhoea, liver disorder (Hussain and Hore, 2007); leaves are used in fever and epilepsy (Chhetri, 2007).	A: (Young shoots) A medicinal tincture made from young shoots has long been in use for the treatment of headache, giddiness, feeble and falling pulse, diarrhoea and severe biliousness (Panda, 2007).	Sedative, antispasmodic, antitumor, antifertility, anticancer, antimicrobial, anti-implantation, antiovulatory, cardiac-depressant, CNS depressant, antiulcerogenic, anti-inflammatory, antipyretic, diuretic (Anonymous, 2000).
29. *Terminalia bellirica* (Gaertn.) Rox. (Combretaceae) Barra (N), ba-ru-ra (T), Balela (U).		
The fruits are used in stomach upsets (Chhetri, 2007); stomach dysfunction (Rai and Sharma, 1994).	T: Decoction taken in eye disease, digestive disorders (Kunzang, 2001); skin becomes thick and heard with pimples on it, psoriasis, patches on skin devoids of pigment (Lhawang *et al.,* 1995). U: (Bark, fruit, seed) disease of gastro-intestinal tract and bronchitis, benign tumours (Anonymous, 1997).	Purgative, blood pressure depressant, antifungal, antihistaminic, activity against viral hepatitis and vitiligo, antiasthmatic, broncho-dilatory, anti-spasmodic, antibacterial, CNS stimulant, amoebicidal, antistress and endurance promoting activity (Anonymous, 2000).
30. *Terminalia chebula* Retz. (Combretaceae) Harra (N), Selim Pot (L), Katukkai (S), a-ru-ra (T), Halelaj Aswad (U).		
The fruits are used in tonsillitis (Chhetri, 2007); pharyngitis and other throat complications (Rai and Sharma, 1994).	H: (Semi mature fruits) Cardiac diseases, Palpitation, Haemorrhages, Spermatorrhoea (Anonymous, 2007). S: (fruit) Jaundice, Eye diseases, Hyper tension, Laxative, Ascitis, Poison (Anonymous, 2008). T: Fever, swelling of stomach, indigestion, jaundice, tumours, dysentery (Kunzang, 2001). U: (Fruit before ripen) Beneficial in Paralysis (Anonymous, 1987).	Antimicrobial, antifungal, antibacterial, antistress, antispasmodic, hypotensive, indurance promoting activity, antihepatitis B virus activity, hypolipidaemic, inhibitory activity, against HIV-1 protease, anthelmintic, purgative (Anonymous, 2000).
31. *Urtica dioica* Linn. (Utricaceae) Sisnu (N), Surang (L)		
The leaves in high blood pressure (Chhetri, 2007); whole plant is use in bone fracture and dislocation, diarrhoea, cough, child delivery (Pradhan and Badola, 2008).	A: (BicchuBooti; leaf) Diarrhoea (Sharma, 1969). H: (Fresh plant in flower) Agalactia, Allergic reactions, Bee-strings, Gout, Erythema, Hives, Leucorrhoea, Renal colic, Whooping cough (Anonymous, 2007).	Antirhenmatic, astringent, anthelmintic, antiasthmatic, antidiarrhoeal, diuretic, stimulant and tonic (Despande, 2006)

Contd...

Table 14.1–*Contd...*

Ethnomedicine at Sikkim	*Use in Alternative Medicine in India*	*Pharmacological Activity*
32. *Zingiber officinale* Roscoe in Trans. (Zingiberaceae) Aduwa (N), Shunthi (A), Cukku (S), bcha'-sga, saga, sga-skya (T), Zanjabeel (U).		
The rhizome is used in cough, fever and throat pain (Pradhan and Badola, 2008); appetiser (Biswas, 1956).	A: (rhizome) Appetiser, anti-inflammatory, anti-cancerous, cough and cold (Anonymous, 2000). H: (rhizome, roots) Albuminuria, Halitosis, Diarrhoea, Dysentery, Hepatitis, Food poisoning, Nasal catarsh, Colic, Back ache, Dyspepsia (Anonymous, 2007). S: (rhizome) Indigestion, cough, gastritis, burning sensation of oesophagus, loss of appetite, headache, Painful ulcer of duodenum (Anonymous, 2008). T: Phlegm, blood pressure irregularities, kideney disese (Kunzang, 2001). U: (rhizome) Indigestion, dyspepsia, flatulence, colic, vomiting, spasm, asthma (Ahmed *et al.*, 2005)	Anti-inflammatory, hypolipidaemic, antiathero-sclerotic, antiulcer, antipyretic, Cardiovascular, analgesic, anti depressant, hepatoprotective, intropic (Anonymous, 2000).

Plate 14.1

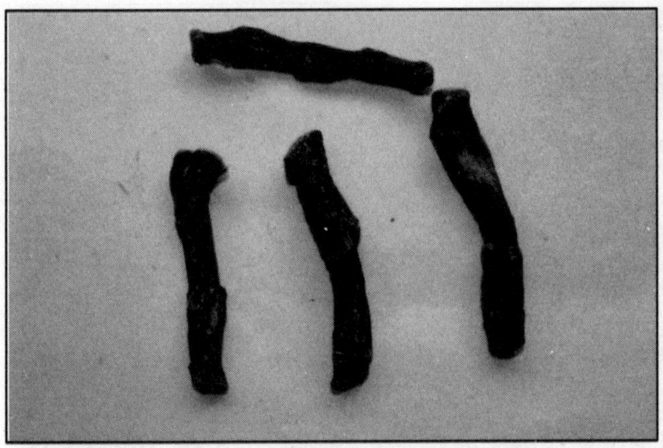

Acorus calamus

Alstonia scholaris

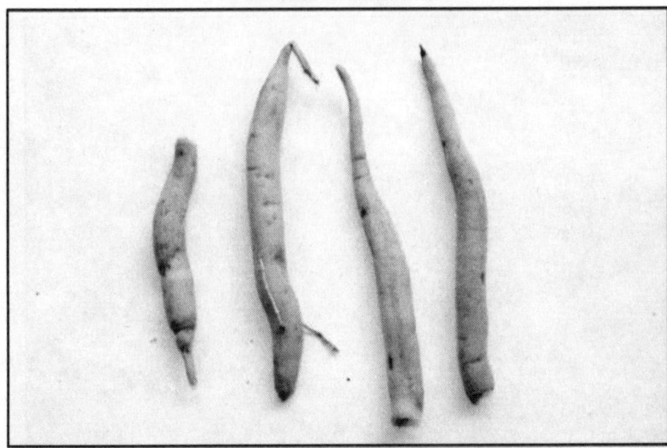

Asparagus racemosus

Contd..

Plate 14.1–*Contd...*

Bauhinia variegata

Bergenia ciliata

Cannabis sativa

Contd...

Plate 14.1–*Contd...*

Celastrus paniculatus

Centella asiatica

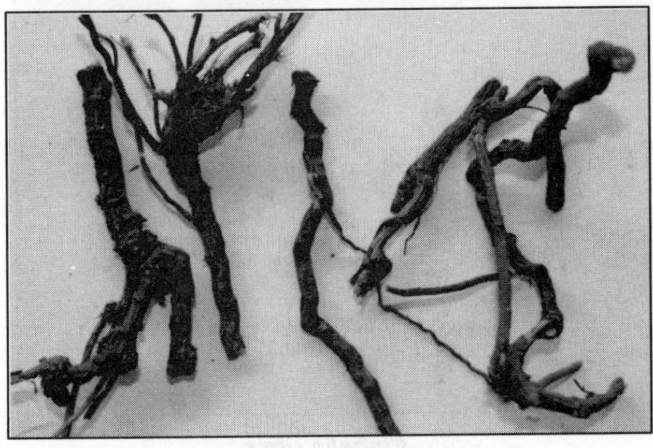

Cissampelos pareira

Contd...

Plate 14.1–*Contd...*

Costus speciosus

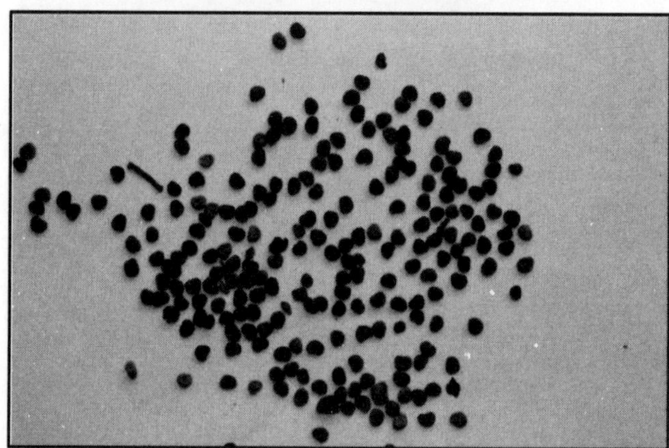

Datura metel

Embelia ribes

Contd...

Plate 14.1–*Contd...*

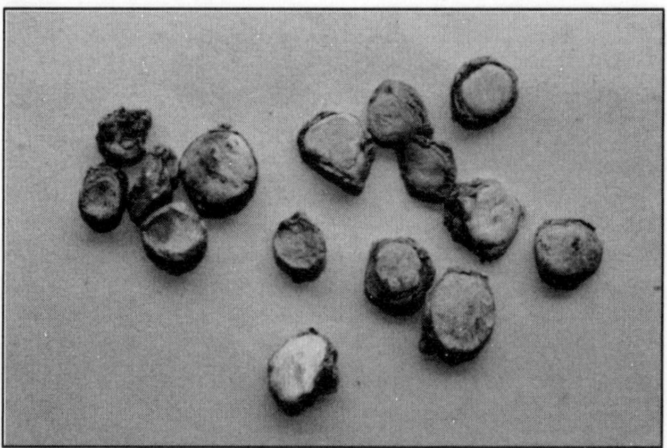

Hedychiums picatum

Holarrhenaanti dysenterica

Lycopodium clavatum

Contd...

Plate 14.1–*Contd...*

Mallotus philippinensis

Mentha viridis

Oroxylum indicum

Contd...

Plate 14.1–*Contd...*

Plantago major

Picrorhiza kurroa

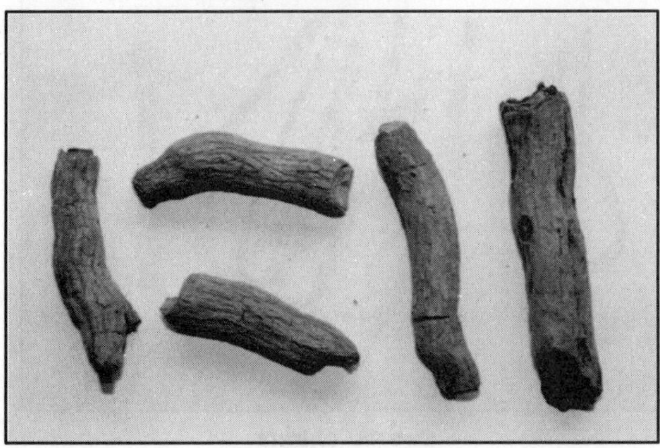

Rauvolfia serpentina

Contd...

Plate 14.1–*Contd...*

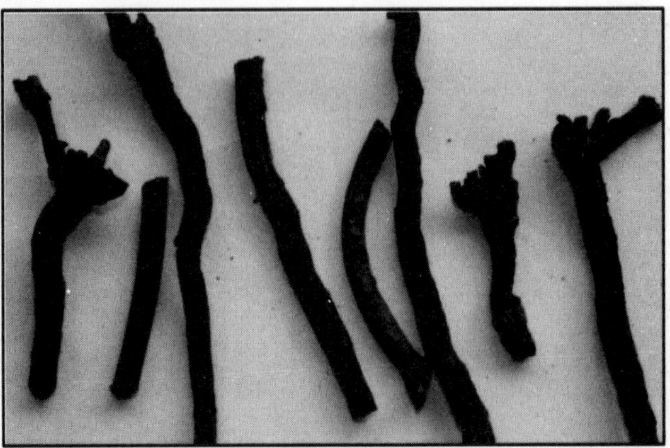

Rubia cordifolia

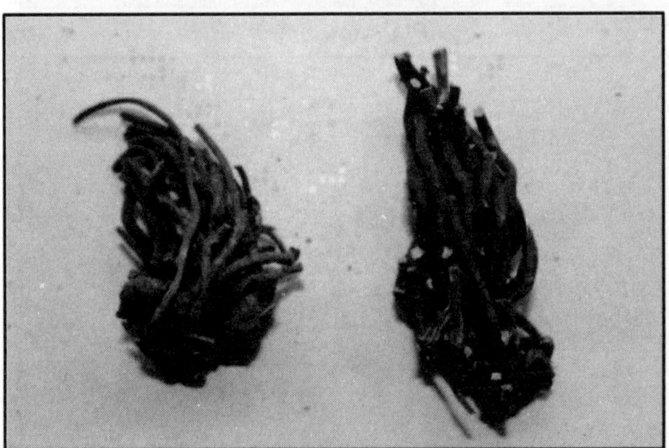

Sinopodophyllum hexandrum

Swertia chirata

Contd...

Plate 14.1–*Contd...*

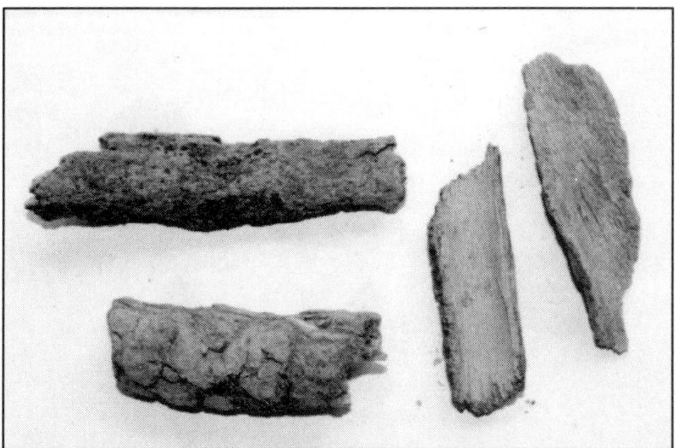

Symplocos racemosa

Taxus wallichiana

Terminalia bellirica

Contd...

Plate 14.1–*Contd...*

Terminalia chebula

Embelia ribes, Holarrhena antidysenterica, Mallotus philippinensis, etc. Then, further research may lead to the development of a miraculous modern drug. Recent survey shows that the percentage of natural products in modern drug armamentarium is considerable, estimates varying from 35 per cent to 50 per cent (Holmstedt and Bruhn 1995).

References

Ahmed F, Nizami Q and Aslam M 2005. *Classification of Unani Drugs,* Maktaba Eshaatul Q'uran Delhi.

Anonymous 1987-97. Standardisation of single drugs of Unani Medicine, Vol 1-3, Central Council for Research in Unani Medicine, New Delhi.

Anonymous 2000. Database on Medicinal Plants used in Ayurveda Central Council for Research in Ayurveda and Siddha, New Delhi.

Anonymous 2006. Standardisation of single drugs of Unani Medicine, Vol 4, Central Council for Research in Unani Medicine, New Delhi.

Anonymous 2007.A handbook of medicinal plants used in Homoeopathy, Vol 1, Central Council for Research in Homoeopathy, New Delhi.

Anonymous 2008. The Ayurvedic pharmacopeia of India, Department of Ayurveda Yoga and Naturopathy Unani Siddha and Homoeopathy, New Delhi.

Anonymous 2008. The Siddha pharmacopoeia of India, Part 1, Vol 1, Department of Ayurveda Yoga and Naturopathy Unani Siddha and Homeopathy, New Delhi.

Arif-ullah Khan and Anwarul Hassan Gilani 2009. Antispasmodic and bronchodilator activities of *Artemisia vulgaris* are mediated through dual blockade of muscarinic receptors and calcium influx. *J Ethnopharmacoly* **126**: 480-486.

Arora R, Chawla R, Puri S C, Sagar R, Singh S, Kumar R, Sharma A K, Prasad J, Singh S, Kaur G, Chaudhary P, Qazi G N and Sharma R K 2005. Radioprotective and antioxidant properties of low-altitude *Podophyllum hexandrum* LAPH. *J Environ Pathol Toxicol Oncol* **24**: 299-314.

Arumugam P, Ramamurthy P, Santhiya S T and Ramesh A 2006. Antioxidant activity measured in different solvent fractions obtained from *Mentha spicata* L: An analysis by ABTS + decolorization assay. *Asia Pac J Clin Nutr* 119-124.

Atta A H and Mouneir S M 2005. Evaluation of some medicinal plant extracts for antidiarrhoeal activity. *Phytotherapy Res* **19**: 481 – 485.

Biswas K 1956. *Common Medicinal plants of Darjeeling and Sikkim Himalayas.* Supdt Govt Press, Kolkata.

Biswas K and Chopra R N 1982.*Common Medicinal Plants of Darjeeling and Sikkim Himalayas.* Periodical Experts Book Agency, Delhi.

Boericke W 2007. *New Manual of Homeopathic Materia Medica with Repertory.* B Jain Publishers P Ltd, Delhi.

Caballero T 1997. Across reactivity between mugwort pollen *Artemisia vulgaris* and hazelnut *Abellananux* in sera from patients with sensitivity to both extracts. *Clin Exp Allergy* **27:** 1203-11.

Chhetri D R 2007. Medicinal plants scenario in Darjeeling Himalayas: Conservation and cultivation as alternative crop. *Indian Forester* May: 665-678.

Chiang L C, Chiang W, Chang M Y and Lin C C 2003. *In vitro* cytotoxic antiviral and immunomodulatory effects of *Plantago major* and *Plantago asiatica*. *Am J Chin Med* **31**: 225-34.

Chiang L C, Chiang W, Chang M Y, Ng L T and Lin C C 2002. Antiviral activity of *Plantago major* extracts and related compounds *in vitro*. *Antiviral Res* **55**: 53-62.

Dash S S, Maiti A and Rai S K 2003. Traditional uses of Plants among urban population of Gangtok. **In:** Ethnobotany and Medicinal Plants of India and Nepal. V Singh and A P Jain (Eds), Scientific Publishers, Jodhpur, 313-324.

Dash VB 1995. Encyclopaedia of Tibetan Medicine, Sri Satguru Publications, Delhi.

Deshpande D J 2006. *A Handbook of Medicinal Herbs*, Agrobios, Jodhpur.

Ghotage N S and Ramdas S R 2008. *Plant Used in Animal care.* Anthra, Pune.

Gilani A H, Yaeesh S, Jamal Q and Nabeel Ghayur 2005. Hepatoprotective activity of aqueous-methanol extract of *Artemisia vulgaris*. *Phytotherapy Res* **19**: 170-172.

Giri A and Narasu M L 2000. Production of podophyllotoxin from *Podophyllum hexandrum*: A potential natural product for clinically useful anticancer drugs **34**: 17-26.

Gomez-Flores R, Calderon C L, Scheibel L W, Tamez-Guerra P, Rodriguez-Padilla C, Tamez-Guerra R and Weber R J 2000. Immuno-enhancing properties of *Plantago major* leaf extract. *Phytotherapy Res* 14:617 – 622.

Gupta A K and Tandon N 2004. *Reviews on Indian Medicinal Plants,* Vols. 2-3. Indian Council of Medical Research, New Delhi.

Heinrich M 2000. Ethnobotany and its role in drug development. *Phytotheraphy Res* **14**: 479-488.

Holmstedt B R and Bruhn J G 1995. Ethnopharmacology-A challenge in Ethnobotany. *In*: Schultes R E and Reis S V (eds) Evolution of a Discipline. Chapman and Hall, USA.

Hson-Mou Chang and Paul Pui-Hay 1986. Pharmacology and Applications of Chinese Materia Medica, 2 Vols, World Scientific, Singapore.

Hussain S and Hore D K 2007. Collection and conservation of major medicinal plants of Darjeeling and Sikkim Himalayas. *Ind J Traditional Knowle* **6**: 352-357.

Khare C P 2004. Indian Herbal remedies rational western therapy ayurvedic and other. Springer Verlag, Berlin Heidelberg, Germany.

Klayman D L 1985. Qinghaosu artemisinin: An antimalarial drug from China. *Science* **228**: 1049-1055.

Kletter C, Kriechbaum M, Szambor P, Qusar N, Holzner W and Kulbelka W 1995. *J Tibetan Med and Astro Inst* **1**: 1-21.

Krishna B and Singh S 1987. Ethnobotanical observations in Sikkim. *J Econ Tax Bot* **9**: 1-7.

Kunwar R M, Uprety Y, Burlakoti C, Chowdhary C L and Bussmann R W 2009. Indigenous Use and Ethnopharmacology of Medicinal plants in Far-West Nepal *Ethnobot Res Applications* **7**: 5-28.

Lewis W H 2003. Pharmaceutical discoveries based on ethnomedicinal plants: 1985 to 2000 and beyond. *Eco Bot* **57**: 126-134.

Lhawang Namgyal Q and Namgyal T 1995. The Knowledge of Skin Care in Tibetan Medicine: as explained in Upadesa. *Tantra* **1**: 47-59.

Mandal S K, Biswas R, Bhattacharyya S S, Paul S, Dutta S, Pathak S, Khuda-Bukhsh A R 2010. Lycopodine from *Lycopodium clavatum* extract inhibits proliferation of HeLa cells through induction of apoptosis via caspase-3 activation. *European J Pharmacol* **626**: 115-122.

María Elena Núñez Guillén José Artur da Silva Emim Caden Souccar and Antonio José Lapa 1997. Analgesic and Anti-inflammatory Activities of the Aqueous Extract of *Plantago major* L. *Pharmaceutical Bio* **35**: 99-104.

Mehr-e-Alam Khan 2004. *Some Common Unani Medicinal Plants.* Central Council for Research in Unani Medicine, New Delhi.

Mkaddem M, Bouajila J, Ennajar M, Lebrihi A, Mathieu F and Romdhane M 2009. Chemical composition and antimicrobial and antioxidant activities of *Mentha longifolia* L and *viridis* essential oils. *J Food Sci* **74**: 358-63.

Nadkarni A K 1976. *Indian Materia Medica,* Vol 2, Popular Prakashan Pvt Ltd, Bombay, India.

Orhan I, Küpeli E, Sener B and Yesilada E 2007. Appraisal of anti-inflammatory potential of the clubmoss *Lycopodium clavatum* L. *J Ethnopharmacol* **109**: 146-50.

Orhan I, Terzioglu S and Sener B 2003. Alpha-onocerin: an acetylcholinesterase inhibitor from *Lycopodium clavatum*. *Planta Med* **69**: 265-267.

Ozaslan M, Didem Karagöz I, Kalender M E, Kilic I H, Sari I and Karagöz A 2007. *In vivo* antitumoral effect of *Plantago major* L extract on Balb/C mouse with Ehrlich ascites tumor. *Am J Chin Med* **35**: 841-851.

Panda A K 2007. *Status of Local Health Traditions of Sikkim Himalaya*. National Training workshop on folk medicine and local health tradition of Sikkim, National Research Institute Ayurveda, Gangtok, Sikkim.

Pandey V N 1991. *Medico-Ethno-Botanical Exploration Sikkim Himalayas*. Central Council for Research in Ayurveda and Siddha, New Delhi.

Pires J M, Mendes F R, Negri G, Duarte-Almeida J M and Carlini E A 2009. Antinociceptive peripheral effect of *Achillea millefolium* L and *Artemisia vulgaris* L: Both plants known popularly by brand names of analgesic drugs. *Phytother Res* **23**:212-9.

Pradhan B K and Badola H K 2008.Ethnomedicinal plant use by Lepcha tribe of Dzongu valley bordering Khangchendzonga Biosphere Reserve in North Sikkim, India. *J Ethnobiol Ethnomed* **4**:22.

Rai L K and Sharma E 1994. *Medicinal Plants of Sikkim Himalaya Status uses and potential*. G B Pant Institute of Himalayan Environment and Development, Gangtok.

rGyud-bzhi and Vaidurya Sngog-pog 1981. Man-nr-rgyud Series 4 Library of Tibetan Works and Archives of His holiness the Dalai Lama, Dharmshala, India.

Sharma P 1969. *Dravyaguna-Vijnana*, Chaukhambha Publishers, Varanasi, India.

Tigno X T, de Guzman F and Flora A M 2000. Phytochemical analysis and hemodynamic actions of *Artemisia vulgaris* L. *Clin Hemorheol Microcirc* **23**: 167-75.

Turel I, Ozbek H, Erten R, Oner C A, Cengiz N and Yilmaz O 2009. Hepatoprotective and anti-inflammatory activities of *Plantago major* L. *Indian J Pharmacol* **41**:120-124.

Ven Rechung and Rinpoche Jampal Kunzang 2001. *Tibetan Medicine*. Sri Satguru Publications, Delhi.

Vogel H G 1991. Similarities between various systems of traditional medicine considerations for the future of ethnopharmacology. *J Ethnopharmacol* **35**:179-190.

Medicinal Plants: Aspects and Prospects (2014) *Pages* 218–232
Editors: Mukesh Kumar, Anjali Khare and C.P. Shukla
ISBN: 978-81-7622-309-6
Published by: BIOTECH BOOKS, NEW DELHI

Chapter 15

Bio-Prospecting of Key Stone Species and Traditional Knowledge of Tribals Community at Vindhyan Region, Uttar Pradesh

C.P. Shukla[1], A.K. Shukla[2] and Kumud Rai[3]

[1]LVP Shodh Sansthan, Koraon, Allahabad – 212 306
[2]Department of Botany, Kripalu Mahila Mahavidyalaya, Kunda – 230 204, Pratapgarh
[3]Department of Zoology, Nehru Gram Bharti Vishwavidyalaya, Allahabad – 212 306

ABSTRACT

A set of species whose impacts on its community or ecosystem are much larger and more influential than its abundance is called Keystone species. It exerts pronounced influence on community structure, micro climate, soil, water and mineral resources as well as physical environment as a whole. The critical impact on community and environment is more important than their dominance or abundance. Even micro-organisms can also act as Keystone species if they are capable of modifying the environment either by the addition of their products or by spoiling the available resources. The Koraon Range, a part of Allahabad Forest division, is managed by Vindhyan region of Uttar Pradesh state. Therefore, it has been given special attention for the protection of the native flora and fauna of the area. Coupled with that, there were selection felling, exploitation of plant resources and forest land by tribals and also neighboring villagers, frequent fires, and so on, lead to severe degradation of the vegetation of the region. Several species of the dry deciduous forests of the area have also been exploited, both by the tribals (Kol, Mushar, Belbans, Bhuiya, Ghosia etc.) and their associated villagers for products of market demand. These products also include fire wood, which imbalances the natural plant populations not only of tree species but also associated shrubs and herbs. When the availability of plant produces dwindled, tribals started to give away their lifestyle, fully and become depended on plant resources, even though they continued to stay inside the forest area. This dual lifestyle of tribals in the range further aggravated the degradation process. Glaring example of this can be demonstrated by the cultivation and processing of Lemon grass (*Cymbopogon flexciosus*) for its oil. The Koraon range is important to conserve the indigenous flora on one hand and the traditional knowledge of tribals on the other. In fact, such species can be considered as the 'Key-stone' species in the context of biodiversity rehabilitation using traditional knowledge *i.e.* both social and ecological senses. *Albizzia lebbeck, Anogeissus latifolia, Cordia myxa, Diospyros cordifolia, Diospyros tomentosa, Garcinia gummi-gutta, Hopea parviflora, Terminali achebula* and *Strobelus asper* are the key-stone species identified during the study with details of their habitat, plant parts exploited, product details and also the main reason by which the species are rare or endangered in the range.

Keywords: Key-stone species, Biodiversity, Vindhyan region, Traditional knowledge.

Introduction

The rich biodiversity of India and her equally rich cultural diversity present unique repertoire. Ayurveda, the ancient science of right livelihood, with its herbal medicine component is an illustration of India's traditional knowledge. Apart from the codified knowledge streams like Ayurveda and Siddha there are several unwritten systems and knowledge streams in India, particularly with the large mosaic of tribal communities. This wealth of the country needs to be appropriately utilized for the betterment of the people and the health of the ecosystem.These have potential to benefit the communities that own the traditional knowledge, the country and also the whole human kind (Venkataraman and Swarnalatha, 2006).

It is not in healthcare alone that India has such potential; it is also in food, agriculture, livestock, native industries and other avenues of life, all of which have tremendous prospects if properly trapped. Being of little monetary or other benefit, to the knowledge holders, traditional knowledge is often undervalued and rapidly eroded. With new technologies and global trade regimes, the traditional knowledge of India is also in the danger of getting misappropriated by others and the country is losing its precious resource (Borders 2007) while there are international protocols like the United Nations Convention on Biodiversity (UNCBD). There are also legally binding trade related agreements which do not honour such rights. These have innate discrepancies and are even mutually incompatible. Side by side, considering biodiversity as unlimited, the native populations plunder it, particularly in the case of medicinal plants where there is a spurt in demand. The large number of industries that utilize medicinal plants across India like, commercial manufacturers of Ayurvedic drugs, continue their indiscriminate use of precious medicinal plants from the wild, which exhausts the plant stock, as there are no effective regulatory systems (Shukla 2009). Export of raw drugs is also going up every year. Though in other countries the commercial users are legally bound to produce the required raw drugs, there is not such regulation in India as on date, except in the case of a few rare plants and even this is not enforced with rigor. Thus traditional communities that depend on the medicinal plants are affected, which caused many precious plants to become endangered. This problem is as serious as inroads made by the external threats. The loopholes in enforcement facilitate unscrupulous manufactures to exploit even, those plants that are classified as severely endangered (Babu and Arora 1999). Traditional medicine for non-commercial uses like home remedies finds it impossible to locate certain medicinal plants in the wild now, which were available plentifully earlier.

Statutory provision to prevent commercial exploitation of resources in the wild must be strictly enforced. Rural healers who provide services at economical terms, or even free, to the poor shall be affected first as they have to compete with the huge companies, only by legally demanding the large manufactures to cultivate their own. The traditional knowledge (TK) holders retaining their knowledge can become redundant for want of resources. Since conservation of biodiversity has direct bearing on human survival, beyond the immediate profit motive, it is necessary to streamline the utilization. Freely available public domain knowledge and the freely accessible biodiversity are the bottom lines of indigenous societies but in the modern context these are facing threat as those wanting to generate huge profits go for destructive and monopolistic uses (Khanna 2006). There are more and more commercial products in the global market which rarely give credit to the traditional knowledge holders.

To the indigenous communities and the practitioners of TK it is least important whether the exploiters are native companies or international ones. Thus there is need for policy frameworks within the country and outside. Meticulous care has to be taken to ensure that the indigenous communities are benefited from the utilization and this shall help to give value addition to their local resources, till now considered as free. More importantly, this will ensure conservation as a peoples need as they shall stand to benefit materially from the resources (Krishnamurthy 1993). While the organized sectors make profit from these very same knowledge systems and resources, the original owners are languishing. The Geographical area of Vindhyan region (especially Koraon forest range) is 600 km^2 and it is located at 25°15' to 35°30' N latitude and between 81°45' to 82°15' E longitude. It is also important to quote here that the knowledge of tribals regarding plants has descended from one generation to another, as domestic practices.

Materials and Methodology

Primary data on the ethnic groups dwelling in the region, their population structure, present and pasts way of life, details on the present occupation, income, species grown and commercial practices, etc. The tribals were also asked as to how they extract, process and use such plants or their products and also the local name by which they identify them in the field. The data thus generated were recorded species-wise and specimen samples were collected either in the flowering or fruiting stage to prepare herbarium specimens and establish their correct identity.The specimens, collected were processed and made into herbarium specimens as per the methodology laid down by standard method (Jain and Rao 1996) and deposited the Herbarium of Lok Vanaspati Paryavaran Sodh Sansthan, Allahabad. In the laboratory, the specimens were tentatively identified with the help of relevant literature like floras, monographs, revision, etc. and such identifications were confirmed with the help of authentically identified specimens available in Botanical survey of India, Allahabad. Later, the specimens were labeled with field data recorded during field surveys and up-to-date botanical names were worked out as per the International Code of Botanical Nomenclature (2008). Local names were also provided for each of the species, which were noted during the field surveys. The following work has been based on the ethnobotanical information and details of traditional knowledge gathered from the tribals and data generated on habit, habitat and distribution pattern of each of the taxon on the investigated area.

Vegetational Growth

The Indian subcontinent has been under the influence of human culture for a very long time, as such the climax formation has been altered for the purpose of human settlement. The characteristics and major vegetation type of the area is tropical dry deciduous forests. The rainfall is about 100-160 cm/year. There are also vegetation along the banks of rivers and branches traversing the area.

Tropical Dry Deciduous Forests

The forest type is met with almost through out the area except for patches of formations belonging to the other vegetational types mentioned before. The forest types have also got transformed into scrub jungle in certain areas mentioned under the category. Tree species common in the forest formation at Vindhyan region (specially Koraon area) are *Anogeissus latifolia, Acacia nilotica, Albizia lebbeck, Azadirachta indica, Butea monosperma, Ceiba pentendra, Diospyros cordifolia, Diospyros melanoxylon, Dalbergia sissoo, Phyllanthus officinalis, Terminalia bellerica, Terminalia arjuna, Pterocarpus marsupium, Cordia myxa, Ficus religiosa, Ficus glomerata, Ficus benghalesis, Grewia tiliaefolia, Hardwickia binata* etc. Shrubby species like *Carissa carandus, Grewia havercens, Grewia orientalis,* Herbs like *Acalypha indica, Bacopa monieri, Boerhavia diffusa* and *Tribulus terrestris* are also common. The forest types are seasonal, annual or perennial growth (Figure 15.1).

Figure 15.1: Glimpse of tropical dry deciduous forest

Thorny Scrubs

The vegetation type is dominated by ground flora with a few tree species *viz. Acacia leucophloea, Tamarindus indica, Azadirachta indica,* shrubs like *Euphorbia antiquirum, Grewia villosa* and *Grewia orientalis* also occupy the ground, partially or fully covered with the growth of the species *viz. Euphorbia hirta, Phyllanthus niruri* and *Euphorbia antiquorum* also herbs like *Acalypha indica, Cleome viscosa, Cleome monophylla* and *Tribulusterres trissurrive* seasonally there. Perennial climbers like *Asparagus racemosus, Gymmena sylvester* etc. are also seen straggling or climbing on blusher or shrubs like *Euphorbia* and *Opuntia* species, common to the vegetation type. Thorny scrub vegetation is very prevalent in the region.

Riparian Vegetation

The vegetation types flourish along the valley close to rivers and rivulets of the region and merges with the tropical dry deciduous forest or scrub jungle on its opposite side, about 20-60 cm from the seasonal or perennial water sourses (Figure 15.2).

Figure 15.2: A View of riparian vegetation

Cordia myxa, Garcinia gummi-gutta, Hopea parviflora, Mangifera indica, Manikara hexandra, Mimusops elengi, Pongamia glabra, Syzigium cumini, Terminallia arjuna etc. are the trees common along the river and rivulet banks forming the dominant component of the vegetation type. Climbers like *Bacopa monnieri, Boerhavia diffusa, Croton sparciflorus, Solanum nigrum* and *Scoparia dulcis* grow in the vegetation type, especially in damp areas or as under growth in the forest type. Shrubs like *Cassia tora* and *Hemidesmus indicus* and herbaceous plant like, *Alstonia scholaris, Helictere sisora, Homonoia riparia* and culums of Bamboos, Canes and few members of grasses and sedges are also characteristics to the vegetation type.

Grasslands

The grasslands are not climax formations but have developed secondarily after the destruction of forests. In most cases they are maintained in the present serial stage due to biotic influence. They are spread in all the major bioclimatic regions of the country. Grasslands region is secondary nature formed by repeated fire and other disturbances that affected the tropical dry deciduous forests in the part. The grasslands are not very extensive but very common here and are bordered by the tropical dry deciduous forest or shrubby vegetation. Emerging tress *viz. Anogeissus latigolia* and *Pterocarpos acerifolium* are rarely seen in the grasslands and the ground flora is dominated by *Cymbopogon flexuosus* and *Phoenix louririi* etc. The grasslands areas either shrink or expanded depending upon the level of intrepidity of adverse factors affecting the vegetation type and also bordering its formations. It may be noted that the flora and vegetation of the range were subjected to severe degradation in the past, when the area was managed only as forest. Low rainfall, rocky substratum, floor soil, drought fire selection and clearing, felling, collection of non-wood products and fire wood both by the tribals and also nearby villagers and introduction of exotics like *Acacia* and *Eucalyptus* were major degrading factors, which contributed to the depletion of the natural vegetation and flora of the area. However, with the establishment of the area most of the external interferences contributing to degradation of the vegetation were restricted to a maximum possible extent.

Ethnobotany of 'Kol' People

Settlements are mostly in the lower elevations of the area. In each of the settlement, there are 12-30 huts with a forest area of about 2-5 ha allotted to each settlement for cultivation of crops of their choice, without removing the trees growing there. The tribe keeps cattle and sheep which they let loose for grazing in the nearby forest areas and the cattle are also often provided with fodder lopped from nearby standing forest trees. The tribals also keep dogs, which accompany them while they go to forest areas for collection of firewood, hunting, fishing or for gathering other products of their need. Attached to some of the tribal settlements are primary schools and temples for worship. The huts in the settlements are mostly with two rooms, one used for cooking and the other for sleeping. The huts are mostly made of locally available materials like bamboos, canes, branches of trees, etc. with roofing leaves. The walls of the huts are having skeleton made of bamboos and canes, often covered with mud. They use vessels of clay and stainless steel. Cattle are left in open areas nearby the settlement. Each hut is occupied by a tribal family composed of husband, wife, parents and children. During night elderly parents sleep in the 'Satrum' along with boys above the age of ten years. Husband and wife perform the function of the collection of forest products, hunting, and fishing, collection of firewood, lemon grass distillation, and labour etc. and earn their livelihood (Figure 15.3).

Tribals store food grains, firewood, seeds, etc. for use in the lean period. They purchase cloths, ornaments, chapels, medicines, provisions, etc. from outside markets. Implements and other accessories are made in the settlement with wood materials collected from the forests.The total tribal community includes men, women and children. They speak to each others in Desi, Hindi, Awadhi and Bhozpuri language.

Figure 15.3: Dweling place of tribal community

In the past, tribal community solely depended on the forest resources for their livelihood, which they were free to exploit and the forest officials, who considered them as a part of the ecosystem, imposed no restriction. The tribal's conventional crops were finger millets, minor forest products of plant and animal origin, and so on, to meet their basic requirements of food, shelter and medicine. At present the major tribals are agriculturists and labours. Major species grown by them include *Arachis hypogea* (Groundnut), *Oryza sativa*, *Citrus* sp., *Eleusine coracana*, *Ipomoea batata*, *Musa paradisiaca*, *Amaranthus* sp., *Brassica* sps., *Dolichos lablab*, *Cajanus cajan*, *Psidium guajava* and *Annona squamosa* etc.

Tribal being settled near to the civilized habitations are at present more non-traditional in their lifestyle. Therefore, they depend on several products available in the market and cultivate only lesser number of conventional crops in the allotted areas and also depend less on natural forest products at present. Therefore, traditional approach to day to day life is comparatively less profound in the case of Kol tribal as compared to other who still retains such an attitude towards their living, making them more dependent on forest area and products. This transformation of tribal folk dwelling within forest areas is partly due their proximity to the villages nearby and also due to lack of sufficient forest based resources for their livelihood. For example, due to insufficient water availability to cultivate conventional crops, they grow and process products like lemongrass oil.

Lifestyle and Cultivation Practices

At present the folk men of tribal are mainly labours employed as forest watchers, fire line workers, etc. The women folk engage themselves in forest nursery work and other supporting jobs along with men. They also do agriculture centered on the crops like *Dolichos lablab, Eleusine coracana, Zea mays* etc. Also, they grow lemon-grass (*Cymbopogon flexuosus*) on a fairly large scale in nearby forest areas of the settlements and distil it for the oil. During leisure and lean periods, tribals engage in fishing, honey collection and also rarely hunting of small animals for food. Being exposed to the nearby village people and also forest officials, they feel mingle and converse with even strangers and do marketing in the nearby towns. Children either go to nearby schools nearby or study in a few outer colonies.

Plants of Medicinal, Economic, Horticultural and Food Value

Table 15.1 represents an alphabetical list of 71 plants which are traditionally used by tribals. They use the plants as medicine, food, detergent, hut-making, fodder, firewood, rope, teeth and vessel cleaning dog and fish poison, religious plants, etc. These are included in day to-day life of the tribals. There are also products like Cabbage, Gooseberry, Soap-bark, Broom-material, Soap-nut etc., which were collected and marketed earlier or sometimes at present. The removal of such products, very much related to the regeneration of the species, has contributed substantially to the impoverishment of their natural population in the area.

Table 15.1: Ethnobotanical species of Vindhyan region

Sl.No.	*Botanical Names*	*Habit*	*Part(s) Used*	*Tribal Use(s)*
1.	*Acacia nilotica* (L.)	Trees	Bark	Medicine
2.	*Acacia tora* Roxb.	Lianas	Bark	Detergent
3.	*Acalypha fruticosa* Forsk.	Herbs	Leaves	Medicine
4.	*Albizia amara* (Roxb) Boiv.	Trees	Leaves, Timber	Hair-Wash, Firewood
5.	*Albizia lebbeck* (L.) Benth	Trees	Timber	Implements
6.	*Alstonia venenata* R. Br.	Shrubs	Latex	Adhesive
7.	*Anogeissus latifolia* Guill. et Perr.	Trees	Timber, Bark	Firewood, Medicine
8.	*Asparagus racemosus* Willd.	Climbers	Rood tuber	Food
9.	*Azadirachta indica* A. Juss.	Trees	Leaves	Medicine
10.	*Bacopa monnieri* (L.) Watts.	Herbs	Plants	Medicine
11.	*Bambusa arundinacea* (Retz.) Willd.	Shrubs	Stems	Hut
12.	*Bauhinia racemosa* Lamk.	Trees	Timber, Bark	Firewood, Rope
13.	*Boerhavia diffusa* L.	Herbs	Twigs	Food, Medicine
14.	*Calotropis gigantea* (L.) R. Br.	Shrubs	Latex	Medicine
15.	*Cassia alata* L.	Shrubs	Leaves	Food
16.	*Cassia fistula* L.	Trees	Leaves	Medicine
17.	*Cassia occidentalis* L.	Shrubs	Leaves	Food
18.	*Catharanthus roseus* (L.) G. Don	Herbs	Leaves	Medicine
19.	*Chloroxylons wietenia* DC.	Trees	Bark	Medicine
20.	*Clemoegy nandra* L.	Herbs	Twigs	Food

Contd...

Table 15.1–*Contd...*

Sl.No.	Botanical Names	Habit	Part(s) Used	Tribal Use(s)
21.	*Cleome monophylla* L.	Herbs	Twigs	Food
22.	*Cleome viscosa* L.	Herbs	Twigs	Food
23.	*Commiphora caudata* (W. et A.) Engl.	Trees	Twigs, Bark	Fooder, Medicine
24.	*Cordia domestica* Roth	Trees	Fruits	Adhesive, Food
25.	*Cordia gharaf* (Forssk.)Ehrenb.	Trees	Fruits	Food
26.	*Dalbergial lanceolata* L.f	Trees	Timber	Firewood
27.	*Dalbergia latifolia* Roxb.	Trees	Heart wood	Fumigation
28.	*Datura metel* L.	Shrubs	Leaves	Medicine
29.	*Datura stramonium* L.	Shrubs	Leaves	Medicine
30.	*Dioscorea wightii* Hook. f.	Climbers	Tuber	Food
31.	*Diospyros cordifolia* Roxb.	Trees	Timber, Leaves	Firewood, Poison
32.	*Diospyros ebenum* Koen. ex Retz.	Trees	Timber	Pestle
33.	*Diospyros ovalifolia* Wt.	Trees	Timber	Firewood
34.	*Ehretia canarensis* Miq.	Trees	Timber	Implements
35.	*Phyllanthus officinalis* Gaertn.	Trees	Fruits	Medicine, Food
36.	*Euphorbia hirta* L.	Herbs	Branches	Fodder
37.	*Ficus racemosa* L.	Trees	Fruits	Food
38.	*Ficus religiosa* L.	Trees	Leaves	Medicine
39.	*Garcinia gummi-gutta* (L.) Roxb.	Trees	Fruits	Food, Sale
40.	*Grewia flavescens* Juss.	Shrubs	Fruits	Food
41.	*Grewia orbiculata* Rottl.	Trees	Fruits	Food
42.	*Grewia orientalis* L.	Trees	Fruits	Food '
43.	*Grewia tiliaefolia* Vahl	Trees	Fruits	Food
44.	*Grewia villosa* Willd.	Trees	Bark, Fruits	Fibre, Food
45.	*Gymnema sylvester* R. Br.	Climbers	Leaves	Medicine
46.	*Hardwickia binata* Roxb.	Trees	Wood, Bark	Firewood, Hut
47.	*Helicter sisora* L.	Shrubs	Fruits, Bark	Medicine, Rope
48.	*Hopea parviflora* Bedd.	Trees	Timber	Firewood
49.	*Jasmincus pidatum* Rottl.	Climbers	Stems	Drum-stick
50.	*Mallotus tetracoccus* Roxb.	Trees	Leaves	Container
51.	*Mangifera indica* L.	Trees	Fruits	Food, Sale
52.	*Mimusops elengi* L.	Trees	Fruits	Food
53.	*Mollugonu dicaulis* Lamk.	Herbs	Leaves	Meditine
54.	*Opuntia dellenii* Haw.	Shrubs	Fruits	Food
55.	*Oxalis corniculata* L.	Herbs	Plants	Cleaning
56.	*Phoenix loureirii* Kunth	Shrubs	Fruits, Leaves	Food, Broom

Contd...

Table 15.1–*Contd...*

Sl.No.	*Botanical Names*	*Habit*	*Part(s) Used*	*Tribal Use(s)*
57.	*Phyllanthus amarus* Schum. et Thonn.	Herbs	Plants	Medicine
58.	*Physalis minima* L.	Herbs	Fruits	Food
59.	*Pongamia pinnata* (1.) Pierre	Trees	Bark	Medicine
60.	*Santalum album* L.	Trees	Leaves	Fodder
61.	*Sida acuta* Burm. f.	Herbs	Leaves, Stems	Hair-wash, Broom
62.	*Sida rhombifolia* L.	Herbs	Roots	Medicine
63.	*Streblus asper* Lour.	Trees	Leaves	Curd-making
64.	*Strychnos potatorum* L.f.	Trees	Leaves, Fruits	Fish-poison
65.	*Syzygium cumini* (L.) Skeels	Trees	Fruits, Bark	Food, Fish-poison
66.	*Tectona grandis* L.f.	Trees	Timber, Leaves	Hut, Pot
67.	*Terminalia belerica* Retz.	Shrub	Fruits	Sale
68.	*Tribulus terrestris* L.	Herbs	Plants	Medicine
69.	*Tridax procumbens* L.	Herbs	Plants	Medicine
70.	*Wrightia tinctoria* R. Br.	Trees	Latex	Curd-making
71.	*Ziziphus oenoplia* (L.) Mill.	Shrubs	Fruits	Food

At present tubers of *Asparagus racemosus, Dioscorea bulbifera*, stem and leaves of *Boerhavia diffusa, Cassia alata, Cleome viscosa, Cleome monophylla, Ficus racemosa, Garcinia gummi-gutta, Grewia tiliaefolia, Mitrephora heyneana, Opuntia dillenii, Solanum nigrum* etc. are eaten or rarely marketed by the tribals from their natural source. This supplements their food requirements from cultivated crops; those purchased from the markets and also contribute to their income from other sources like labour. However, the removal stress of such plant products on the biodiversity of the area is not very profound. But, the negative impact on the natural flora of the area is in still worse in the case of large scale removal of firewood of species like *Albizia odoratissima, Hardwicki abinata, Carrisa spinosa, Anogeissus latifolia, Cordia myxa, Diospyros melanxylon, D. ovalifolia, Hopea parviflora*, etc. used both for domestic purposes and for distillation of Lemongrass oil. This negative process on the conservation of the plant diversity of the area can only be inverted by enrichment planting of such species in the area. Moreover being tree species that too of tropical dry deciduous forest areas, indiscriminate removal of such plants from their natural habitat have both direct and indirect impacts especially on the associated flora and their artificial introduction can only help in the rehabilitation of them and their associated species.

Plants of Botanical Interest

As mentioned before, 71 flowering plants are found to be related to the traditional life of tribal people staying in Vindhyan region. The occurrence of such species in and nearby forest areas has been given in Table 15.2, elucidating their availability and distribution pattern for whole of the area. It may be noted here that out of the 71 species, 25 taxa are restricted to one settlement area, 48 to two and the remaining 42 species are wider in their distribution, available in three or all the areas.

Table 15.2: Distribution of ethnobotanical species in the vicinity of different tribal settlements

Sl.No.	Botanical Names	Habit	Distribution
1.	*Acacia nilotica* (L.)	Trees	Common
2.	*Acacia tora* Roxb.	Trees	Common
3.	*Acalypha fruticosa* Forsk.	Lianas	Common
4.	*Albizia amara* (Roxb) Boiv.	Lianas	Common
5.	*Albizia lebbeck* (L.) Benth	Herbs	Common
6.	*Alstonia venenata* R. Br.	Herbs	Common
7.	*Anogeissus latifolia* Guill. et Perr.	Trees	Common
8.	*Asparagus racemosus* Willd.	Trees	Common
9.	*Azadirachta indica* A. Juss.	Trees	Rare
10.	*Bacopa monnieri* (L.) Watts.	Shrubs	Rare
11.	*Bambusa arundinaceae* (Retz.) Willd.	Trees,	Common
12.	*Bauhinia racemosa* Lamk.	Trees	Very rare
13.	*Boerhavia diffusa* L.	Herbs	Fairly common
14.	*Calotropis gigantean* (L.) R. Br.	Climbers	Fairly common
15.	*Cassia alata* L.	Shrubs	Common
16.	*Cassia fistula* L.	Trees	Planted
17.	*Cassia occidentalis* L.	Shrubs	Rare
18.	*Catharanthus roseus* (L.) G. Don	Trees	Common
19.	*Chloroxylon swietenia* DC.	Herbs	Common
20.	*Clemoe gynandra* L.	Climbers	Rare
21.	*Cleome monophylla* L.	Shrubs	Rare
22.	*Cleome viscosa* L.	Shrubs	Common
23.	*Commiphora caudata* (W. et A.) Engl.	Climbers	Common
24.	*Cordia domestica* Roth	Shrubs	Rare
25.	*Cordia gharaf* (Forssk.)Ehrenb.	Shrubs	Fairly common
26.	*Dalbergia lanceolata* L.f	Trees	Common
27.	*Dalbergia latifolia* Roxb.	Shrubs	Common
28.	*Datura metel* L.	Herbs	Rare
29.	*Datura stramonium* L.	Shrubs	Common
30.	*Dioscorea wightii* Hook. f.	Trees	Rare
31.	*Diospyros cordifolia* Roxb.	Trees	Common
32.	*Diospyros ebenum* Koen. ex Retz.	Climbers	Common
33.	*Diospyros ovalifolia* Wt.	Herbs	Common

Contd...

Table 15.2–*Contd...*

Sl.No.	Botanical Names	Habit	Distribution
34.	*Ehretia canarensis* Miq.	Herbs	Common
35.	*Phyllanthus officinalis* Gaertn.	Herbs	Common
36.	*Euporbia hirta* L.	Habit	Distribution
37.	*Ficus racemosa* L.	Trees	Rare-
38.	*Ficus religiosa* L.	Trees	Fairly common
39.	*Garcinia gummi-gutta* (L.) Roxb.	Trees	Fairly common
40.	*Grewia flavescens* Juss.	Trees	Rare
41.	*Grewia orbiculata* Rottl.	Shrubs	Rare
42.	*Grewia orientalis* L.	Shrubs	Rare
43.	*Grewia tiliaefolia* Vahl	Lianas	Rare
44.	*Grewia villosa* Willd.	Trees	Common
45.	*Gymnema sylvester* R. Br.	Climbers	Fairly common
46.	*Hardwickia binata* Roxb.	Trees	Fairly common
47.	*Helicter sisora* L.	Trees	Very rare
48.	*Hopea parviflora* Bedd.	Trees	Fairly common
49.	*Jasmincus pidatum* Rottl.	Shrubs	Rare
50.	*Mallotus tetracoccus* Roxb.	Trees	Rare
51.	*Mangifera indica* L.	Trees	Common
52.	*Mimus opselengi* L.	Herbs	Rare
53.	*Mollugo nudicaulis* Lamk.	Herbs	Common
54.	*Opuntia dellenii* Haw.	Trees	Rare
55.	*Oxalis corniculata* L.	Trees	Rare
56.	*Phoenix loureirii* Kunth	Trees	Rare
57.	*Phyllanthus amarus* Schum. et Thonn.	Trees	Rare
58.	*Physalis minima* L.	Trees	Rare
59.	*Pongamia pinnata* (1.) Pierre	Shrubs	Rare
60.	*Santalum album* L.	Shrubs	Rare
61.	*Sida acuta* Burm. f.	Shrubs	Rare
62.	*Sida rhombifolia* L.	Trees	Fairly common
63.	*Streblus asper* Lour.	Shrubs	Fairly common
64.	*Strychnos potatorum* L.f.	Climbers	Fairly common
65.	*Syzygium cumini* (L.) Skeels	Trees	Fairly common
66.	*Tectona grandis* L.f.	Shrubs	Rare
67.	*Terminalia belerica* Retz.	Trees	Very rare
68.	*Tribulus terrestris* L.	Climbers	Fairly common
69.	*Tridax procumbens* L.	Trees	Fairly common
70.	*Wrightia tinctoria* R. Br.	Shrubs	Very rare
71.	*Ziziphus oenoplia* (L.) Mill.	Shrubs	Common

Table 15.3: Keystone plant species in the context of biodiversity rehabilitation using traditional knowledge

Botanical Names	Habit	Part	Product	Reason for Rarity
Acacia tora	Trees	Bark	Soap-bark	Destructive collection
Albizia amara	Trees	Wood	Firewood	Overexploitation
Albizia odoratissima	Trees	Wood	Firewood	Overexploitation
Anogeissus latlfolia	Trees	Wood	Firewood	Overexploitation
Antiaris toxicaria	Trees	Bark	Mat	Overexploitation
Canarium strictum	Trees	Resin	Fumigant	Destructive collection
Cordia domestica	Trees	Wood	Firewood	Overexploitation
Cordia gharaf	Trees	Wood	Firewood	Overexploitation
Diospyros cordifolia	Trees	Wood	Firewood	Overexploitation
Diospyros ovalifolia	Trees	Wood	Firewood	Overexploitation
Phyllanthus officinalis	Trees	Fruits	Food	Destructive collection
Garcinia gummi-gutta	Trees	Fruits	Food	Poor regeneration
Hopea parviflora	Trees	Wood	Firewood	Overexploitation
Ochlandratra vancorica	Shrubs	Stems	Hut, Basket	Overexploitation
Phoenix loureirii	Shrubs	Leaves	Brooms, Hut	Overexploitation

Reasons for Undertaking the Present Work and its Importance

Allahabad forest division especially Koraon forest range has been managed like any other reserve forest area of Uttar Pradesh. No special attention has been given for the protection of the native flora and fauna of the area. Coupled with that, selection felling, exploitation of forest resources and forest land by the tribals and also neighboring villagers has been carried out there, leading to severe degradation of the vegetation of the region. Severe grazing by cattle reared by the tribals and villagers nearby, shifting cultivation by the tribals introduction of cash crops like Lemongrass, *Bombax* and many other food and cash crops also aggravated the degradation process and at present the vegetation of Vindhyan region is rather a relict of what existed in the past. Several species of the dry deciduous forests of the area were also excessively exploited in the past both by the tribals and their associated villagers, for products of market demand including firewood, which impoverished the natural plant populations not only of tree species but also of associated shrubs and herbs. When the availability of such forest products dwindled, tribals started to give away their traditional lifestyle, fully depended on forest resources, even though they continued to stay inside the forest area. This dual lifestyle of tribals in the area further aggravated the degradation process and a glaring example of this is demonstrated by the cultivation and processing of Lemongrass for its oil. The tribals engaged forest land for its cultivation after removal of the native flora and also exploited firewood trees on a very large scale from neighboring forest areas for distillation of the oil, which vigorously continues even at present.

Key-stone Plant Based Industries

Many other forest products of market value like soap bark (*Acacia tora*), resin collected from the bark of *Canarium strictum*, fruits of *Garcinia gummi-gutta*, *Phyllanthus officinalis*, etc, were also excessively exploited in the past, which continues even at present, depending on the availability of such products and their market demand also had its negative impact on the natural populations of the trees in the Sanctuary and also their associated species. Therefore, in any attempt to use traditional knowledge for the rehabilitation of the biodiversity of the area, it is essential to replenish the natural populations of such species heavily depended on by the tribals. This will help to conserve the indigenous flora on one hand and the traditional lifestyle of the tribals, on the other. In fact, such species can be considered as the 'Key-stone species' in the context of biodiversity rehabilitation using traditional knowledge both social and ecological senses. Table 15.3 represents an alphabetical list of the major key-stone species identified during the study with details on their habit, part exploited product details and also the main reason by which the species are rendered rare or endangered in the area.

Over-exploitation, destructive collection procedures, excessive collection of regenerating parts like fruits, roots, etc. are the major reasons for the impoverishment of the natural populations of the species over a long period of time and the process of such plants getting rare in the area was accelerated by the overall degradation of the forest ecosystem, stemmed from various inherent and external factors mentioned earlier. It is very evident that, the tribals living in the area are very poor. They exploit marketable forest products or generate the products like lemongrass oil of market value in order to earn their livelihood, supplementing what they get as wages from labour. This being a socio-economic problem can only be solved by the measures taken to improve their economic conditions and also by educating those on the need to conserve the bio-diversity of the area and in turn their traditional way of life, which are inter-related. Other than the key-stone species identified, which are mainly arborescent in habit playing a key role in the ecosystem structure and functioning, and whose continued survival is essential for the existence of associated flora like shrubs and herbs.There are about another 45 species which are exploited by the tribals for their domestic consumption or for marketing. They are mostly food, medicinal, hut and implements making plants like *Asparagus racemosus* (root tuber as food), *Gymnema sylvestre* (leaves as medicine) *Phyllanthus amarus* (whole plant as medicine), *Tribulus terrestris* (whole plant as medicine) etc. many of which are also rare in the area at present due to over-exploitation. Details on such species have already been shown in Tables 15.1 and 15.2.

Biodiversity Rehabilitation

In the rehabilitation programmes, such species of economic, consumption or utilization potential also deserve attention. In fact, what is essential for the biodiversity restoration of the area is a system approach to rejuvenate the ecosystem as such with maximum indigenous floral and faunal elements which automatically will sustain the traditional lifestyle of the tribals also. A unique instance of wild pigs, digging and eating away the roots of *Capparis spinosa* devastating the natural populations of this endemic species in the area has also been observed. Protection from all such causes of degradation, rehabilitation using key-stone and other species linked with the day-to-day life of the tribals and tribal education on the need for the conservation of the biodiversity of the area and retention of their traditional way of life can facilitate restoration characteristic to the sanctuary. This will also help to improve the quality of the ecosystem as such, which is also essential for species survival.

Dependence on the native flora of the area is mainly for the products like firewood, medicine, food, rope hut and implements making. A total of 71 ethnobotanical species being exploited by the tribals have been identified from the area. This includes species which produce marketable products like (*Phyllanthus officinalis*), Soap bark *(Acacia tora)*, Resins *(Canariums trictum)*, Broom material *(Phoenix loureirii)* and Camboge (*Garcinia gummi-gutta*). Over-exploitation, destructive extraction methods poor regeneration due to lack of sufficient number of propagules (as they are removed as products) and degraded ecological conditions of the area are the major factors for the rarity of such plants and their associates in the area. Exploitation of firewood, mainly for the distillation of lemongrass is the most important factor in operation at present which has led to the drastic impoverishment of the floral diversity of the sanctuary. Also, extensive cultivation of lemongrass in forest areas after removing the natural vegetation, devastated the native flora substantially. In order to rehabilitate the floral diversity of the area, major ethnobotanical and firewood species identified as key-stone species have to be regenerated to replenish their natural populations. This will also bring in several other species associated with them. Cultivation has to be restricted and tribals have to be educated on the need for sustained utilization and regeneration of the species related to the traditional lifestyle. With their active involvement in rehabilitation programs, the biodiversity of the area can be achieved and the ethnobotanical information generated will facilitate the rehabilitation process as the data source for species selection, propagation and sustained utilization.

References

Babu Sarat G V and Arora S 1999. Hot spots of biodiversity, Ministry of Environment and Forest. (http# envfor.nic.m/news/Aug 99/biodiv.html).

IAPT 2008. International Code of Botanical Nomenclature. Germany

Jain S K and Rao R R 1976. *Handbook of Field and Herbarium Methods*. Today and Tomorrow's Printers and Publ., New Delhi, p. 33-58.

Khanna K K 2006. Changing pattern in the flora of Allahabad district, U.P.: A case study of monitoring. *Phytotaxonomy* **6:** 49-52

Krishnamurthy T 1993. *Minor Forest Products of India*. Oxford and IBH, New Delhi, p. 679.

Shukla C P 2009. Biological resources and traditional therapeutical knowledge: A review. *In:* Kumar A and Govind D (eds.) *Biodiversity to Biotechnology*. Narosa Publ. House, New Delhi, p. 135-174.

Venkatraman K and Swarna Latha S 2006. Intellectual property rights, traditional knowledge and biodiversity of India. *J Intellect Prop Rights,* **13:** 326-335.

Medicinal Plants: Aspects and Prospects (2014) ***Pages 233–240***
Editors: **Mukesh Kumar, Anjali Khare and C.P. Shukla**
ISBN: 978-81-7622-309-6
Published by: **BIOTECH BOOKS, NEW DELHI**

Chapter – 16

Anti-Urolithiatic Plants of District Bijnor, Uttar Pradesh

Deepika[1], Deepak Kumar[1], Alka[2] and P. Kumar[2]

[1]Department of Botany, R.S.M. Degree College, Rampur, Ghoghar, Moradabad – 244 002
[2]Department of Botany, Hindu College, Moradabad – 244 001

ABSTRACT

Across the world human societies have invented different ways for curing illness but all of these are chiefly plant based. The present investigation has been carried out for documentation of native medicinal plants exploited for medicinal purposes from district Bijnor, Uttar Pradesh. The study area is rich repository of economically important plants. The survey was conducted by holding interviews and using questionnaires among the people of local area having traditional knowledge of medicinal plants. The survey has resulted the compilation of 15 indigenous taxa used as urolithiatic medicines.

Keywords: Repository, Traditional knowledge, Urolithiatic medicines.

Introduction

A floristic survey of ethno-medicinal plants used in district Bijnor was conducted to access the potentiality of plant species. It has been observed that the plants belonging to 24 families of angiosperms are used as anti-urolithiatic agents for local remedies. The information on medicinal plants use is based on the exhaustive interviews with local healers practicing in traditional system of medicines. The importance of ethno-medicine is established for various economic uses of plants among the primitive human societies, which is equally beneficial to modern man. It has also brought into light, numerous little known or unknown medicinal uses of the plants (Jain 1991).

Bijnor district lies in the extreme North-West of Uttar Pradesh in the semi-arid region of upper Ganga-Yamuna Doab. District Bijnor of UP and Panipat and Karnal districts of Haryana lie in the Eastern and Western sides of Ganga and Yamuna respectively. Its northern border is shared by districts Muzaffarnagar and Saharanpur of U.P. and Haridwar of Uttarakhand, while Southern border by district Moradabad and Meerut (U.P.). The Bijnor covers an area of 4008 sq km with maximum length and width of 99 km and 58 km, respectively. The slope of the district is from North to South. It is traversed by the rivers like Ramganga, Kho and Ganga. From North to South, they flow almost parallel. The district may generally be described as an alluvial plain consisting of 3 main geographical tracts-Khadar, Bargar and Bhood zone. In khadar zone, it consists of rural country side having many small human settlements and tiny homelets in interior areas with poor connectivity. Due to a large number of villages and huge rural population with rich traditional utilization of plants for medicinal uses, the district consists of a suitable and significant area for ethno-botanical studies. It, therefore, contains a rich and diverse flora, with the use of a majority of traditional folk medicines. There is an immense need of the studies related to the ethno-botany which has yet not been attempted, and there is complete lack of information about the traditional usage of medicinal plants.

This chapter attempts to present briefly the information about a few medicinal plants being used in this area for their anti-urolithiatic, litholytic and litho-expulsive properties. Related ethno-medicinal data has been collected through interviews, discussion and observations. Many visits have been made to interact with the villagers and people of remote areas particularly those who are familiar with herbal medicines. This chapter covers 30 taxa, belonging to 24 families of angiosperms. The nomadic people recipes were compared with the information available in literature. (Kirtikar and Basu 1945, Anonymous 1948-76; Chopra *et al.*, 1956-69 and Singh 1998). The present information is a maiden attempt and has not been recorded in the available literature as yet.

Ethno-medicinal Plants of Bijnor

Thirty angiospermic taxa having ethno-medicinal properties have been found to utilize as herbal medicines by remote villagers. These plant species have been arranged alphabetically followed by the family, local name and utilization in various ailments. The summary of these plants has been presented as follows:

1. *Acacia nilotica* L. (Varn. Kikar)

Family: Mimosaceae

Young leaves taken orally with water in dhaturog.

2. *Achyranthes aspera* L. (Varn. Latjeera)

Family: Amaranthaceae

Leaves are crushed and applied in scorpion sting.

3. *Allium sativum* L. (Varn. Lahsun)

Family: Liliaceae

Fresh juice of bulb is applied on wounds.

4. *Aloe vera* L. (Varn. Gwarpatha)

Family: Liliaceae

Orally used in diabetes, constipation and stone. Leaf pulp/juice is applied externally on burns, wounds and other skin diseases.

5. *Azadirachta indica* A. Juss. (Varn. Neem)

Family: Meliaceae

Decoction of leaves is taken orally for the treatment of fever and stomach ache. Leaves are used in touch therapy.

6. *Cassia fistula* L. (Varn. Amaltas)

Family: Caesalpinaceae

Decoction of legume is taken orally in stomachache and piles. Stem bark is used in skin diseases.

7. *Cassia occidentalis* L. (Varn. Kasaundi)

Family: Caesalpinaceae

Leaf-pulp is used externally in skin diseases.

8. *Chenopodium album* L. (Varn. Bathua)

Family: Chenopodiaceae

Young tender shoots and leaves are used as green vegetable; they are believed to be appetizer and to relive constipation.

9. *Chrysanthemum coronarium* L. (Varn. Guldaudi)

Family: Asteraceae

The whole plant is boiled and filtrate is taken orally to cure jaundice.

10. *Cleome viscosa* L. (Varn. Hurhura)

Family: Capparidaceae

Fresh leaf extract is applied on the effected parts of ringworms.

11. *Coccinia grandis* L. (Varn. Kanduri)

Family: Cucurbitaceae

Decoction of leaves is useful in respiratory diseases.

12. *Crinum difixembker* L. (Varn. Sudarshan)

Family: Amaryllaceae

Juice is externally used in earache.

13. *Cyperus rotundus* L. (Varn. Motha)

Family: Cyperaceae

Two-three tubers are orally given per day with water for five days for treatment of spermatorrhoea.

14. *Datura metel* L. (Varn. Dhatura)

Family: Solanaceae

Fresh juice is applied externally on the swollen part of the body.

15. *Eclipta alba L. (Varn. Bhangra)*

Family: Asteraceae

The plants are used as hair tonic. Crushed leaves are applied in cuts and wounds.

16. *Euphorbia hirta* L. (Varn. Dudhi)

Family: Euphorbiaceae

Leaves together with stems are crushed and taken orally in dysentery.

17. *Evolvulus alsinoides* L.

Family: Convolvulaceae

Decoction of leaves and roots is taken orally to increase IQ of children.

18. *Ficus religiosa* L. (Varn. Peepal)

Family: Moraceae

Powder of stem bark is used in asthma.

19. *Girardinia zeylanica* (Varn. Bicchua)

Family: Urticaceae

Leaves paste or juice is applied on bleeding of cuts and wounds.

20. *Indigofera tinctoria* L. (Varn. Jitia)

Family: Mimosaceae

Leaves pulp is applied on head during headache.

21. *Jatropha curcas* L. (Varn. Rotonj)

Family: Euphorbiaceae

Fresh leaves decoction is taken orally in Tuberculosis. Oil externally used in skin diseases.

22. *Lawsonia inermis* L. (Varn. Mehandi)

Family: Lythyraceae

Ground leaves are rubbed on affected surface to cure whitlow and decoction of boiled root is used for abortion.

23. *Leucas capitata* Spreng (Varn. Guma)

Family: Lamiaceae

Two spoonful decoction of leaves and flower head mixed with *Momordica charantia* is given once for 7-10 days for the cure of jaundice.

24. *Oxalis corniculata* L. (Varn. Khattibuti)

Family: Oxalidaceae

Fresh leaves are taken orally with water to cure dysentery.

25. *Phoenix humilis* Roxb (Varn. Khajur)

Family: Aricaceae

Decoction of leaves and roots is given for cooling.

26. *Psidium guajava* L. (Varn. Amrud)

Family: Myrtaceae

Decoction of stem bark along with *Azadirachta indica* stem bark is taken orally or used in bathing water in the treatment of malaria, fever and jaundice.

27. *Sida cordifolia* L. (Varn. Samas)

Family: Malvaceae

Leaves and seeds are used in dhaturog.

28. *Tamarindus indica* L. (Varn. Imli)

Family: Caesalpiniaceae

Boiled fruit extract is used for porridges taken in large quantity it helps in milk formation in lactating mothers.

29. *Tephrosia purpurea* Pers (Varn. Sarphonka)

Family: Fabaceae

Decoction of leaves is taken in treatment of respiratory disease.

30. *Vitex negundo* L. (Varn. Samhalu)

Family: Verbenaceae

Decoction of leaves is taken orally in gout rheumatism, arthritis and sciatica.

Ethno-medicinal Plants of District Bijnor Used in the Treatment of Urinary Tract and Kidney Stones

Ethno-medicinal properties of the plants being used in the treatment of urinary tract and kidney stones is given below. It includes plant names, families, local names and their medicinal uses, along with the plant parts used, as has been reported by the local people. Occurrence of urinary tract and kidney stones is a common clinical disorder, which has affected mankind since ancient times. Urolithiasis is an entity, which has high morbidity and low mortality but having serious and significant socio-economic impact. The prevalence of urolithiasis is estimated to be 1-5 per cent. However, its frequency varies with differences in dietary habits of different people, food and water contamination in different geographical areas, and their level of development and environmental pollution, etc. The overall probability of forming stones differs in various parts of the world. Its worldwide prevalence is estimated to be 2-5 per cent while it is 2-13 per cent in developed countries (Zaidi *et al.*, 2006). During ethno-botanical survey of district Bijnor, 15 plant species belonging to 13 families were recorded as effective remedies used by local people to treat and cure stone ailments of urinary tract and kidney. These crude drug preparations inhibit further stone formation and their enlargement, dissolve or break the calculi and stones, expel them and reduce and relieve the suffering from pain. Some of these plants have also been reported earlier, being used in anti-urolithiatic and lithotriptic preparations from different parts of the country. A survey of *Vedic* literature was conducted to elucidate pharmacognostic aspects of herbal crude drugs of plant sources for the cure of urinary tract stones. *Boerhavia diffusa* L. was among the important medicinal plants used for the treatment of stones during *Vedic* period. Decoction of whole plant is employed in the treatment of calculi by tribal people of Saurashtra (Gujarat) while its roots are used in the treatment of urinary stones by tribals of Akola and Sanganner talukas of Ahmednagar, Maharashtra.

List of Antiurolithiatic and Urolithiasis Ethnomedicinal Plants of District Bijnor

1. *Abutilon indicum* L. (Varn. Kanghi)

Family: Malvaceae

Leaf juice taken twice daily for two weeks is effective for the treatment of urinary tract and kidney stones.

2. *Aerva lanata* L. (Varn. Chaya)

Family: Amaranthaceae

Whole plant decoction of Chaya, along with castor (*Ricinus communis* L.) root and Gokhuru (*Tribulus terrestris* L.) fruits is given twice a day for two weeks to cure stones. Root decoction is also utilized.

3. *Boerhavia diffusa* L. (Varn. Bishkapra)

Family: Nyctaginaceae

Root decoction is taken daily for one month to expel kidney stones.

4. *Bryophyllum pinnatum* Lamk. (Varn. Patharchata)

Family: Crassulaceae

Fresh leaf juice along with 2-3 kalimirch (*Piper nigrum* L.) powder is taken twice a day for 15 days to expel stones.

5. *Crataeva nurvala* Buch-Hain. (Varn. Barna)

Family: Capparaceae

Bark decoction twice a daily for seven days is given in urinary tract infection (UTI) and for removal of stones from urinary tract.

6. *Cynodon dactylon* L. (Varn. Doobghas)

Family: Poaceae

Root decoction is given with honey or misri (clarified and crystallized sugar) twice daily for 3 weeks to cure urolithiasis.

7. *Daucus carota* L. (Varn. Gajar)

Family: Apiaceae

One glass gajar juice is taken regularly for a fortnight to remove stones from urinary bladder and kidney.

8. *Equisetum debile* Roxb. (Varn. Jodetodeki ghas)

Family: Equisetaceae

Whole plant juice is given along with 1 g *Piper nigrum* L. twice a day for 7 days for the removal of both urinary tract as well as kidney stones.

9. *Gomphrena celosioides* Martius (Varn. Kasia)

Family: Amaranthaceae

Whole plant juice along with 4 *Piper nigrum* L. and lemon juice twice a day is taken for 10 days to cure urolithiasis.

10. *Musa balbisiona* Colla. (Varn. Kela)

Family: Musaceae

Decoction of *Musa* roots and gulli (axis of maize cob, *Zea mays* L.) is given twice a daily for 7 days in complaints of kidney and urinary tract stones and severe pain.

11. *Ricinus communis* L. (Varn. Arandi)

Family: Euphorbiaceae

Root decoction along with half a gram sunthi (dried and powdered rhizomes of *Zingiber officinale* Rose.), one pinch of *heeng* (*Ferula asfoetida* L.) and common salt is taken twice a daily for 7 days to treat kidney stones.

12. *Solanum surattense* Burn. (Varn. Neeli Kateli)

Family: Solanaceae

Root powder along with *Bari Kateli* (*Solanum indicum* L.) root powder is given with curd daily for two weeks to expel kidney stones.

13. *Trianthema portulacastrum* L. (Varn. Lalsubuni)

Family: Ficoidae

Fresh leaf juice is given twice a day for a week in case of stone problems of both urinary tracts as well as kidneys.

14. *Tribulus terrestris* L. (Varn. Gokhuru)

Family: Zygophyllaceae

Fruits and root decoction thrice a day is taken regularly for a fortnight to help in expelling kidney stones.

15. *Zea mays* L. (Varn. Makka)

Family: Poaceae

Decoction of styles obtained from female inflorescence or immature cobs are given twice daily for 7 days to expel stones from kidney.

Conclusion

Bijnor is very rich in floristic wealth. The assessment and conservation of biodiversity is of paramount importance. There is a need to set up of a co-ordinated programme for biodiversity measurement and monitoring, at least in major forest types. It is also desirable to keep the regulatory regime and concerted efforts need to be made to attract peoples participation in big way. People should be educated and trained to take benefits from conservation of the medicinal flora in the region.

References

Anonymous 1948-1976. The wealth of India, Raw Materials, CSIR, New Delhi.

Chopra R N, Chopra I C and Verma B S 1956. Supplement to glossary of Indian medicinal plants. CSIR, New Delhi.

Jain S K 1991. Dictionary of Indian Folk Medicine and Ethno-botany. Deep Publications, New Delhi.

Kirtikar K R and Basu B D 1945. Indian medicinal plants. 1-4 Volumes (II ed.) Bishan Singh, Mahendra Pal Singh, Dehradun.

Singh D 1998. Studies on the flora of district Etawah. Ph.D. thesis submitted in CSJM Univ. Kanpur.

Zaidi S M A, Jamil S S, Singh R N and Asif M 2006. Clinical evaluation of herbo-mineral Unani formulation in urolithiasis. **In**: Int Conf. Ethnopharmacol Alternative Med., V[th] Annual Conf. Nat. Soc. Ethnopharmacol Abstr., Amla Ayurveda Hospital and Research Centre. Thrissur, Kerala, India.

Medicinal Plants: Aspects and Prospects (2014) ***Pages* 241–249**
***Editors:* Mukesh Kumar, Anjali Khare and C.P. Shukla**
ISBN: 978-81-7622-309-6
***Published by:* BIOTECH BOOKS, NEW DELHI**

Chapter 17

Pharmacognositc Evaluation of *Curcuma longa* Linn.

***A.K. Shukla*[1], *Maneesh Kumar*[2] *and D.B. Singh*[3]**

[1]*Department of Botany, Kripalu Mahila Mahavidyalaya, Kunda – 230 204, Pratapgarh*

[2]*Department of Chemistry, K.N. Govt. P.G. College, Gyanpur – 221 304, S.R.N. Bhadohi*

[3]*Department of Botany, D.D.U. P.G. College, Saidabad – 221 508, Allahabad*

ABSTRACT

Medicinal plants are very precious and considers as the wealth of the nation. All the plants which produce and store certain secondary metabolites of therapeutic use are placed under the category of medicinal and aromatic plants. *Curcuma longa* Linn. (Family-Zingerberaceae), commonly known as 'Haldi' is widely depicted in various smahitas and texts. It is used in Indian system of medicines since the time immemorial. The plants grow well in wild habitat and are also cultivated throughout India. Medicinally, it is regarded as cooling, aromatic astringent and is used to promote digestion. A paste of the rhizome is externally useful for cuts, wounds, itching and also in sprains. A detailed pharmacognostic analysis of its rhizome show moisture content (83.22 per cent), total ash (9.97 per cent), acid insoluble ash (1.80 per cent), alcohol soluble extractives (11.45 per cent), water soluble extractives (20.41 per cent), ether soluble extractives (6.35 per cent), sugar (3.14 per cent) and starch (18.42 per cent). It also contains 1.2 per cent volatile oil.

Keywords: Curcuma longa, Indian System of Medicine, Volatile oil.

Introduction

Curcuma longa Linn. is being used in Indian system of medicine sine times immemorial as carminative and stomachic. The plant is found wild in Bengal, Tamil-

Nadu, Konkan and many other parts of India and often cultivated in gardens in rotation with vegetable crops. It is known as Haldi, in Hindi and Bengali, Rajaninisa in Sanskrit and Indian Saffron/Turmeric in English and Zaod-Chobah, dar-zard in Persian. Fresh rhizomes emits the aroma of the green mango, hence various names has been given to this species. Rhizome is expectorant, astringent and useful in diarrhoea. Fresh root is used in perfumery, and as an ingredient in chutneys like ginger. The rhizome is mostly used in pickles and is considered to be stomachic and carminative, used in chronic rheumatism, bruises, sprains and for improving the quality of the blood (Chopra *et al.*, 1956, 1969; Kirtikar and Basu, 1953; Nadkarni, 1954).

A fair amount of chemical extraction has been done on this plant. The rhizomes yield 1.1 per cent essential oil (containing d-a-pinene 1.8 per cent, Ocimene 47.2 per cent, Safrole 9.3 per cent resin, sugar, gum, starch, albiminodis, crude fibre and organic acids. The effect of *Curcuma longa* on the hypercholesterolemia of rabbits show that the ethereal extract of *Curcuma longa* rhizomes has a pronounced blood cholesterol lowering action in these animals. Although the drug is fairly important and has good economics, but only a little pharmacognostical work have been carried out so far in details. The present work documents the detailed pharmacognostical information, which can be utilized by the industries for the authentication and further use of this drug for mankind (Anonymous, 1950).

Materials and Methods

The plant material had been procured from farmer's field near Sant Ravidas Nagar (Bhadohi). The rhizomes were preserved in 70 per cent ethyl alcohol for histological studies. After microtomy routine procedure of staining (with safranin and fast green) and mounting was followed and photographs were taken with Nikon F70X camera. Physico-chemical and phytochemical studies like total ash, acid insoluble ash, tannins, sugars and starch were done from the shade dried powdered material as per the recommended procedures (Anonymous, 1984; Peach and Tracy, 1955)

Taxonomic Description of the Plant

Rhizomes large (4-53×4cm) and light yellow inside; sessile tubers thick (5-10×2-3cm), cylindric or ellipsoid, branched, horizontal; roots fleshy and without root tubers; pseudostem 30-35cm long; leaves 4-6; petiole 5-10cm long; lamina of lower leaves much smaller(18×8cm) than the upper (45-60 ×14-15cm); shape of lamina varies from oblong lanceolate to acuminate; base equal, pinnately veined, lower surface puberulous, upper glabrous, tip hairy; inflorescence lateral or central; peduncle 20-22 cm long, covered by 5-6 sheaths; spike 12-18cm long; 4-5 coma bracts in each bract longer than the bracts, fused only at the base, light violet; fertile bracts 15-18cm long; calyx truncate, 1 cm long, deeply cleft on one side, 3-lobes at the tip minutely pubescent; corolla tube funnel-shaped, 3 cm long, pale yellow, minutely pubescent; lobes unequal, dorsal lobe larger, 1.5 ×1.2cm, hooded at the apex; lateral c. 1.4 ×1 cm long, tip rounded, glabrous, labellum somewhat elliptic, c. 1.8 × 1.5cm, 3-lobed, middle emarginated, recurved, pale yellow with a median dark band, glabrous; lateral staminodes c. 1.5 ×

0.9cm apex slightly incurved, glabrous, pale yellow, stamen white, thecae parallel, slightly convergent, glabrous, epigynous glands two, linear, 6 mm long, tip acute; ovary trigonous, 3 mm long, tricarpellary, syncarpous with many ovules, densely hairy; style long, filiform; stigma closely apprised within the anther lobes; fruit setting not seen.

Macroscopic Characters of the Rhizome

Rhizome laterally flattened, longitudinally wrinkle, 2 to 6 cm long, 0.5 to 2 cm in diameter, branched, remnant of scaly leaves arrange circularly giving the appearance of growth rings; cut pieces 1.5 to 3.5 cm in diameter, circular, punctuate scars on the surface, branching sympodial, horizontal; roots long, unbranched, tapering, thread like, yellowish brown; rhizome buff coloured with short and smooth fracture; raw mango like smell and sweet in taste (Figure 17.1).

Microscopic Characters of the Rhizome

Transverse section of rhizome is circular in outline. Epidermal cells are rectangular-oval in shape covered with thick cuticle, long unicellular trachoma's present. Followed to these, 4 to 7 layered suberized cork cells, interrupted by lysigenous oil glands are present. A wide cortex having irregularly scattered vascular bundles are present. Each vascular bundle is enclosed within a prominent fibrous sheath, inner limit of cortex marked by endodermis, and pericycle followed by vascular bundles devoid of bundle sheath, arranged in a ring; schizogenous canals and abundant oil cells with suberized walls are also found in central region. Most of the parenchymatous cells are filled with oval-ellipsoidal starch grains,sometimes polygonal in shape, 10 to 60µm, simple, hilum cirucular or a 2 5 rayed cleft, lamellae distinct and concentric. Vascular bundles in the central cylinder are similar to those in the cortex, scattered, closed, collateral, surrounded by thick walled bundle sheath. Secondary wall thickening reticulate, fibers thin walled with lignified narrow central lumen (Figure 17.1).

In transverse longitudinal sections, cortex cells irregular in shape with variable size. Vessels elongated with spiral or annular thickenings and with no clear end-walls. Tracheids with bordered pits are observed. On maceration, vessels (737.863×14.665µm) with annular and spiral thickenings are observed. Tracheids with bordered pits measuring 432.635 × 13.290 mm are also clearly discernable.

Powder Study

Powder light yellow, sweet in taste with raw mango like odour. Shows fragments of storied cork, xylem vessels with reticulate thickenings, lignified xylem fibres, oil cells, patches of parencymatous cells, filled with oval-elliposidal starch grains, sometimes polygonal in shape, 10 to 60µm, simple, hilum circular or a 2 to 5 rayed eleft, lamellae distinct and concentric (Figure 17.1).

The behavior of the powdered drug with different chemical reagents has been shown in the Table 17.1.

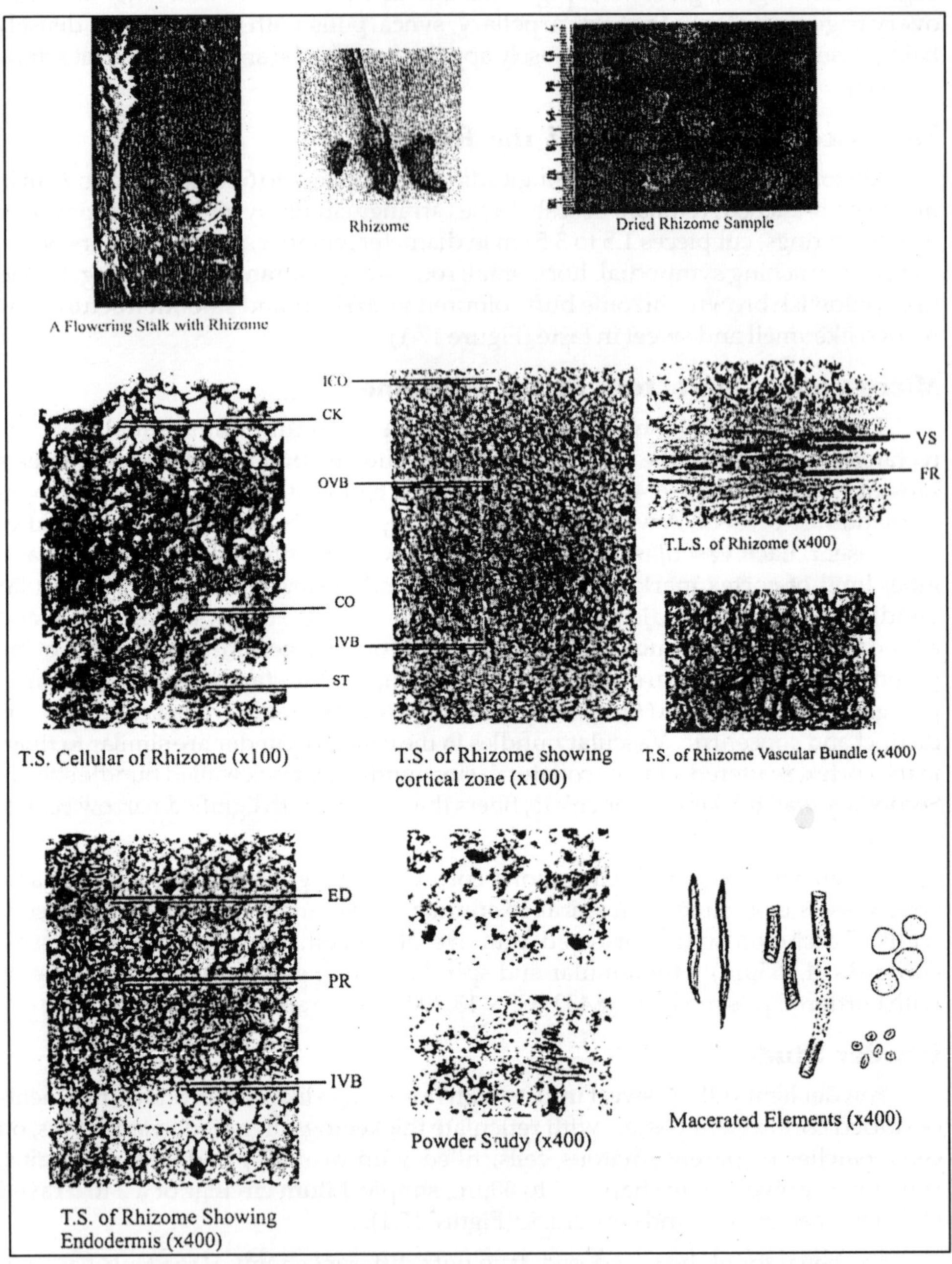

Figure 17.1: Macroscopic characters of *Curcuma longa* rhizomes

Table 17.1: Fluorescence analysis

Sl.No.	*Treatment*	*Day light*	*UV-254nm.*	*UV-366nm.*
1.	Powder (P) as such	Light Brown	Green	Fluorescence Light Green
2.	P+Nitro-celluose in amylacetate	Light Orange Brown	Dark Green	Fluorescence Brown
3.	P+N, NaOH in water	Light Brown with Yellow Tinge	Green with yellowish Tinge	Green
4.	P+1N, NaOH+Nitro-cellulose in acetate	Yellowish Green	Dark Green	Yellowish Green
5.	P+1NHCL+Nitro-celluiose in amylacetate	Orange	Dark Green	Light Green
6.	P+1N NaOH in Methanol	Light Brown with Orange Tinge	Light Green Purple Margins	Yellow
7.	P + 50 per cent KOH	Cream	Green	No Change
8.	P + NHCL	No change	No change	No change
9.	P+50 per cent H_2SO_4	Brownish Green Tinge	Bright Green	Green
10.	P+50 per cent HNO_3	Light Brown	No change	No change
11.	P+ Conc. HNO_3	Brownish Yellow	Greenish Yellow	Brown
12.	P+ Acetic acid	Yellow	Bluish Green	Yellow
13.	P+Conc. H_2SO_4	Brown	Bright Green	Bright green
14.	P+ Iodine water	Navy blue	Bluish with Green tinge	Greenish Blue

Physico-chemical Studies

Different physico-chemical values obtained are recorded in Figures 17.3 and 17.4.

On microscopical examination, rod shaped starch grains and fibres are observed in the rhizome. Similarly, number of curcumin containing cells, which are yellow in colour and also very low in the rhizome.

Physico chemical values *viz.* percentage of moisture, total ash, acid insoluble ash, alcohol and water-soluble extractives have also been observed. The total ash and acid insoluble ash, which are considered to be an important and useful parameter for detecting the presence of inorganic substance like silicate ion, were found 9.97 per cent and 1.80 per cent, respectively. Similarly, the alcohol and water-soluble extractives, which are indicators of the total solvent soluble components, are 11.45 per cent and 20.41 per cent, respectively. Likewise the essential oil, which is an important parameter for identification and authentication has to be 1.2 per cent.

Successive Soxhlet extraction from non-polar to polar solvents *viz.* hexane, chloroform, acetone, alcohol and water were also carried out. It is interesting to note that *C. longa* rhizome possessed an exceptionally low amount of acetone extracts *i.e.*

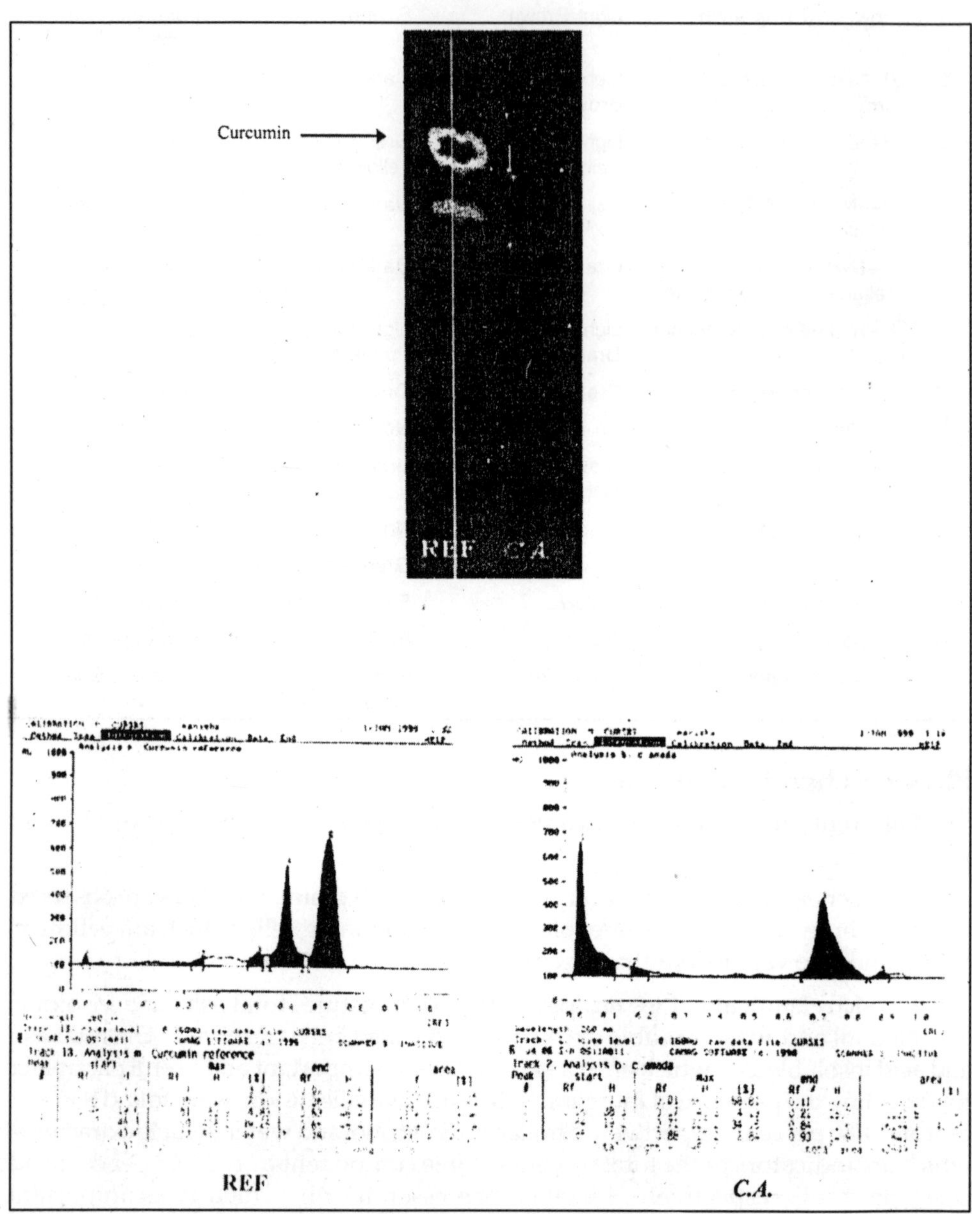

Figure 17.2: HPTLC profile of *Curcuma longa* rhizomes

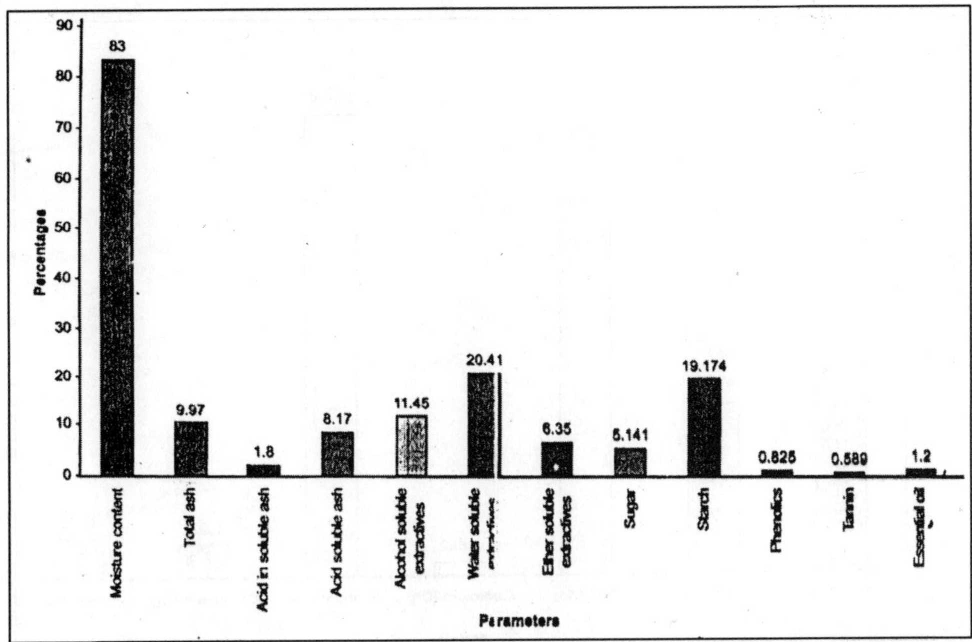

Figure 17.3: Physico-chemical studies in the rhizome of *Curcuma longa*

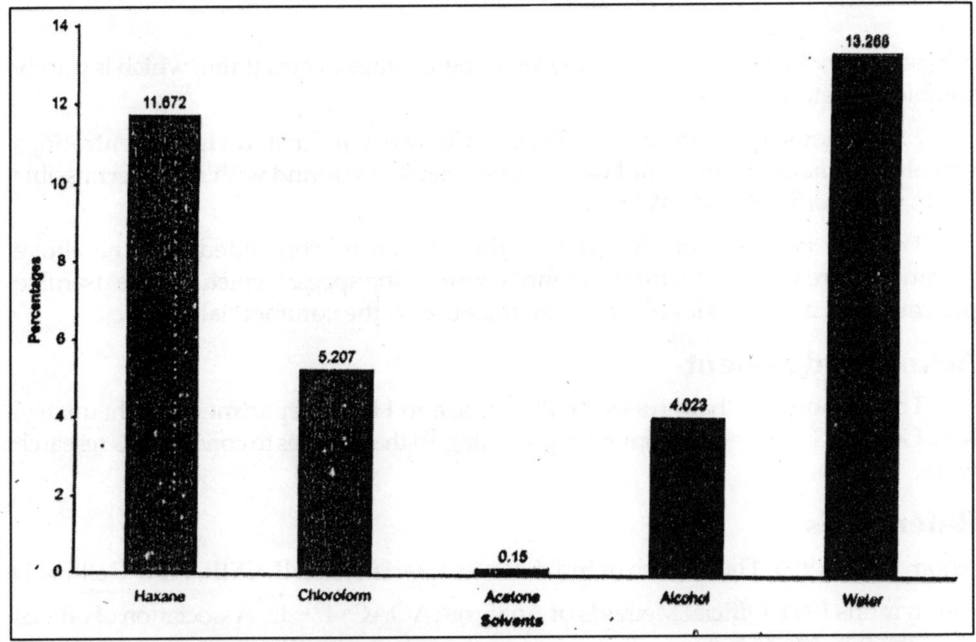

Figure 17.4: Successive soxhlet studied in the rhizome of *Curcuma longa*

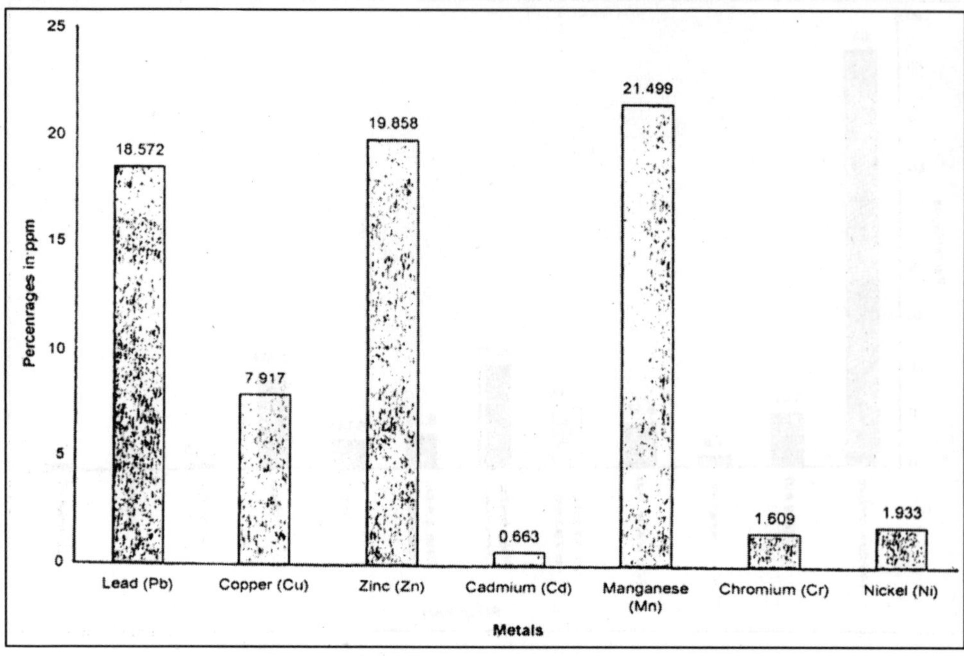

Figure 17.5: Rate of heavy metals concentration

0.15 per cent which may be due to the lesser percentage of curcumin, which is purely soluble in acetone.

The various heavy metals *viz.* Pb, Cd, Co, Mn, Cu, Zn and Hg concentrations was also estimated in the samples and all the metals are found within the permissible limits as prescribed by the WHO.

Thus on the basis of aforesaid studies, it can be concluded that the above parameters are very useful for the identification of the species which may be useful to pharmaceutical industries for the authentication of the commercial samples.

Acknowledgement

The authors are thankful to the Principle and Head Department of Chemistry, K.N. Govt. P.G. College, Gyanpur, for providing all the facilities to conduct this research work.

References

Anonymous 1950. The Wealth of India: Raw materials, vol. II, CSIR, New Delhi 401.

Anonymous 1984. Official Methods of Analysis (AOAC) 4[th]Edn. Association of official chemists Inc. U.S.A.

Chopra R N, Nayar S L and Chopra I C 1956. Glossary of Indian Medicinal Plants. PID, CSIR, New Delhi.

Chopra R N, Chopra I C and Verma B S 1969. Supplement to the Glossary of Indian Medicinal Plants. PID, CSIR, New Delhi.

Kirtikar K R and Basu B D 1933. Indian Medicinal Plants. LM Basu Pub Allahabad, **4**: 2422.

Nadkarni A K and Nadkarni K M 1954. *Materia Medica.* Dhootpapeshwar Prakasan Ltd, Mumbai, 412-13.

Josansen D A 1940. *Plant Microtechnique.* McGraw Hill Book Co, New York and London.

Peach K and Tracy M V 1955. *Modem Methods of Plant Analysis*, Springer, Heidelberg, 3rd and 4th Vol.

Medicinal Plants: Aspects and Prospects (2014) ***Pages* 250–278**
***Editors:* Mukesh Kumar, Anjali Khare and C.P. Shukla**
ISBN: 978-81-7622-309-6
***Published by:* BIOTECH BOOKS, NEW DELHI**

Chapter 18

Quality Control of Therapeutics: An Approach to Development and Evaluation of Herbals

Sandeep Tiwari[1], Ajay Shukla[2] and J.P. Singh[3]

[1]CSM Group of Institutions, Institute of Pharmacy, Iradatganj – 212 306, Allahabad
[2]Department of Botany, Kripalu Mahila Mahavidyalay, Kunda – 230 204, Pratapgarh
[3]Swami Kalyan Dev Gardens and Training Institute, Nara – 251 001, Muzaffarnagar

ABSTRACT

Potentials of herbals range from parts of plants, through simple extracts to isolate active constituents. Plants and plant derived products as a source of food are being utilized from time immemorial. There has been a resurgence of interest on such products as source of medicine in the last few decades as the herbal products has occupied a major part in curing different human ailments, but what about the quality of those products? Certain of these Herbals have been known and are being used by man for many centuries, while others are still being isolated and evaluated. Dealing with Herbals as functional foods is not always easy to define in the context of their quality and purity. Quality Herbals have become one of the totems in this era of phytotherapy. The Herbals have got enormous commercial potentials throughout the globe. Along with the increased interest in this field there has been an explosion in the amount of literature on the subject and quality control so much so the evaluation of their safety and efficacy is of utmost essential in this respect world over. In the herbal boom world wide it is estimated that high quality phytoceuticals will provide safe and effective food as well as phyto-medication. Development and evaluation of Herbals with an approach to their safety and efficacy will without question become a significant landmark.A milestone of achievements, has been set in the development of phytoceuitcals as food as well as medicine and its wide application of human health. Pharmacological evaluation of various natural products used in different food systems as well as alternative systems of medicine is becoming increasingly potential, so the approach for the development of safety, efficacy and quality control of herbal medicine are getting momentum which are being made through the regulation of different countries. In this respect, exploration of the Herbals through international coordination with the development of operational methodologies consisting of wide array of standard operating procedures on screening, evaluation, quality control and standardization based on the efficacy of phytomedicine is the need of the day for development of medicine as well as food supplements from natural resources.

Keywords: Herbals, Food supplements, Toxicity, Nutraceuticals.

Introduction

Plants are considered to be medicinal if they possess pharmacological activities of possible therapeutic use. These activities are often known as a result of millennia of trial and error, but they have to be carefully investigated if we wish to develop new drugs that meet the criteria of modern treatment. The goals of research in this field are especially:

- ✰ The identification of the active principles of medicinal plants, and investigation of the extracts in order to ensure that they are safe, effective, and of constant activity.
- ✰ The isolation of these active principles and the determination of their structure, in order that they may be synthesized, structurally modified, or simply extracted more efficiently.

The methodology of research into medicinal plants must be rigorous. Often, simple technical errors undermine the value of research on natural products. There are many who believe that a little rapid research is sufficient to confirm the reputation of a plant, and who then attempt to proceed from there towards lucrative industrial production. The review is intended to present elements of both the methods and theoretical backgrounds of different aspects on quality safety and efficacy of Herbals in every aspects of their use as functional foods and drugs. More emphasis has been made on different aspects of quality control and standardization by virtue of new techniques. Scientists and interested lay persons can no longer disregard the scientific and clinical evidences supporting herbal utility simply by claiming ignorance of data available so forth in this field. The examples and evidences represented in this review strongly supports the quality parameters as well as different safety profiles, which will facilitate to improve the concepts of the Herbals in one hand and will increase use of desirable standardized plant parts in the other. Various efforts have been made globally to highlight various developments taking place in the field of Herbals with the steady growth on the subject, the significant modern approaches that will be described through the review include all aspects of their evaluation with chromatographic parameters based on marker compound analysis, WHO guidelines, GAP regulations and different other official monographs.

Herbals so much so the herbal medicines are prepared from a variety of plant materials *viz.* leaves, stems, roots, bark and so on. They usually contain many biologically active ingredients and are used primarily for treating mild or chronic ailments. Herbs can be prepared at home in many ways, using either fresh or dried ingredients. Herbal teas and infusions can be steeped to varying strengths. Roots, bark or other plant parts can be boiled into strong solutions called decoctions. Honey or sugar can be added to infusions and decoctions to make syrups. Herbal remedies can also be purchased in the form of pills, capsules or powders, or in more concentrated liquid forms called extracts and tinctures. They can be applied topically in creams or ointments, soaked into cloths and used as compresses, or applied directly to the skin as poultices. Across the spectrum of alternative medicine, the use of herbs is varied: Naturopathic medicine, traditional Chinese medicine, and Ayurvedic medicine all differ in how diseases are diagnosed and which herbal remedies are prescribed (Mukherjee, 2003). Out of these, the Chinese herbal medicine or the Traditional Chinese Medicine (TCM) has a potential usage similar to the Indian systems of medicine.

Herbals versus Conventional Food and Medicine

While many conventional drugs or their precursors are derived from plants, there is a fundamental difference between administering a pure chemical and the same chemical in a plant matrix. It is this issue of the advantage of chemical complexity, which is both rejected by orthodoxy as having no basis in fact and avoided by most researchers as introducing too many variables for comfortable research. Herein lays the fundamental difference between the phytotherapist, who prefers not just to prescribe chemically complex remedies but often to administer them in complex formulations, and the conventional physician who would rather prescribe a single agent.

Synergy is an important concept in herbal pharmacology. In the context of chemical complexity, it applies if the action of a chemical mixture is greater than the arithmetical sum of the actions of the mixture's components: the whole is greater than the sum of the individual parts. A well-known example of synergy is exploited in the use of insecticidal pyrethrins. A synergist known as piperonylbutoxide, which has little insecticidal activity of its own, interfaces with the insect's ability to break down the pyrethrins, thereby substantially increasing their toxicity. This example emphasizes what is probably an important mechanism behind the synergy observed for medicinal plant components: increased or prolonged levels of key components at the active site. In other words, components of plants which are not active themselves can act to improve the stability, solubility, bioavailability or half-life of the active components (Mukherjee 2001). Hence a particular chemical in pure form might have only a fraction of the pharmacological activity that it has in its plant matrix. This important example of synergy therefore has a pharmacokinetic basis. Synergy can also have a pharmacodynamic basis. One example is the antibacterial activity of major components of lemongrass essential oil. While geranial and neral individually elicit antibacterial action, the third main component myrcene did not show any activity. Sennoside A and sennoside C from senna have similar laxative activities in mice, however, a mixture of these compounds in the ratio 7:3 (which somewhat reflects the relative levels found in senna leaf) has almost double the laxative activity (Kisa *et al.*, 1981).

Chemical complexity leading to enhanced solubility or bioavailability of key components has been the theme of a number of scientific studies. Lots of evidences have proved these basic issues (Eder and Mehnert 1998). The isoflavone glycoside daidzin given in crude extract of *Pueraria lobata* achieves much greater concentrations in plasma than equivalent doses of pure daidzin (Keung *et al.*, 1996). Ascorbic acid in a citrus extract was more bioavailable than ascorbic acid alone (Virison and Bose 1988). Co-administration of procyanidins from *Hypericum perforatum* significantly increased the *in vivo* antidepressant effects of hypericin and pseudohypericin. This effect was attributed to the observed enhanced solubility of hypericin and pseudohypericin in the presence of procyanidins and indicates that pure hypericin and pseudohypericin have considerbaly less antidepressant activity than their equivalent amounts (Butterweck *et al.*, 1998).

The Extent of Plant Kingdom in Medicine and Food

The relationship between man and plants has been very close throughout the development of human cultures. Throughout the history, Botany and Medicine were, for all practical purposes, synonymous fields of knowledge, and the shaman, or witch doctor usually an accomplished botanist represents probably the oldest professional man in the evolution of human culture. At no time in the development of mankind, however, has there been more rapid and more deeply meaningful progress in our understanding of plants and their chemical constituents than during the past quarter century. And this is curious, especially in view of the somewhat earlier deprecation in pharmaceutical chemistry of any emphasis on plants. The gradual sophistication of phytochemistry in the last half of the nineteenth century and the exaggeration of hope for specific remedies from vegetal sources for any and all ills set up a counter-current, a tendency to disparage any data concerning the potential value of physiologically active natural products.

Eighty per cent of the world's population relies on traditional medicine to maintain its health (Weragoda 1980). In recognition of the political, economic, and social barriers slowing the delivery of modern biomedical health care to most of the world's population, the World Health Organization (WHO) has embarked upon an ambitious program to evaluate herbal medicines (WHO 1978). This project ultimately hopes to circumvent the problems of developing and distributing appropriate pharmaceuticals by encouraging the cultivation and use of locally adapted medicinal plants with proven empirical value. Ethnobotany can contribute to this strategy in two ways. First, Eco-ethnological studies may provide models for profitable and environmentally sound multiple use land management programs. Second, ethnobotanists can invoke the considerable economic potential of as yet undiscovered or undeveloped natural products,of an estimated 75,000 edible plants, for example, only 2500 have ever been eaten with regularity, a mere 150 have entered world commerce, and a scant 20, mostly domesticated grasses, stand between human society and starvation. To diversity of this resource base is goal of ethno- botany, and numerous promising crops that can be exploited in ecologically sound ways have already been identified(Balick 1985). Possibly, the greatest economic potential of ethnobotany lies in the area of folk medicine. Annually worldwide sales of plant-derived pharmaceuticals currently total over $20 billion, and a great many of these drugs were first discovered by traditional healers in folk contexts (Farnsworth 1982). Ethnobotany routs for drug development (Figure 18.1). The gifts of the shaman and the sorcerer, the herbalist and the witch, include such critical drugs as pilocarpine, digitoxin, vincristine, emetine, physostigmine, atropine, morphine, and reserpine (Farnsworth 1988). The forests of tropical America have yielded scopolamine, cocaine, quinine and d-tubocurarine.

The startlingly effective drugs that have come from this decade or two of discovery are scattered throughout the plant kingdom. They range from muscle relaxants from South American arrow poisons, antibiotics from moulds, actinomycetes, bacteria, lichens and other plants; rutin from a number of species; cortisone precursors from sapogenins of several plants, especially from *Strophanthus* and *Dioscorea*; hypertensive agents from Veratrum; cytotoxic principles from *Podophyllum, Vinca* and other sources; khellin from *Ammivisnaga*; reserpine from *Rauwolfia*; hesperidin from the *Citrus* group; bishydroxycoumarin from *Melilotus* and sundry others - not to mention the numerous psychoactive structures of potential value in experimental psychiatry, some new, some old, from many cryptogamic and phanerogamic sources. Not only have new drugs from vegetal sources been discovered, but new methods of testing and refined techniques have led to the finding of novel uses for older drugs.

Despite the enormous availability of medicines and, above all, of pharmaceutical specialties, plants have a potential place in current therapy and there are renewed interest in using plants in therapy such as is the case of *Artemisia* (Klayman 1985), a source of anti malarial. Traditional medicine depends on a number of plants that are currently used in scientific medicine although they have not yet been improved upon. Such is the case of *Digitalis purpurea* L. Many other drugs exist to which therapeutic effects have been attributed. As is well known, synthetic chemistry has until now had little success in obtaining drugs effective in the treatment of various viral diseases; even though immunotherapy has achieved great success, we still do not have vaccines for all viral diseases. It is possible that plants may be useful to treat these diseases. An example from Ecuador is Lanigua (*Margyricarpus setosus* Ruiz and Pavon), the roots of which, in infusion, are used in the symptomatic treatment of measles. Other examples are nachag (*Bidens humilis*), the flowers of branches of which, in infusion, are used in the treatment of infectious hepatitis, or the latex of several species of Euphorbiaceae, especially of the genera *Croton* and *Euphorbia*, which in topical form are used in the treatment of common warts. Furthermore, numerous plants are known for certain antineoplastic effects (Cassady and Douros 1980), modifiers of fertility (Moreno and Schwartzman 1975), and other effects (Perdue and Hartwell 1969).

The controlled studies are needed, for example, by the double blind system, to confirm the therapeutic effect of these plants of traditional medicine in therapy. Nonetheless, in folk medicine these plants are employed with apparently favourable results, and above all without causing detectable unfavorable side effects. An inventory of medicinal plants compiled by the World Health Organization (WHO 1978) and

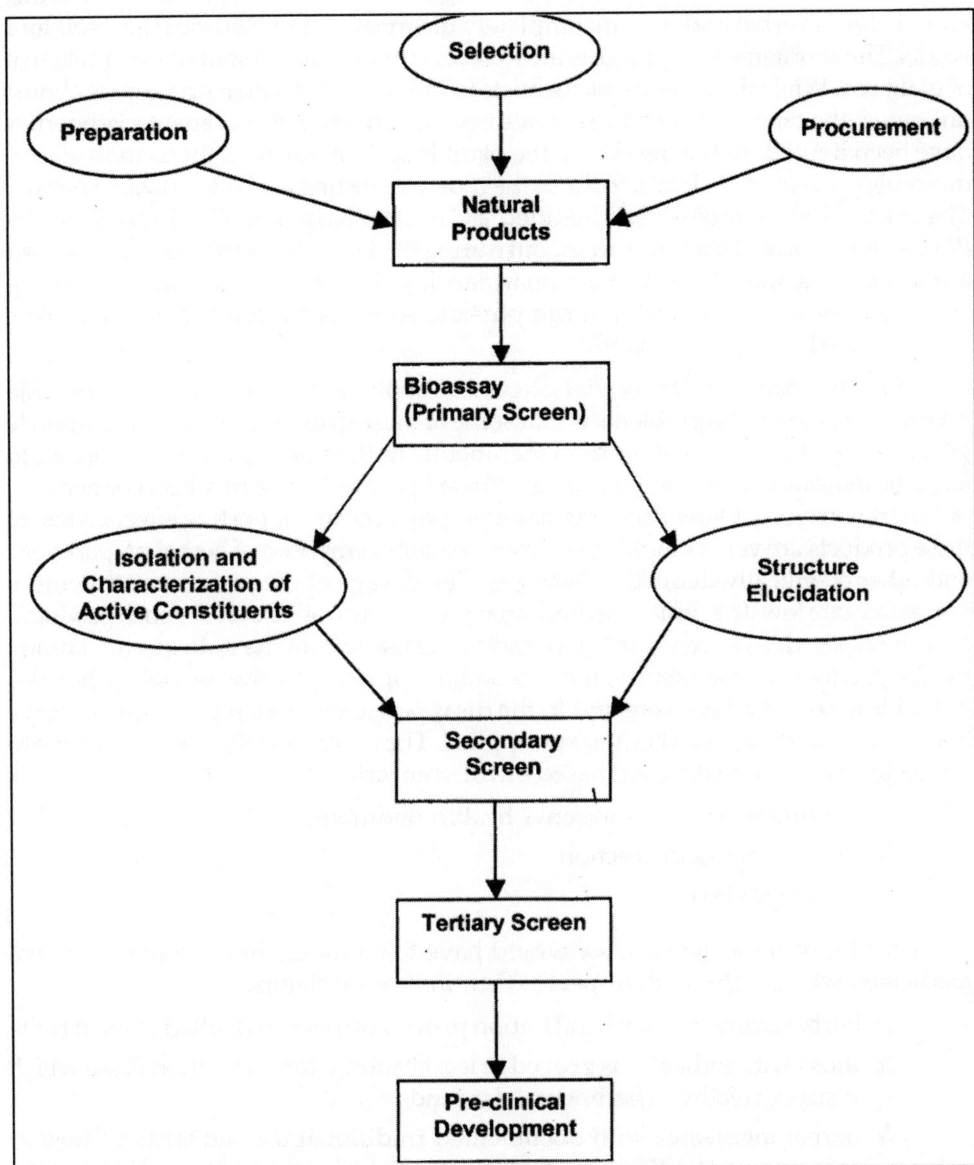

Figure 18.1: Various strategies for the discovery of drugs from natural resources

encompassing only ninety member countries gave the large figure of 20,000 species, of which only 250 were of wide-spread use or had been analyzed to identify their main active chemical compound(s). That sample, even though a partial one, reveals the enormous empirical traditional knowledge about medicinal plants. Most of this knowledge is verbal and only incompletely incorporated in historical and folklore works. The aboriginal knowledge is the fruit of centuries and, in some cases, millennia of plant use. While the capacity of chemists to modify a molecular structure is almost unlimited, the capacity to invent or create new structures with therapeutic properties have been limited. In the meantime the plant kingdom offers us thousands of new molecules (Evans *et al.*, 1982), fruits of the most interesting and often bizarre nature. The study of those molecules identified as "active compounds" is indispensable. Phytochemical investigations carried out during the 1970s and 1980s have discovered a number of alkaloids and other pharmacologically active substances that are currently being studied and that can possibly serve as models for new synthetic compounds (Barz and Ellis 1980).

There has been a controversial discussion about herbal medicines sold outside pharmacies. One of the problems is that some of these drugs can only be sold outside pharmacies if they claim other than therapeutic indications. This legislation led to fantastic indication claims as for example "blood purifier', "to fortify heart or nerves", or "heart nutrition". Other products claim prophylactic or supporting effects. Most of these products are very complex fixed combinations composed of herbal preparations and other chemically defined substances. The dosage of the active constituents is normally quite low. It is difficult to find scientific evidence of efficacy for such products. In some cases, this is even true for a' traditional use within the indications claimed for the product. Consequently, the evaluation approach by the official authorities would lead to a negative vote and to the disappearance of such products from the market, even if the weakest criteria be applied. The corresponding products have to be labeled as traditionally used based on different criteria as follows:

- for amelioration of subjective health conditions
- to support organ function
- for prophylaxis

Because of these aspects, we would have to consider three groups of herbal medicines, which differ with respect to their indication claims:

- herbal medicines with indication proved by new controlled clinical trails
- those with indications proved at least by long- term traditional use which is supported by experimental data and
- herbal medicines with documented traditional use but without further assessment of efficacy requiring a special labeling on the package of the finished drug

Phytoceuticals from Plants

Several factors have contributed to the revival of interest in plant-derived products as follows:

- ☆ There is an undisputed clinical efficacy of several natural product anticancer drugs. The early discovery of vincristine and vinblastine from *Vinca rosea* (the Madagasar periwinkle) was followed by other agents including the aryl lignametoposide derived from *Podophyllum hexandrum* (mayapple), and the taxoids' from *Taxus brevifolia* (Pacific yew) and *Taxus baccata* (European yew).
- ☆ Compounds with less direct therapeutic potential may offer new molecular templates for the design of more effective drugs *e.g.* the development of atracurium and related muscle relaxants from the alkaloids of curare, the South American dart poison obtained from *Chlorodendron tomentosum* (Waigh 1988).
- ☆ Natural products can offer an alternative to established therapy because they act at a different stage in the disease', and be useful in combination therapy. The search for synthetic molecules active against human immunodeficiency virus (HIV) has resulted largely in reverse transcriptase inhibitors, but investigations into plant extracts have produced a wide range of chemical compounds with various modes of action that result in viral non-proliferation (Kinghorn and Balandrin 1993).
- ☆ Plants have proved invaluable as inexpensive sources of "feedstock" molecules that can be readily transformed into drugs. Thus, development of the steroid-based oral contraceptives would have been virtually impossible without plentiful supplies of compounds from the processing of the steroidal components of plants such as yams (*Dioscorea bulbifera*) and sisal (*Agavea americana*).

The renaissance of interest in plant products has also been stimulated by the use of plant extracts in chronic conditions for which conventional medicine is perceived to offer very little comfort, *viz.*, Chinese herbs for severe atopic eczema. Other conditions are hard to define clinically, but "herbal" medicines claimed to alleviate them enjoy huge sales *viz.*, ginseng as a tonic and for fatigue. Studies of some of these plants have yielded compounds with unique activity *viz.* ginkgolides, specific platelet activating factor antagonists that were obtained from the Chinese tree *Ginkgo biloba* (Kleijnen and Knipschild 1992). A standardized extract of ginkgo leaves is one of the most frequently prescribed medicines in Germany and is taken to alleviate cerebral ischemia (Braquet *et al.*, 1991). Amid all these developments it is sobering to realize that less than 10 per cent of the estimated 250,000 flowering plant species in the world have been examined scientifically for their potential in medicine more than 60,000 species of higher plants and this accelerating loss of species due to destruction of habitat, as well as loss of knowledge of traditional uses of plants with the attrition of indigenous cultures, strikes a note of urgency into the quest for the new compounds of therapeutic interest.

How should we select plants for serious study? One approach is to consider why it is that plants produce biologically active secondary metabolites, often with highly complex chemical structures. A common view is that such compounds are mainly concerned with survival (Waterman 1992). This feature can have a positive dimension such as the.attraction of pollinators or a more negative aspect such as a feeding deterrent, general toxicity to predators or prevention of growth of competing organisms in the immediate environment. Although chemical protection against the common classes of plant pathogens (viruses and fungi) and insect predators has implications for development of drugs and pesticides, it is difficult to explain the vast array of activity that other secondary metabolites exert on various animal species. Consequently, the search for new drugs has had to rely on other methods, the two most important being random screening of plants for chemicals or activity and ethno pharmacological investigations. By the middle of the 20th century, certain groups of chemical compounds such as alkaloids were known to be responsible for the biological activity observed in drug-producing plants. This observation led to the screening of other plants for similar compounds that could be detected by simple chemical tests. Few major advances ensued, but this approach furnished many novel structures whose potential as drugs is still not exhausted. The detailed chemical investigation of plants related botanically to those that have already yielded interesting compounds is a logical approach. A good example is the introduction of species of *Valeriana* other than the traditional *V. officinalis*, commonly known as valerian, as sources of the tranquilizing valepotriates (Houghton 1988). Standardized extracts have been widely used in mainland Europe as an alternative to drugs such as benzodiazepines.

The random screening of plants for food and medicine by our prehistoric ancestors is probably the basis of the botanical pharmacopoeia that exists in virtually all cultures. Some early tests for pharmacological activities were done with animal models, but numerous methods have not been introduced for testing of biological activities. The bioassays were made possible by the introduction of molecular biological techniques, with the development of receptor-ligand binding assays. Over twenty-five of these receptors are now used routinely by the pharmaceutical industries. This approach is also used by some large government programmes *viz.*, the National Institute of Health screen for anti-HIV agents in this way. Development of new test systems may reveal move towards activities and compounds that could explain the traditional use of a plant previously thought to be inactive. Thus the general immuno-stimulant effect of plants such as *Echinacea* species, used as a tonic and to prevent infections, was shown only with the advent of tests that could detect an increase in immune response (Bauer 1988). Ethno pharmacology is the scientific study of plants used by a cultural group for medicinal purposes. As a name, discipline ethno-pharmacology is new, but its principles have long been followed in the interchange of information between different cultures *e.g.*, the introduction of drugs such as Senna into western Europe after contact with the Islamic Middle East at the time of the crusades. During the twentieth century the investigation of plants used for generations has resulted in some important therapeutic advances *e.g.*, the introduction of the first tranquilizer, reserpine, from the Indian plant *Rauwolfia serpentina*, sold in Indian markets as a treatment for insanity. Moreover, new activities and chemical substances are still being reported for plants with a long history of medicinal use, the beneficial cardiovascular effects of garlic being one example (Warshafsky *et al.*, 1993).

Less well-known are some new therapeutically useful compounds that have been isolated from such plants. The Indian plant *Coleus forskholii* is a source of forskolin, which has unique activator properties on adenylatecyclase and is being developed as a treatment for glaucoma, and artemisinin, from *Artemisia annua*, is a promising antimalarial (Phillipson *et al.*, 1993). Artemisinin contains an unusual endoperoxide moiety in its structure and minor modifications have yielded several clinically useful drugs such as artemether (Anonymous 1994). The whole area of antiprotozoal drugs from plants is especially interesting since only in recent years have *in vitro* tests enabled bioassay-guided fractionation of plants extracts. Research programs in India and China are active in the investigation of their local medicinal plants, but little ethnopharmacological research has been carried out for the ethnic groups who live in areas of maximum biodiversity, where many plants of interest are likely to occur (Grindley 1993). A related area of interest is the scientific study of "nutraceuticals" – *i.e.*, plants used for food but also taken regularly as a preventive measure against disease. The role of compounds such as flavonoids and others with activities such as antioxidation and free radical scavenging is largely unknown but could be important in the prevention of chronic inflammatory diseases and cancer (Kinghorn 1993). What is the future for plant-based agents? There are many possibilities for research, but priority should be given to tropical infectious and chronic diseases for which current mediations have severe drawbacks, and to the scientific appraisal of plant-based remedies that might be safer, cheaper, and less toxic for self- medication than existing prescription medicines.

Quality Assurance for Efficacy of Herbals

Ideally the exact geographical source of an herbal material so much so the condition under which it has been grown, harvested, dried and stored should be known. The chemical treatment such as pesticides or fumigants used during harvesting or storing should be known. However, in many cases since the herbal raw materials are obtained from varied geographical and commercial sources, the above conditions may not be always known. For these reasons therefore, appropriate level of testing additional to the monograph criteria of any official book should be carefully assessed based on various factors, including the nature of the material, knowledge of its batch history and test results from previous batches. Quality assurance of herbal products may be ensured by proper control of the herbal ingredients and by means of good manufacturing practices. Some herbal products have many herbal ingredients with only small amounts of individual herbs being present. Chemical and chromatographic tests are useful for developing finished product specifications. Stability and shelf life of herbal products should be established by the manufacturers. There should be no differences in standards set for the quality of different dosage forms; such as tablets or capsules of herbal remedies as well as from those of other pharmaceutical preparations (Mukherjee 2002).

In UK, for the licensed herbal remedies the European Scientific Cooperative for Phytotherapy (ESCOP) monographs are important developments. In India the majority of the herbal remedies available are being marketed for a long time, in fact, for many products it may be before the Drugs and Cosmetic Acts 1948. The condition in other developing countries for the sale and production of herbal products are similar to UK. The review of herbal remedy in UK showed that majority of the licensed herbal remedies available are being marketed even before the existence of Medicines Act 1968 or any such drug licensing system. In September 1971, a registration exercise issued all medicinal products already on the market including herbal with a Product License of Right (PLR) and no scientific assessment was undertaken. For any order to be issued with PLRs for their products, pharmaceutical companies had simply to provide details of the products and evidence that the products had been marketed prior to 1971. This procedure was applied to all medicinal products including herbal remedies, and in total, some 39,000 PLRs were granted. It was obvious that at some future date all PLR products (including herbal remedies) would have to be assessed by the Licensing Authority (LA) for their quality, safety and efficacy in the same manner as those products which have applied for a product license after 1971. During the UK review of herbal remedies holding a PLR, the Licensing Authority agreed to accept bibliographic evidence of efficacy for herbal remedies which were indicated for minor, self-limiting conditions. No evidence would be required from new clinical trials provided the manufacturers agreed to label their products as a traditional herbal remedy for the symptomatic relief of any diseases and to include the statement "if symptoms persist consult your doctor". The licensing authority considered it inappropriate to relax the requirement for proof of efficacy for herbal remedies that are indicated for more serious conditions. Thus evidence was required from controlled clinical trials for herbal remedies which are indicated for conditions considered to be inappropriate for self-diagnosis and treatment.

Indicative Substances for Quality Assurance of Herbals

In order to rationalize the use of natural products in different forms more particularly the extracts in therapy as is being used now-a-days, a need-based and novel concept of biomarkers is getting momentum. Unfortunately, the development of this paradigm is not an easy task due to the complex interplay of many variables in this approach. The bio-markers in combination with other chemical entities via poorly understood mechanisms of synergy or antagonism are supposed responsible for the efficacy of the standardized extracts for any particular therapeutic area or disease (Mills and Bone 1999). The bioactive extract, being the composite mixture in terms of classes and groups of organic compounds in conjunction with many extraneous materials (both organic and inorganic), could be better understood in terms of this concept. However, defining of the biomarker has to be very specific and a lot of insight has to go into it before declaring any distinct molecule as a biomarker. Moreover, a good number of biomarkers are desirably required to achieve the elusive goal for the rational use of any given standardized extract.

Thin Layer Chromatography (TLC) is a simple and reliable aid for supporting the macro and microscopic identification of a plant drug and also for identification of an extract and of any medicinal dosage forms which may be prepared from it. If no tested authentic plant drug or extract material is available, indicative substances are also chromatographed. They can be selected from among the constituent substances of the drug plant, irrespective of whether these are active or accompanying substances. They should however be characteristic of the drug plant under investigation and easily demonstrable. Analytical data other than those obtained by thin layer chromatography can also be used for identification and quality assurance (Anonymous 1990). However, it is not sufficient to demand that identically coloured zones or spots from the investigated solution appear in the chromatogram at the level of the indicative substances. Rather, the colour and Rt value of the zones appearing in addition to the indicative substance spots must be described.

It is interesting that German Pharmacopoeia permits the use of so called external indicative substances which are not contained in the drug plant itself, 'Arnica flowers, may be quoted as an example. A comparative solution of routine in methanol is also applied to the thin layer plate in addition to the solution under investigation, which is prepared by extraction of the drug powder with methanol. This flavonoid (routine), which does not occur in Arnica flowers, is used to recognize any possible adulteration by marigold (*Calendula*) flowers, as it is stated: 'No zones with a yellowish to green colour should appear in the chromatogram of the investigated solution at the same level or below the zone in the chromatogram of the comparative solution' (Anonymous 1993).

Marker Analysis

A marker can be defined as a chemical entity in the plant material that may or may not be chemically defined and serves as a characteristic fingerprint for that plant. In other ways through various analytical techniques like TLC and HPLC we can visualize the presence of this compound in the plant and also quantify it to ascertain the limits. A bio-marker on the other hand is a group or chemical compounds which are in addition to being unique for that plant material also correlative with biological efficacy.

The development of parameters for quality control of Herbals is a big task involving biological evaluation for a particular disease area, chemical profiling of the raw material and laying down specifications for the finished product. Therefore, the word "standardization" should encompass the entire field of study from birth of a plant to its clinical application. It is a vast study and involves various disciplines that only interdisciplinary laboratories can undertake. The industry however cannot wait for this to happen and looks forward to easy and quick answers. To solve their problems, first and foremost task is the way of selection of the right kind of plant material which is therapeutically efficacious. The difficulty can be elucidated with the example of *Withania somnifera*. This plant is reported to have three chemo types depending upon the presence of class of closely related steroidal lactones like Withnaolides, Withaferin A etc. In spite of this classical chemo type distinction, selection of the right chemo type to which all clinical effects are attributed is still elusive. Therefore, the need arises to lay standards by which the right material is selected and incorporated into the formulation. The content of Withanolides and Withaferin A and other definite biologically active compounds may be variable depending upon the geography, genome, time of collection of plant material, etc. Therefore, there is a need for setting up the limits for the content of these multiple components in the clinically acceptable material. So here arises the concept of multiple markers and biomarkers to define the right material.

The assay part of standardization is chemical and biological profiling by which the chemical and biological effects could be assessed and curative values are established. Safety for use could also be assessed through this parameter. In biological assays, the drug activity is evaluated through a pharmacological model. For example insulin again a natural product is expressed in international units. These units correlate with the activity of a unit weight of Insulin to lower bloodglucose levels. Similarly, efficacy of Digitalis, a cardiotonic drug, can be effectively evaluated by biological assay of its action on cardiac muscles. The force of contraction of these muscles on administration of known amount is an indicator of its biological activity. However, biological testing is highly subjective in nature due to lot of variations and secondly there are strict regulatory controls on animal testing. Therefore, development of standardization procedures for quality control through biological assays shall be running against time.

Chemical profiling on the other hand is a versatile technique and can be made to good use in standardization. Fingerprints could be generated with known analytical procedures. Fingerprinting in essence is chemoprofiling which means establishing a characteristic chemical pattern for the plant material or its cut or fraction or extract. It is important to understand that a plant extract consists of established classes of chemical compounds. These include the primary metabolites, secondary metabolites and inorganic salts and metals. Primary metabolites are compounds like carbohydrates, proteins, lipids which are essential for the plant physiology as such but are formed as byproducts in the biochemical pathways. These include very interesting and useful classes of compounds like alkaloids, flavonoids, coumarins, terpenoids, anthocyanins, etc. Many modern drugs have come from these secondary metabolites like morphine from *Opium somniferoum,* reserpine from *Rauwolfia serpentina,* vinblastine and vincristine from *Vinca rosea*. In fact this is thepoint of divergence between the modern system of medicine and classical system of medicine like Ayurveda with the former laying emphasis on pure compounds. However coming to the mainline of our topic we can utilize these secondary metabolites for the identification of plant material as our knowledge of chemistry has advanced sufficiently and though sophisticated analytical techniques we can measure these compounds qualitatively and quantitatively. But the catch lies here for two reasons. First, if in a plant material we can measure the presence of a unique secondary metabolite it is not sufficient to say with certainty that the plant material is of the desired quality. As an example, Vasicine is a very characteristic alkaloid found in this plant, but merely establishing the presence of this compound and measuring is not a sufficient indicator of plant quality as vasicine as a pure entity has no biological activity attributed to Vasaka. The second important factor is that there are no databases for characterizing compounds obtained from specific plant materials. The example of the use of some herbal drugs with varying marker compounds in therapy in traditional system of medicine has been given in Table 18.1.

Marker testing is no way a substitute for other tests like Physico-chemical, chemical, macro and microscopically characterization etc. Nevertheless, it is an efficient procedure to ensure the identity and purity of herbal drugs (Mukherjee 2001).

Table 18.1: Some marker compounds present in therapeutically active ayurvedic herbal drugs in India

Name of the herb	*Marker Compound*	*Therapeutic Activity*
Andrographis Paniculata	Andrographolides 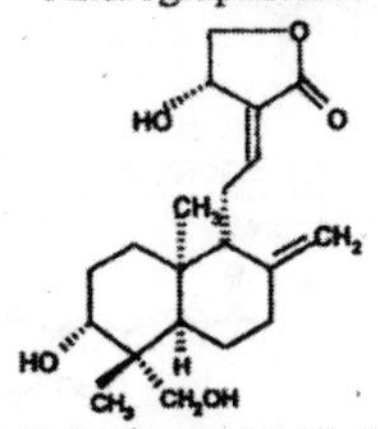 $C_{20}H_{30}O_5$ Mol. Wt. 350.46	Hepatoprotective
Acorus calamus	β-Asarone 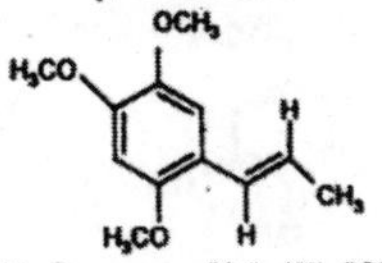 $C_{12}H_{16}O_3$ Mol. Wt. 208.26	*CNS active*
Adhatoda vasica	Vasicine 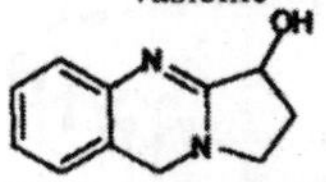 $C_{11}H_{12}N_2O$ Mol. Wt. 188.23	*Bronchodilator, expectorant*
Bacopa moniri	Bacoside A_3 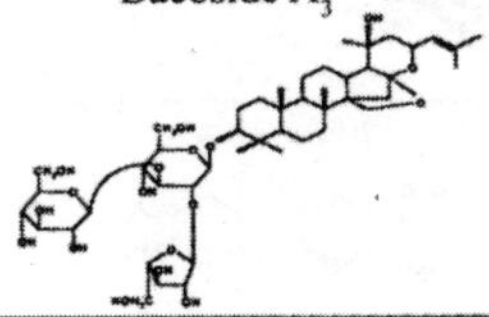	*Brain Tonic*
Baberis aristata	Baberine 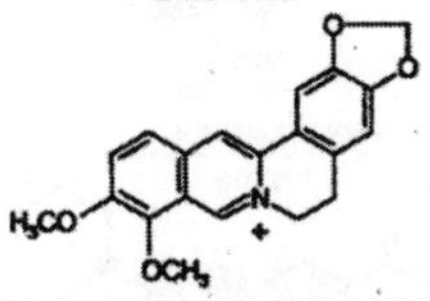 $C_{20}H_{18}N_4$ Mol. Wt. 336.37	*Diaphoretic, antiinflammatory*
Boswellia serrata	Boswellic acid 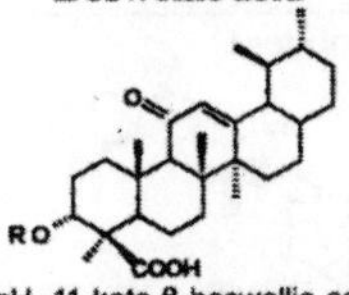 R=H, 11-keto-β-boswellic acid R=CH_3C0, acety-11-keto--β-boswellic acid	*Anti-arthritic*

Contd...

Table 18.1–*Contd...*

Commiphora mukul	Guggulsterone-Z 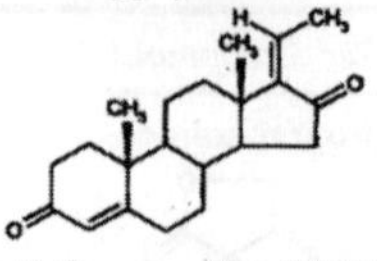 $C_{21}H_{28}O_2$ Mol. Wt. 312.45 GUGGULSTERONE-Z	*Hypolipidemic*
Curcuma longa	Curcumin 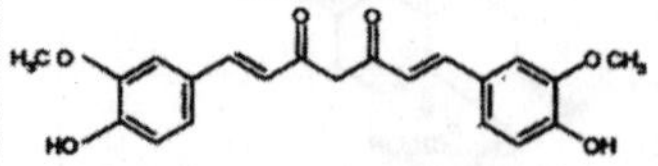 $C_{21}H_{20}O_6$ Mol. Wt. 368.39	*Anti-inflammatory*
Emblica officinalis	Gallic Acid 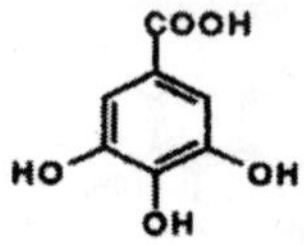 $(C_7H_6O_5)$ Mol. Wt. 170.12	*Carminative, cerebral and G.I. tonic, antioxidant, anti-scorbutic*
Glycyrrhiza glabra	Glycyrrhzin 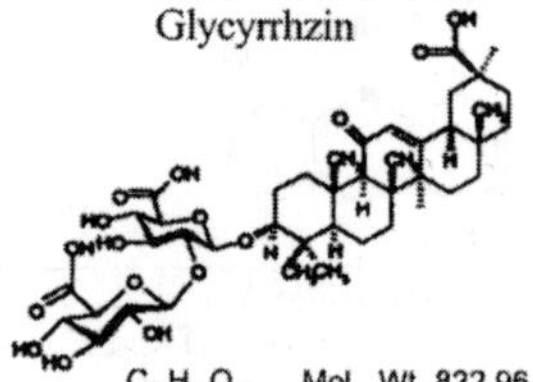 $C_{42}H_{62}O_{16}$ Mol. Wt. 822.96	*Anti tussive*
Phyllanthus niruri	Phyllanthin 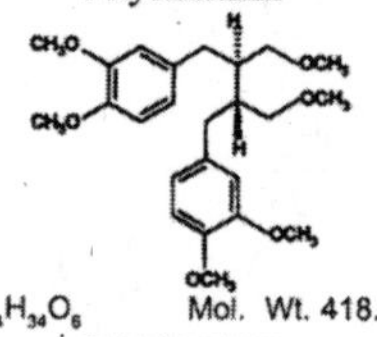 $C_{24}H_{34}O_6$ Mol. Wt. 418.53 PHYLLANTHIN	*Hepatoprotective*
Piper longum	Piperine 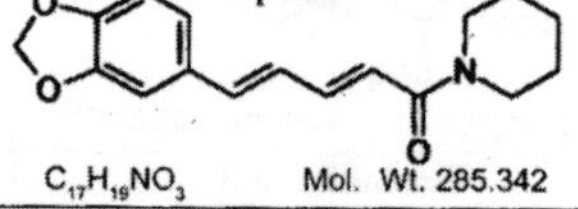 $C_{17}H_{19}NO_3$ Mol. Wt. 285.342	*Cough and cold*
Withania somnifera	Withanolides 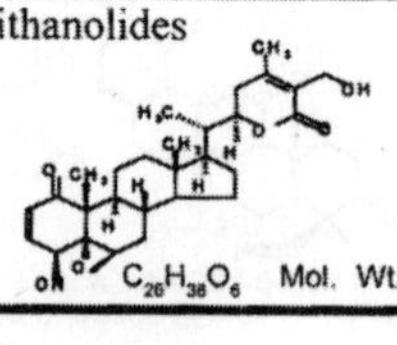 $C_{28}H_{38}O_6$ Mol. Wt. 470.61	*Adaptogen*

There are several constrains in adapting this technique for regular testing of the herbs and herbal formulations in the traditional system of medicine as follow:

- ☆ Non availability of library of marker compounds isolated from herbs in their pure form . Lack of proper communication and the willingness of quality assurance department to undertake tedious testing procedures.
- ☆ Non availability of a system where standard or properly identified herbs are available for the industry.
- ☆ Procurement of costly instruments and also its validation and proper maintenance.

Good Manufacturing Practices (GMP) in Traditional Systems of Medicine

In the production of traditional medicine, the concept of Good Manufacturing Practices (GMP) is an emerging trend to produce quality products. Each and every aspects are studied including:

- ☆ Collection, transport and storage of raw materials
- ☆ Unit operation for herbs
- ☆ Process validation
- ☆ Sanitation and hygiene
- ☆ Packing and finished goods checking
- ☆ Storage of finished products

Hazard Analysis and Critical Control Point (HACCP)

It is another aspect in herbal drug production. This is accepted on an analogy that manufacturing of herbal drugs is close to manufacturing of food products as both have large number of similarities. The basic principle of HACCP is "Doing only what is needed". It is now a strategy on production of traditional medicinal formulations, that production be based on thorough analysis of the prevailing conditions during purchase, production, quality control, packaging and marketing (Narayana *et al.*, 1997). In all the traditional systems of medicine the quality assurance aspects are considered as an integral part and they need to be exercised to the full extent (Anonymous 1992). There are several factors which make the quality assurance of traditional medicinal formulations (TMF) more hazardous like:

- ☆ Due to geographical and seasonal variation the active constituents varies
- ☆ The dosage of different Ayurvedic and other TMF varies from person to person depending on the intensity of diseases.
- ☆ Most of the TM formulations have to be administered with some specific vehicles like ghee (cooking fat), honey, milk etc.

Efficacy of Herbals as Food and Medicinal Products

Phytomedicines and neutarceuticals consist of many chemical constituents with complex pharmacological effects on the body. They are used continuously for many decades or centuries, often in ways that differ from those of conventional medical prescribing. Research and development in phytotherapy has suffered through lack of patent protection and the diversity and relatively small-scale of the industries involved compared to the rest of the pharmaceutical industry. The differing regional uses of traditional herbal remedies present extra difficulties for the harmonization of quality control procedure around the world. Therapeutic efficacy and clinical trials are two most important criteria for the development of herbal drugs.

Although preliminary assessments of efficacy can be obtained through the results of *in vitro* testing and experiments on animals, authorities licensing new medicines for public use require evidence of their effect on human beings. Only carefully planned clinical trials that clearly minimize experimental bias are able to satisfy these requirements. Most herbal remedies can call on a tradition of popular use, which has in practice, allowed manufacturers to submit relevant bibliographic evidence in reviewing their earlier licenses of right. Nevertheless this has been a reluctant concession by licensing authorities and in reviews of licenses; there is the prospect that additional evidence may be required.

A clinical trial is any systematic study of medicinal products in human subjects whether in patients or non-patient volunteers in order to discover or verify the effects of and/or identify and adverse reaction to investigational products, and or study their absorption, distribution, metabolism and exertion in order to ascertain the efficacy and safety of the products under normal conditions of use. Therapeutic advantages must outweigh potential risks (Rules governing medicinal products in the European community Part-IV, 1989). All test procedures shall correspond to the state of scientific progress at the time and shall be validated through suitable procedures and the results of the validation studies shall be provided (Rules governing medicinal products in the European community part II, 1989) to be concerned.

The particulars of each clinical trial must contain sufficient detail to allow an objective judgment. In general, clinical trials shall be done as 'controlled clinical trials' and if possible, randomized; any other design shall be justified. Inclusion of a large number of subjects in a trial must not be regarded as an adequate substitute for a properly controlled trial. Clinical statements concerning the efficacy or safety of a medicinal product under normal conditions of use which are not scientifically substantiated cannot be accepted as valid evidence. In order to agree with the labeled and advertised claims for the efficacy of medicinal products, regulators will be required to see the results of several "pivotal" trials for each application and indication, usually double blinded, random-assignment, and controlled against placebo or other standard medicine.

For researching over-the-counter (OTC) label indications for discrete medicinal products, in which individual responses to remedies are not the critical issue, the double-blind controlled clinical trial is clearly the most appropriate method. In addition as the "patient" is in many cases not being diagnosed professionally and is determining his or her own treatment and prognosis, self-assessment questionnaires are often an appropriate measure of progress. These are not expensive to administer. Also as this is often "outpatient" medicine research costs can be saved as close clinical supervision need not be always necessary throughout the trial.

In pursuing good clinical research in herbal medicine the problems usually encountered with are as follows:

1. Herbal medicine in the west can boast few teaching hospitals or research institutes, or support from public resources. Industrial investment has been limited to a few larger manufacturers used to working in a pharmaceutical culture. In most parts of the sector the necessary infrastructure is lacking. Neither can the costs of undertaking research studies easily be justified commercially; it is difficult to patent herbs and the size of the market for any individual product is only occasionally comparable to that for any patentable conventional drug.
2. The indications often claimed for phytomedicines include many without robust outcome measures. As many are destined for the self-medication OTC market they are by definition directed at lesser degrees of morbidity where hard measures are elusive. By contrast most synthetic OTC medicines on the market have "switched" from prescription status and have acquired their efficacy evidence on harder clinical indications and in hospital or similar settings. Without hard or acceptably validated outcome measures, with more variable and lower grade symptoms among the patient population, with a greater likelihood of self-limiting or other spontaneously changing conditions, clear treatment effects are thereby harder to establish. The result is often the need to recruit large patient samples and to devise particularly artificial exclusion criteria to constrain sample variability.
3. Herbs are complex medicines, occupying an unusual position as being medicines with many of the characteristics of vegetables. Being a complex of pharmacologically active chemicals the whole package will have different properties from that of any single constituent acting alone. Knowing the action of the latter will not itself be predictive on the effect of the former, particularly if the experimental evidence is based on work done on laboratory animals. It is therefore rare to find the satisfactory pre-clinical evidence often required by ethics committees for approvals of major clinical studies.

Further limitations of the controlled clinical study are noted by the practitioner, the phytotherapist. Such practitioners will often emphasize more strongly than in conventional medicine the individuality of their treatments and often mix a number of individual herbs in a prescription. They point out that conventional clinical trials involve the homogenization of the patient population so that only an average effect is confirmed. Clinical trial data will help with, but still not answer, the basic question "is the drug going to be good for this patient?" It is also likely that genuinely important benefits for a minority of the population will be overlooked.

There are also certain cases where satisfactory blinding will always be difficult. In particular it will always be impossible to blind for the effects of bitters or other prominent testing agents on oral consumption.These have played important parts in the claims of traditional herbal medicine. There will as always be a role for the good single-blind study, especially if other elements are rigorously controlled. Nevertheless, the controlled clinical trial is a notably flexible instrument. With an adequate investment structure, a significant section of the herbal medicinal market could be clearly evaluated in the public interest. The increasing rate of publication of rigorous clinical trials on herbal products in the scientific literature highlights the potential for a stronger evidence base in conventional terms.

Traditional views of herbal remedies emphasis their primary influence on transient body functions *viz.*, diaphoretics, expectorants, circulatory stimulants, diuretics, digestive stimulants, laxatives and so on. In other words, contrary to common belief, many herbs may have almost immediate results on the body. It is possible to devise methods by which such effects can be detected. Activity on biological markers, physiological functions and tissue or fluid constitution can be monitored directly in healthy or morbid populations and could provide much useful information on the effects on the body of herbal products. With advances in non-invasive monitoring technologies, it is possible to conceive of important trials in human subjects, both in observational and controlled studies. Against such ideas, it has been pointed out that measures of efficacy tend to be based on a medicine's effects on morbidity or mortality. "Surrogate parameters" may be acceptable but would restrict possible indications as labeled claims. It would need to be argued in each individual case, with the support of other medical evidence, that a verified effect in changing some physiological parameter would be likely to have an effect on morbidity or mortality.

Herbal Therapies

The history of herbal treatments is marked by the enormous diversity of local traditions, with much more variety than consistency. When these are distilled further and then translated into modern terminology, it is possible to identify a few features that recur consistently. These appear to encapsulate essential characteristics of the material, universal archetypes of the effects plants have on humans. They reflect also the therapeutic priorities of earlier ages and, at the least, need adaptation to be incorporated into a phytotherapy appropriate to modern needs. They are valuable, however, because they probably represent the roots of herbal therapeutics, those characteristics of the remedies that are the most potent. They also draw stark contrasts with the most potent. They also draw stark contrasts with the approaches that have emerged with the development of modern technological medicine. They are antidotes to the modern tendency to view herbs solely as milder versions of modern drugs. In essence, they should underpin any rational phytotherapy.

Although there is indeed considerable published research, much reviewed in this text, there is still a relative paucity of top-quality controlled clinical studies of the whole herb that clearly separate its effect from placebo, and most good clinical studies show only a modest additional effect. Herbal practitioners will attest to consistent therapeutic performance and will also have accounts of dramatic responses from their patients, where genuinely powerful pharmacological effects have followed consumption of herbal prescriptions. They will also point to the essential 'neutering down' of clinical trials data as denying precisely that which they value most about their therapy the individualization of treatment and response.

This is enough to convince most practitioners that herbal medicine is a serious alternative to conventional drug treatment with few of the adverse effects. Any move in the standoff between the skeptics and the believers is, however, likely to include a full discussion of the placebo effect.

Among both conventional and complementary healthcare practitioners, there is an understandable feeling that cures that have nothing to do with the treatment so skillfully provided are something of an embarrassment and even a challenge to one's choice of vocation. Thus the placebo effect is one of the least discussed phenomena in clinical medicine. Yet it is by far the most powerful factor of all.

Clinical Research in Herbal

Compared with the experience of most modern drugs, the human use and approval of most herbal remedies is awesome. The requirement by the medical and scientific establishment for research to 'prove' that herbs are effective is not found among the population at large. It is apparent that most ordinary people are content to rely on their impressions of the world to get by in it. Judging by the substantial markets for herbal products in the developed world, and others one the vast use in traditional cultures, a great many people have already found herbal medicines useful.

However, the public do want to be assured that someone is looking after them. They therefore assume the questions are being asked by those who ought to do it. The physician and the regulators are charged with the job of making sure that medicine is safe and effective. Knowledge within traditional medicine, however, has generally been in the form of received wisdom molded to the individual needs and prowess of each practitioner. Such means of acquiring healing skills seem temperamentally suited to most practitioners' herbal researchers even today. Their interest in inquiry for its own sake, with secure truths up for constant possible refutation, is understandably secondary to their concern to survive in practice. In the case of herbal medicine, adherents understandably tend to give it the benefit of doubt. To perform with good clinical research in herbal medicine, some problems as follows needed to be answered:

- To produce results carrying sufficient statistical weight is expensive and laborious (each trial has to be casted in research salaries plus logistical expenses). Herbal medicine in the West can boast no teaching hospitals or research institutes, nor funding by government or a wealthy industrial sector. The necessary infrastructure is lacking. Neither can the costs of undertaking research studies easily be justified commercially. It is difficult to patent herbs and the size of the market for any individual product is not comparable to that for any patentable conventional drug.

- Herbs are complex medicines, occupying an unusual position in being medicines with many of the characteristics of foods. Being a complex of pharmacologically active chemicals, the whole package will have different properties from that of any single constituent acting alone. The action of the latter will not predict the effect of the former, particularly if the experimental evidence is based on the work done on laboratory animals.

☆ The application of herbs and their effect on the body are not always the same as usually understood for conventional medicines. As has been suggested above, herbal medicines may be used more to evoke healing responses in the body than to attack symptoms; this generates a different research question.

Any rationale for herbal medicine is likely to be based on the activity of many plant chemical constituents. There are fundamental technical questions raised in building a rational case for herbal therapeutics more particularly in what ways are plant constituents likely to interact, in the context of the gut and body tissues, to affect bioavailability and activity for which the obvious interactions are between tannins, alkaloids, saponins, minerals, complex carbohydrates etc. Beside this how do changes in pharmaceutical preparation affect the bioavailability and activity of plant constituents is an important factor *e.g.* do alcoholic extracts have significantly different actions from the aqueous extracts generally dominant in traditional practice. With new biochemical monitoring technology it is feasible that some of the parameters could be obtained non-invasively in healthy human subjects which however have been addressed traditionally with animal and organ culture experiments.

It can be recalled that herbal therapy aims to support vital functions of human body. If the intention of a trial was to assess the effects of herbal medication applied in approximately therapeutic doses adjusted for body weight and metabolism, there could be little complaint that the animals would be harmed and they are likely to actually benefit. Feasible trials might include monitoring the effects of posited 'adaptogens' on life expectancy, stamina and reproductive capacity, the effects of antimicrobial remedies on normal resistance to disease among large populations and observing changes in digestive and urinary performance as judged by changes in excretion, appetite and weight. As the common laboratory animal with a metabolism closest to the human being, the rat would probably be the creature of choice. It might also be valuable to have the results of careful observations of the use of herbs in veterinary practice, where many of the limitations attending patient-practitioner interaction can be minimized. Nevertheless, in modern times political sensitivities to any animal research are so great that few herbal researchers are likely to consider the modest net contribution of such trials to a useful understanding of the effects of herbal remedies on human beings as justifying the effort.

In reviewing the prospects for validating herbal medicine in a sceptical age, the works on herbal drugs are overwhelming. The large number of medicines in use, the complexity of their pharmacology and the low financial gearing of the industry rule out an early overhaul by scientific scrutiny. This led the public to rely largely on the precedent of tradition in choosing to use herbal remedies. The suppliers of herbal medicines and the legislators overseeing the supply have a common interest in working together to improve the level of knowledge available as best they can. The industry and profession may thus consider adopting an integrated, but rigorous andrevalidatable research programme that reflects the special character of the material. The projects should be able to be conducted in small clinics and laboratory facilities and without prohibitive capital investment. To screen out the time-wasting project, protocol design and project oversight should be performed by university specialists. In this endeavors, the herbal industry and profession could expect co-operation from regulatory authorities. They should have clear encouragement for investigation that is not necessarily identical to that appropriate to the proving of new synthetic drugs and a declared willingness to include expert testimony from within the industry and profession in deliberations on the fate of herbal medicines. An undertaking of this nature will provide a strong incentive for the suppliers to invest time and finance in the necessary infrastructure for research activity.

Safety in Herbals

Major differences in the assessment of quality, safety and efficacy would hinder free circulation of herbal medicinal products which may represent a risk for consumers. The complexity of herbal drug preparations and the interpretation of bibliographic data on safety and efficacy reflecting the experience gathered during long-term use, are best addressed by involving specific expertise and experience. Safety and efficacy of complex biological products, such as herbal medicinal products, are directly linked to pharmaceutical details such as the way of production and the specification of extracts.

A consistent quality of herbal drugs may need more detailed information on aspects of agricultural production. The selection of seeds, conditions of cultivation and harvesting represent an important aspect in producing a reproducible quality of herbal drugs. Ongoing discussions on Good Agricultural Practices (GAP) for medicinal plants should be monitored regularly.

Safety Guidance for Herbals

The European Agency for the Evaluation of Medicinal Products (EMEA) discussed criteria for the pre-clinical assessment of herbal medicinal products and a guideline "Non-clinical testing of herbal drug preparations with long-term marketing experience – Guidance to facilitate mutual recognition and use of bibliographical data". It takes the available experience into account that has been gained with the use of well established herbal drug preparations in humans, and defines conditions where additional studies may be required if an acceptable level of safety cannot be demonstrated by bibliographic data and experience in humans.

Fixed Combinations

The use of fixed combinations is one characteristic aspect of phyto-therapy. There must be an adequate and proportionate justification of the fixed combination considering both the indications claimed for the product and the risks due to the ingredients. Each active ingredient must contribute to the overall safety and/or efficacy of the fixed dose combination. Irrational fixed combinations should be avoided. Combination products form an important part of the phyto-medicines market. They are widely accepted by patients and the medical profession. On the other hand, they represent a serious problem within the quantification procedure.

The German drug laws prescribed that if the product contains more than one active constituent, it shall be substantiated that every active constituent contributes to the positive judgments of the product. But this can't be the same for all combination products (Shaw *et al.*, 1997). In some instances the combination herbal products are more useful which should be exploited further for their quality specification. In ancient Indian systems of medicine these combination products used to play the major role in therapy. So considering these aspects the combination products have great potentials and they should be explored further for fixing the specification of their standardization parameters as they posses some extra advantages as follows:

- ☆ The different constituents of a combination may influence the different symptoms of a single syndrome.
- ☆ A combination of different constituents with the same active principle or with different active principles leading to the same effect should be allowed in order to achieve cumulative efficacy.
- ☆ If combining it with other substances can reduce the dosage for a single substance, the combination should be acceptable.
- ☆ The indication of a single compound may be different from that of the combination, as long as it enhances the efficacy of the total combination.
- ☆ Further criteria could be an improvement in tolerance and compliance, a simplification of the dosage scheme or avoidance of pharmaceutical incompatibilities.
- ☆ Instead of proof for every single compound, proof of the efficacy of the product in total should be accepted.
- ☆ A differentiation has to be made between combinations with a 'fictive registration' and a new drug application.

The prophylactic use of phytomedicines plays an important role, especially in the distribution of outside pharmacies. It is extremely difficult to show scientific evidence for prophylactic indications because they cannot be supported by clinical trials; one has to rely on plausible interpretations of case histories by experienced practitioners. This kind of experience however is more prominent in Indian traditional medicine.

Toxicity in Herbals

Apart from determining efficacy, the FDA is also charged with determining the safety of drug products, and not all Herbals/herbal remedies are harmless. In this context the reference to the incidence of 1991 and 1992 in Brussels, Belgium, in which 30 women treated with a Chinese herbal slimming preparation died from renal failure caused by the presence of aristocholic acid in it can be taken into account. One of the herbs had been incorrectly identified as a nontoxic herb (Marwick 1995). So the importance on controlling the correct identification of herbal preparations should be taken into account from the very beginning. In addition to the problem of incorrect plant identification, some mixtures may be toxic, particularly if they are misused. The message consumers get is that these products are natural, and they assume that natural is safe. It is very much needed to continually review and assess the safety of

Herbals, with an emphasis on surveillance of the use of these products to identify unknown hazards or risks and address them expeditiously. Importance should be given to continuous surveillance and of actively requesting information rather than just collecting reports and even this can be considered as national program.

Potential and Prospect of Herbal-Drug Interactions

Drug interactions may have little or no effect in-patients and some are harmless. For those drug-drug interactions which are potentially harmful, the effects may only occur in a small proportion of patients (Mehta 1998). However, serious interactions can occur and these may be life threatening. Some drugs have a small therapeutic ratio (toxic dose and therapeutic dose) *viz.*, phenytoin, and a series of other drugs require careful control to be maintained of their dosages, *viz.*, anticoagulants, antihypertensives, antidiabetics etc. These drugs are the ones, which are most frequently involved in interactions. Patients with impaired renal or liver functions, or elderly, are at most risk from drug interactions and may be more prone to adverse effects. Healthcare professionals providing pharmaceutical medicines for the treatment of disease should be aware that concomitant use of herbal remedies so much so the consumers purchasing herbal remedies or herbalists treating patients should be aware of potential herb-drug interactions. Education and continuing education program for healthcare professionals should include information on drug-drug, drug-food and drug-herb interactions. With all drug interactions continued pharma-vigilance is necessary to detect potential drug-herb interactions, especially when the incidence of such reactions is low (Anderson 1985).

There is only limited information available in the literature concerning the interaction of herbal medicines and conventional medicines. The findings of a five year toxicological study on traditional remedies and food supplements used in the UK have been reported by the Medical and Toxicological Unit of Guy's and St. Thomas Hospital, London (Shaw 1997). Their assessments were based on reports to the National Poisons Information Service, London, which provides emergency information to medical professionals and from a follow-up questionnaire, clinical consultations, toxicological analyses and botanical identifications. These findings highlighted allergic reactions, liver problems and heavy metal poisoning from some Asian remedies, the largest single group of enquiries on herbal products involved sedatives, generally containing valerian passiflora, producing drowsiness, gastro-intestinal disturbances and liver abnormalities (D'Arcy 1993). Herbal laxative preparations reportedly caused abdominal pain and diarrhoea in some patients. There were relatively few cases involving interactions between herbal remedies and pharmaceutical medicines and none of them were proved conclusively. Four possible types of interaction were reported, papaya extract/warfarin (increased plasma levels of warfarin), devil's claw/warfarin, ginkgo/thiazide diuretics (hypertension) and evening primrose oil/anaesthetics (seizure) (Maclennan *et al.*, 1997).

It might be stated that there are fewer problems posed to patients by interactions between herbal remedies and conventional pharmaceutical drugs. Nevertheless, before dismissing this out-of-hand, the potential for such interactions should be evaluated. Consideration of the potential for herb-conventional pharmaceutical drug interactions have been re-viewed previously and a selected summary of some possible interactions should be studied (Newall *et al.*, 1996). Some herbs contain constituents, which have the potential to interact with pharmaceutical drugs used to treat a wide range of disorders, *e.g.* the gastro-intestinal tract, cardiovascular, central nervous and endocrine systems. In general, the herb may either potentiate or antagonize effective drug treatment whilst in some instances, existing toxicities of conventional pharmaceutical drugs may be enhanced (Fahishi 1996).

Adverse Effect of Herbal Medicine

A commonly heard argument in favor of herbal medicines is that these products have a longstanding history of traditional use, resulting in considerable experience with and knowledge about their wanted and unwanted effects. Of course the traditional experience is a powerful tool for the identification of adverse effects which occur in the majority of users and develop rapidly after the start of therapy (Park *et al.*, 1992). A classical example is the induction of anticholinergic symptoms, such as palpitations, dryness of the mouth and dilatation of the pupils, by herbal medicines rich in belladonna alkaloids. Such reactions are pharmacologically predictable and dose-dependent, which implies that they can be anticipated, and that they could be prevented by dose reduction.

Obviously, traditional experience can bring such dose-dependencies to light and it may help to detect ways of processing, which reduces the likelihood of acute problems. However, not all adverse reactions occur immediately after the therapy has been started. The importance of delayed reactions was recently underlined by a retrospective study covering clinical safety trials. The occurrence of muscular weakness due to hypokalemia in long term users of herbal anthranoid laxatives (Anonymous 1995). It will be even more difficult for herbal prescribers and their clients to recognize all the adverse drug reactions and more particularly those which are not related to the principal pharmacological properties of a drug and they do not improve when the dose is reduced or the drug has to be withdrawn completely. These reactions are often immunologically mediated but some have a non-immunological basis, such as genetic cause (Bateman and Chaplin 1988). An example of a herbal medicine which has been repeatedly associated with such type reactions is the Japanese Kampo medicine Sho-saiko-to, consisting of *Bupleuri radix*, *Ginseng radix*, *Glycyrrhiza radix*, *Pinelliae tuber*, *Scutellaria radix*, *Zingiber rhizoma*, and *Zizyphi fructus*. This formula has been used in China since the Han Dynasty (*i.e.* about 100 AD), and it was generally considered to be devoid of serious side effects. There is now a respectable series of case reports from Japan, however, which associate its use with allergic pneumonitis and/or hepatitis (Kubo *et al.*, 1986, Kawasaaki *et al.*, 1994). Shosaiko-to is by no means the only Oriental herbal medicine which is associated with rare but serious cases of hepatitis (Perharic *et al.*, 1995, De Smet, 1997). In recent years, there have also been various reports of hepatotoxic reactions to Western herbs, such as wall germander (*Teucrium chamaedrys*) (De Smet 1997), skullcap (*Scutellaria* or *Teucrium* sp.) (De Smet 1993) and chaparral (*Larreatri dentata*) (Gordon *et al.*, 1995). To which extent herbal medicines play a role in such cases, is currently unknown. Doctros should definitely keep this possibility in mind, however, when they examine patients with an unexplained hepatic disease (Anonymous 1995).

Another category of adverse reaction that may be readily overlooked is those types, which consists of certain delayed effects, such as eratogenicity and carcinogenicity (Park *et al.*, 1992). Some years ago, there was a ragic outbreak of fibrosing interstitial nephritis in Belgian women who had been treated with a slimming preparation that supposedly included *Stephania tetrandra* but in reality contained *Aristolochia fangchi* (De Smet 1995). This latter plant drug contains aristolochic acids, which are not only nephrotoxic but also extremely potent rodent carcinogens (De Smet 1992).

Conclusion

Healing plants are the nature's gift for everyone and preservation of this precious gift with their maximum utilization in every aspect of drug development from natural resources is the need of the day. Keeping the interest of the consumers for ensuring quality and standard of Herbals we must have to follow various regulations and guidelines including. GAP, GHP, GSP, GMP and so on. To maximize the utilization of these natural resources there should be an international co-ordination on various regulations of different countries. With this international co-ordination the Herbals should be explored further in the context of modern development.

The assessment of herbal drug production has shown that the cultivation, harvesting and primary processing of plant material which is used in manufacturing health products and herbal medicines is often insufficiently monitored. Because the quality of such preparations is to a large extent determined by the quality of the raw materials, cultivation and harvesting of the same plays a major role.

References

Akerale O, Heywood V and Synge H 1991 Conservation of medicinal plants. Cambridge University Press, Cambridge.

Anderson L A and Phillipson J D 1985. Herbal medicine, education and the pharmacist. *Pharm J* **236:** 303-305.

Anonymous 1994. Herbal hope. *New Scientist* **142:** 11.

Anonymous 1995. Drug development from natural products. *Phar J* **255:** 430-431

Anonymous 1990. Good Clinical Practice for Trials on Medicinal Products in the European Community. CPMP Working Party on Efficacy of Medicinal Products.

Anonymous 1992. Quality Control Methods for Medicinal Plant Materials. Geneva: World Health Organization 492-559.

Anonymous 1993. Council Regulation of 22nd July 1993: Laying down Community procedures for the authorization and supervision of medicinal products for human and veterinary. Medicinal Products.

Bateman D N and Chaplin S 1988. Adverse Reactions. *I Br Med J* **296:** 791-794.

Balick M J 1985. Useful plants of Amazonia: A resource of global importance. In: Prance G T and Lovejoy T A (ed.). Amazonia, Key Environment Series. Pergamon Press, New York.

Barz W and Ellis E 1980. Natural Products as Medicinal Agents. In: Bel J L, Reinhard E. (eds.). Hippokrates, Stuttgart.

Bauer R, Jurcic K, Puhlmann J and Wagner H 1988. Immunologische *in vivo* and *in vitro* Untersuchungenmit Echinacea-Extrakten. *Drug Res* **38:** 276-281.

Braquet P and Hosford D 1991. Ethnopharmacology and the development of natural PAF antagonists as therapeutic agents. *J Ethnopharmacol* **32:** 135-49.

Butterweck V, Petereit F and Winterhoff H 1998. Solubilized hyper-antidepressant activity in the forced swimming test. *Planta Medica* **64:** 291-294.

Cassady J and Douros J 1980. Anticancer Agents based on Natural Product Models. Academic Press, New York.

D'Arcy P F 1993. Adverse reactions and interactions with herbal medicines. Part 2. Drug interactions Adverse Drug React. *Toxicol Rev* **12:** 147-162.

De Smet P A G M 1992. Toxicological outlook on the quality assurance of herbal remedies. *In*: De Smet P A G M, Keller K, Hansel R, Chandler E R (eds.). Adverse Effects of Herbal Drugs, Vol 1. Springer Verlag, Hiedelberg.

De Smet P A G M 1993. An introduction to herbal pharmaco-epidemology. *J Ethnopharmacol* **38:** 197-208.

De Smet P A G M 1995. Should herbal medicine-like products be licensed as medicines *Br Med J* **310:** 1023-1024.

De Smet P A G M 1997. Herbal Pharmacovigilance. *In*: Keller K, Hansel R, Chandler RF editors Adverse Effects of Herbal Drugs. Springer Verlag, Hiedelberg

Eder Mand Mehnert W 1998. Bedeutungpflanzlicher Begleitstoffe in Extrakten. *Pharmazie* **53:** 285-293.

Evans D, Nelson J and Taber T 1982. Topics in stereochemistry. *In:* Alinger N L, Eliel E L (eds.). New York.

Fahishi A 1996. Complementary Medicine. Financial Times and Health care Publishing, London.

Farnsworth N R 1982. The consequences of plant extinction on the current and future availability of drugs, Paper presented at AAAS Annual Meeting, 3-8 January, Washington DC.

Farnsworth N R 1988. Screening plants for newmedicines. In: Biodiversity, (ed.) E O Wilson National Academy Press, Washington DC.

Gordon D W, Rosenthal G, Hart J Z, Sirota R and BakerA L 1995. Chaparral ingestion, the broadening spectrum of liver injury caused by herbal medications. *JAMA* **273:** 489-90.

Grindley J 1993. The natural approach to pharmaceuticals. *Scrip* 30-33.

Houghton P J 1988. The biological activity of valerian and related plants. *J Ethnopharmacol* **22:** 121-42.

Kleijnen J and Knipschild P 1992.*Ginkgo biloba. Lancet* **340:** 1136-1139.

Kinghorn A D and Balandrin M F 1993. Human medicinal agents from plants. *Am Chem Soc Symp Ser* **534:** 1-318.

Klayman D L 1985. Qinghaosu (Artemisinin): An antimalarial drug from China. *Sci* **228**: 1049.

Kawasaki A, Mizushima Y, Kunitani H, Kitagawa M and Kobayashi M. 1994. Useful Diagnostic method for drug induced pneumonitis: a case report. *Am J Chin Med* **22:** 329-336.

Keug W, Lazo, O and Kunze L 1996. Potentiation of the bioavailability of daidzin by an extract of Radix puerariae. *Proc Nat Acad Sci* **93:** 4284-4288.

Kisa K, Sasaki K, and Yamuchi K 1981. Potentiating effect of sennoside C on purgative activity of sennoside A in mice. *Planta Medica* **42:** 302-312.

Kubo K, Watanabe F, Sakuma S, Abe A, Kutsukake S, Shimano K, Abe M and Kuribayashi N 1986. Hypersensitive hepatic injury induced by Shosaikoto, a liver supporting herb medicine. *Jap J Nat Med Serv* **40:** 257-60.

Maclennan A, Wilson D H and Taylor A W 1997. Prevalence and cost of alternative medicine in Australia. Mintel International Group, London.

Marwick C 1995. Growing use of medicinal Herbals forces assessment by drug regulators. Medical News and Perspectives. *J A M A* **273:** 607-610.

Mehta D K 1998. British National Formulary, British Medical Association and the Royal Pharmaceutical Society of Great Britain. The Pharmaceutical Press, London.

Mills S and Bone K 1999. Principles and Practice of Phytotherapy. Churchill Livingston, London.

Moreno R and Schwartzman B 1975. 268 plants medicinalesutilizadaspara regular la fertilidad en algunospaises de Sundamerica. *Reproduction* **2:**163.

Mukherjee P K 2001. Evaluation of Indian traditional medicine. *Drug Info J* **35:** 623-632.

Mukherjee P K 2002. Problems and prospects for the GMP in herbal drugs in Indian systems of medicine. *Drug Inf J* **63:** 635-644.

Mukherjee P K 2003. Exploring Herbals in Indian systems of medicine- regulatory perspectives. *Clinical Res Reg Affairs* **20:** 249- 264.

Mukherjee P K, Shau M and Suresh B 1998. Indian herbal medicine. *Eastern Pharmacist* **42:** 21-24.

Mukherjee P K 2003. GMP in Indian system of medicine. *In*: Mukherjee P K, Verpoorte R (ed.) GIMP for Herbals - Regulatory and Quality Issues on Phytomedicine. 1st ed. India: Business Horizons, New Delhi.

Newall C A, Anderson L A and Phillipson J D 1996. Herbal Medicines a guide for health-care professionals'. The Pharmaceutical Press, London.

Park B K, Pirmohamed M and Kitteringham N R 1992.Idipsyncratic drug reactions: a mechanism evaluation of risk factors. *Br J Clin Pharmacol* **34:** 377-395.

Perharic L; Shaw D, Leon C, De Smet P A W and Murray V S G 1995. Possible association of liver damage with the use of Chinese herbal medicine for skin disease. *Toxicol* **37:** 562-666.

Penso G 1980. The role of WHO in the selection and characterization of medicinal plants (vegetable drugs). *J Ethnopharmacol* **2:** 183-185.

Perdue R E Jr and Hartwell J L 1969. The search for plant sources of anticancer drugs. *Morris Arboretum Bulletin* **20:** 35-53.

Phillipson J D, Wright C W, Kirby G C and Warhurst D C 1993. *In*: Downum K R, Romeo J T, Stafford H A (eds.) Tropical plants as sources of antiprotozoal agents. Plenum, New York.

Shaw D, Leon C, Koler S and Murray V 1997. Traditional remedies and food supplements - a five year toxicological study (1991-1995). *Drug Safety* **17:** 342-356.

Onawunmi G O, Yisak W and Ogunlana E O 1984. Antibacterial constituents in the essential oil of *Cymbopogon citratus* (DC). *J Ethnopharmacol* **12:** 279-286.

Virison J A and Bose P B 1988. Comparative bioavailability to humans of ascorbic acid lone or in a citrus extract. *Am J Clin Nutr* **48:** 601-604.

Waigh R D 1988. The chemistry behind attracurium. *Chem Britain* **24:** 1209-1212.

Warshafsky S, Kramer R S andSivak S I 1993. Effect of garlic on total serum cholesterol: meta-analysis. *Ann Intern Med* **119:** 599-605.

Waterman P G 1992. Roles of secondary metabolites in plants. *In:* Chadwick D J, Vhelan J (eds.). Secondary metabolites: their function and evolution, Ciba Foundation Symposium: Wiley, Chichester.

Weragoda P B 1980 Some questions about the further of traditional medicine in developing countries. *J Ethnopharmacol* **2:** 193-194.

World Health Organization (WHO) 1978. World Health Organization Official Records WHO, Geneva.

Medicinal Plants: Aspects and Prospects (2014) *Pages* **279–301**
Editors: **Mukesh Kumar, Anjali Khare and C.P. Shukla**
ISBN: 978-81-7622-309-6
Published by: **BIOTECH BOOKS, NEW DELHI**

Chapter 19

In vitro Micropropagation and Secondary Metabolite Production in Medicinal Plants

Hans R. Dhingra and Suresh C. Goyal

Department of Botany and Plant Physiology,
CCS Haryana Agricultural University, Hisar – 125 001

ABSTRACT

India is one of the twelve centers of mega-biodiversity since all known types of agro climatic, ecological and edaphic conditions are met and thus one among the top repositories of medicinal plants. An increasing reliance on the use of medicinal plants in the industrialized society on medicinal plants is very well evident from extraction and development of drugs; and chemotherapeutics from these plants till date. These gerbil remedies are very popular for the treatment of minor ailments due to their easy availability, affordability, non toxic nature and lack of any side effects. With an ever increasing global inclination towards herbal medicines, there is an obligatory demand not only for huge amount of raw material of medicinal plants but also of right stage when the active principles are available in optimum quantities. Despite the increasing use of herbal medicinal plants, their future is being threatened by complacency concerning their conservation. This chapter deals with *in vitro* propagation of some of the important endangered/threatened medicinal plants of the country. Moreover, since chemical synthesis is at present fraught with problems, the possibility of using plant tissue culture technique for production of secondary metabolite has been explored because the phytohormonal adjuvant not only regulate callusing and morphogenesis but monitors qualitative and quantitative production of steroids, alkaloids and other compounds. This will help in easing out the tremendous pressure on the natural habitat.

Keywords: Medicinal plants, Micropropagation, Productivity.

Introduction

India in general, is rich in plant diversity since all known types of agro climatic, ecological and edaphic conditions are met. Moreover, the geographical location of India is so unique that all types of ecosystem ranging from coldest Nubra valley (-57°C), dry cold desert of Ladakh, temperate and alpine and subtropical regions of North-West and trans-Himalayans, rain forests of Cheerapunjii Meghalaya, wet evergreen humid tropics of Western Ghats, arid and semi-red Rajasthan and Gujarat, tidal mangroves of Sunderban exist. Indian culture has made use of its medicinal plant resources to maintain the health of the society, people and pets alike. This is the resource which is renewable, reliable and promises health and wealth. Out of angiospermic 17,000 plants, the classic system of medicines like Ayurveda, Sidha and Unani make use of about 12 per cent plants in various formulations. Interest in medicinal plants as re-emerging health aid has been fuelled by the rising costs of medicines in maintenance of personnel health and well being. As per UNESCO report (1996) the use of traditional medicines and medicinal plants in most of the developing countries as a normative basis for the maintenance of good health has been widely observed. Increasing reliance on the use of medicinal plants in the industrialized society has been traced to the extraction and development of several drugs and chemotherapeutics from these plants as well as from traditionally used rural herbal remedies (UNESCO 1998). These herbal remedies are very popular for the treatment of minor ailments due to their easy availability, affordability, non-toxic nature and lack of any side effects. It is estimated that approx. 25 per cent of the drugs prescribed contain plant extracts. These medicinal plants are largely restricted to families like Asteraceae, Euphorbiaceae, Lamiaceae, Fabaceae, Solanaceae, Rubiaceae, Poaceae, Acanthaceae, Rosaceae and Apiaceae with predominance of Asteraceae.

The practice of traditional medicines is predominantly spread in India, China, Pakistan, Japan, Thailand and Srilanka. In China, about 40 per cent of the total medicinal consumption is attributed to traditional tribal medicines. Gorman (1992) drew attention to the power of Chinese folk medicinal potions in treating maladies from eczema and malaria to respiratory disorders. In Japan, herbal medicinal preparations are more in demand than mainstream pharmaceutical products. In Africa, extracts of *Phytolacca dodecondra* (Endod) is used as effective molluscicide to control schistosmiasis (Lemma 1991). Other notable examples are *Catharanthus roseus* (yield antitumor agent) and *Ricinus communis* (laxative). Medicinal plants also constitute a valuable source of foreign exchange for most of the developing countries and India is known to export large quantities of phytochemicals like papain, solanesol, quinine sulphate, quininehygrochloride, atropine sulphate and other alkaloids. The use of medicinal plants like *Eupatorium perfoliatum* (honest), *Podophyllum peltatum* (mayapple) and *Panax quinquefolium* -(ginseng) in USA has long been associated American Indians. In the mid-90s, it is estimated that receipts of the more than 2.5 billion dollars (US) have resulted from the sale of herbal medicines.

With an ever increasing global inclination towards herbal medicines, there is an obligatory demand not only of raw material of medicinal plants but also of right stage when the active principles are available in optimum quantities. For instance, diosgenin content of tubers of *Dioscorea floibunda* reaches to its maximum after 3 years of vegetative growth of leader shoot while solasodine content in berries of *Solanum khasianum* is maximum upon color changes from green to yellow. Despite the increasing use of herbal medicinal plants, their future is being threatened by complacency concerning their conservation. More than 90 per cent of the medicinal plants for herbal industries in India and for export are drawn from the natural habitats. Thus, combination of factors like over exploitation, habitat destruction and unsustainable harvesting coupled with illegal trade practices have driven many medicinal plant species to the brink of extinction, thus making the conservation of valuable resources all the more essential. Poor seed set, seed sterility or recalcitrant nature of seeds also comes in the way of their propagation whereby making task of conservation difficult. Introduction of invasive exotic species like *Parthenium, Eichhornia, Lantana,* etc. are also reported to threaten many of the native species. Thus, medicinal plants are under tremendous pressure all across the globe especially in India.

Plant tissue culture (PTC) technique has a great potential in propagating and conserving rare, recalcitrant endangered plant species in the absence of viable conventional propagation methods. Protocols have been developed for their propagation at a rapid pace by the Government and academic institutions. Propagation of medicinal plants can be achieved by three methods. The most common approach is to isolate shoot tips or axillary buds and induce them to grow into complete plantlets (micropropagation). In the second approach, adventitious shoots are initiated on leaf, root and stem segments or on callus derived from these organs. The third system involves induction of somatic embryogenesis in cells and callus cultures. This system is theoretically most efficient as large number of somatic embryos can be obtained once the whole process is standardized. PTC technique holds much promise as a method for producing complex active ingredient (alkaloids, steroids etc.) of the medicinal plants *in vitro* since chemical synthesis as at present fraught with problems. This is so because the phytohormonal adjuvant not only regulate callusing and morphogenesis but monitor qualitative and quantitative production of steroids, alkaloids and other compounds. Singh (2005) reported that addition of 50 mg^{-1} cholesterol to the medium increased sapogenin production significantly over the natural roots of safed musali. This is further substantiated by the fact that foliar application of IAA and NAA (50 mg^{-1}) not only improved whole plant biomass in *Andrographis paniculata* but andrographolide content also (Gudhate *et al.*, 2009). This chapter is devoted to micropropagation of some of the selected threatened/endangered medicinal plants which are in great demand in drug industry. Relevance of PTC technique in the production of secondary metabolites is discussed.

Micropropagation

Micropropagation of medicinal plants remained neglected till complete plants of *Rauwolfia serpentina* were produced from its somatic callus tissue (Chaturvedi 1968, Mitra and Chaturvedi 1970). This was followed by report on plantlet formation from leaf and root explants of *Atropa belladonna* (Zenktler 1971). Large scale rapid production of clonal plants through *in vitro* cultures of single node stem segments and shoot to shoot proliferation in *Dioscorea floreibunda* is often quoted in the success story wherein more than a million fold increases in the rate of multiplication over the conventional method has been possible (Chaturvedi 1975, Chaturvedi and Sinha 1979). The technique of clonal multiplication is imperative in case of medicinal plants which gibe low yield of active principles but are required in huge amounts like *Catharanthus roseus,* 2 tonnes of leaves of which yield only 1 g alkaloid required to treat a leukemia patient for 6 weeks (Wickens 2001) or *Taxus brevofolia,* bark of a 200 year old tree is required for treating one patient of ovarian cancer (Payne *et al.*, 1992).

Micropropagation in general, is achieved through rapid proliferation of shoot apices and axillary buds in culture and for this cytokinins and weak auxins are most widely used growth regulators. Benjamin *et al.* (1987) reported that 2-5 ppm BAP stimulated the development of axillary meristems in shoot tips of *Atropa belladonna.* Lal *et al.* (1988) observed rapid multiplication in *Picrorhixa kurroa* using 1-5ppm kinetin. Further, Barna and Wakhlu (1988) indicated that the production of multiple shoots was higher in *Plantago ovata* on medium adjuncted with 4-6µM kinetin along with 0.005µM NAA. Micropropagation of endangered medicinal plants is dealt family wise below.

Acanthaceae

Adhatoda vasica also known as Malabar nut tree is exploited for extraction of an alkaloid vasicine, an uterotonic abortifacient. Extract of the plant is also used as an effective ayurvedic medicine in the treatment of tuberculosis. Efficient propagation method using shoot tip culture was developed by Nath and Buragohain (2005). Shoot tips collected from mature plants produced multiple shoots with stunted curled leaves upon inoculation on MS medium supplemented with 0.3mg/l thidiazuron (TDZ) and 15 per cent coconut milk. Sub culturing these shoots on MS medium supplemented with 0.5 mg/l BAP resulted in shoots with healthy leaves. High frequency rooting (9.33 roots per shoot) was observed in MS basal medium without growth hormone. The rooted plantlets could be established in the field at 85 per cent success rate after hardening.

Peristrope bicayculalata, is widely used in traditional system of medicine against tuberculosis for its antibacterial properties. Sharma and Vimala Devi (2006) developed conservation strategies to propagate this species *in vitro* via nodal explants. Single node explants produced six shoots of 3.67 cm with 4-5 nodes upon culturing on MS medium supplemented with 0.5 per cent BA and 0.1 mg/l IAA. Rooting was readily achieved upon transfer of the shoots on half strength MS basal medium. Plantlets were acclimatized and transferred to the soil.

Rhinacantus nasutus (popularly called as snake jasmine) is a rich source of naphthoquinones, rhinacantin-A and B, lupeol, β-sitosterol and stigmasterol. Roots are used to treat skin diseases, worm infestation, disorders of blood and poisoning. Nodal explants produced multiple shoots (8) on MS medium supplemented with 2.22µM BAP. The excised shoots successfully rooted on half strength MS medium fortified with 9.2μM IBA. Rooted plantlets have been established at 75 per cent success to the nursery field without any variariations in morphological or growth characteristics (Johnson *et al.*, 2002).

Apiaceae

Anethum graveolence, popularly called as dil, is carminative, stimulant and diuretic. A protocol has been designed for rapid and large scale *in vitro* propagation (Sharma *et al.*, 2004). Inoculation of shoot tips excised from 2- weeks old seedlings on MS medium supplemented with 0.5 mg/l BAP and 0.1 mg/l IBA produced healthy shoots with similar frequency. Rooting of excised shoots was achieved on MS medium with 1.0 mg/l IBA and 0.5 mg/l kinetin. The rooted shoots were hardened and transferred to field with 64 per cent survival rate.

Centella asiatica, Indian pennywort a source of nervine tonic is used in the treatment of leprosy, asthma, bronchitis, dropsy, leucorrhoea, skin disease and urethritis, the wound healing activity of the species is ascribed to a triterpenoid saponin, asiaticoside acid. Nath and Burgohain (2003) cultured shoot tips isolated from mature plant which produced multiple shoots (mean 3.8) on MS medium fortified with 4.0 mg/l BA and 0.1 mg/l NAA. Profuse rooting (46.6 per shoot) of micro shoots is reported on MS medium fortified with 2.0 mg/l IBA with mean root length of 19.7 cm. Well rooted plantlets were acclimatized and successfully established in the field with hardly 1-2 per cent mortality.

Pimpinella tirupatiensis, the forest coriander, roots are generally supplemented to few other ingredients to cure colic and rhematic ailments in cattle. Fruits are used to cure asthma and are considered as an effective remedy for flatulent colic. Culturing of hypocotyls explants taken from 4 week old seedlings on MS medium supplemented with 1.0 mg/l TDZ and 0.5 mg/l NAA produce friable callus which upon subculture on MS medium with 1.0 mg/l TDZ, produced green shoots with white hairy roots. Regenerated plantlets showed normal growth and flowering pattern with 90 per cent success in the soil (Prakash *et al.*, 2001).

Apocyanaceae

Rauwolfia serpentina, commonly known as sarapgandha, a wonder drug plant, yields around 50 indole alkaloids with predominance of reserpine. Reserpine is still today a potent drug for controlling hypertension as well as mental diseases like paranoia and schizophrenia. Another alkaloid, ajamaline also has a similar therapeutic value. Four morphogenetic patterns of differentiation composing of somatic embryogenesis, cauloguenesis, direct plantlet formation and regeneration of shoot buds from root explants has been reported. However, propagation rate has only been moderate (Chaturvedi, 1968). In a later study, an efficient multiplication of clonal plants through shoot to shoot proliferation by employing nodal segments has been reported (Chaturvedi, 1979). The other commercial source of these drug molecules (terpenoids, indole alkaloids) are *R. vomiforia* and *R. tetraphylla* (Banerjee and Sharma 1989). Ghosh and Banerjee (2003) developed a suitable *in vitro* protocol for mass clonal propagation for *R. tetraphylla.* Shoot apices from mature plant cultured on MS medium supplemented with 1.0 mg/l IAA and 5.0 mg/l BAP is reported to produce approximately 15 multiple shoots per explants. Multiple shoots rooted on MS medium fortified with 2.0 mg/l NAA. These rooted shoots exhibited further growth as well as branching of the root system upon transfer to half strength MS basal. The plantlets could be transferred to the soil after stepwise hardening process with 70 per cent success.

Asteraceae

Holostemma annulare, is a rare medicinal plant with tuberous roots, which constitute a major ingredient of the drug 'Jivanthi' in the indigenous system of medicine. Alcoholic root extract possesses antidiabetic property. A rapid clonal propagation system was achieved through chlorophyllous root segments derived from *in vitro* rooted plants by Sudha *et al.* (2000). Root segments when implanted horizontallly on half strength MS medium supplemented with 0.3 mg/l IBA induced solitary shoot buds with small laterals. The regenerated shoots having elongated roots were transferred to the field with 80 per cent survival.

Saussurea lappa, commonly called as costus or Kuth has been successfully micropropagated (Arora and Bhojwani 1989). They observed 3.5 fold shoot multiplication every 3 weeks on MS medium supplemented with 5µM BAP and 3 µM GA_3 by culturing 10-15mm long terminal portion of phyocotyl plumula and bearing the shoots bud. These shoots rooted as MS medium +0.5µM NAA with 90 per cent efficiency.

Spilanthes acmella, also known as toothache plant or paracress, is used as remedy for stammering in children. The hexane extract of dried flower buds contains bioactive N-isobutylamides, which are effective against *Aedes aegypti* larvae. Nodal segments from the young and healthy branches produced multiple shoots (up to 28) upon inoculation on MS medium supplemented with 0.75mg/l BAP after 60 days of culture. Shoots rooted best on MS medium with 0.5 mg/l IBA (48.8 roots). Rooted shoots after hardening have been successfully transferred to field (Deka and Kalita, 2005).

Vernonia cinerea called as sahadevi, is known to contain antihyperglycaemic and anti-ulcer active triterpenoid 'lupeol acetaate' besides other chemical compounds as sterols, flavonoids, sesquiterpene lactones, luteolin 4'-o-glucoside, chlorogenic acid, 3,5-dicaffeoylquinicacid and methyl caffeate. Leaf and nodal explants obtained from field frown mature plant produced multiple shoots upon inoculation on MS medium supplemented with 3.0 mg/l and 2.5 mg/l BAP, respectively. Further shoots proliferation has been reported upon sub culturing of shoots on MS medium containing 2.0 mg/l BAP along with the 1.5 mg/l NAA. Transfer of these shoots to half strength MS medium containing 1.5 mg/l IAA yielded 25-30 roots per shoot. *In vitro* raised plantlets were acclimatized and transferred in the field with 80 per cent survivalist (Seetharam *et al.*, 2007).

Wedelia calendulaceae, the bhringraj, is a perennial herb and its active ingredient is hepatoprotective in action. The leaves are regarded as a tonic and useful in cough, cephalagia and skin diseases especially alopecia. Emmanuel *et al.*(2000) established a suitable protocol for its micropropagation through axillary bud and shoot tip explants on MS medium containing Gamborg vitamins. Explants produced multiple shoots on the above medium containing 8.9 µM BAP and 4.5µM kin. These shoots rooted upon excision on 4.9µM IBA in half strength MS medium.

Asclepiadaceae

Gymnema sylvestre, a potent antidiabetic plant and a vulnerable species is slow growing and perennial woody climber of tropical and sub-tropical regions. It is used in the treatment of asthma, eye complaints, inflammation, family planning and snake bite (Uniyal 1993, Selvanayagam *et al.*, 1995). In addition, it possesses antimicrobial, antihypochloesterolemic (Bishayee and Chatterjee 1994), hepatoprotective (Rana and Avdhoot 1992) and sweet suppressing activities (Kurihara 1992). Komalavalli and Rao (2000) succeeded in obtaining multiple shoots (57.2) from 30 day old seedling axillary node explants on MS +1 mg/l BAP + 0.5 mg/l Kn + 0.1mg/l NAA +100mg/ l malt extract +100 mg/l citric acid. Shifting these shoots to half strength MS medium adjunct with 3 mg/l IBA yielded high frequency of rooting. These plants could be successfully hardened and established in natural soil.

Hemidesmus indicus, Indian sarasspilla is reported to be rich source of saponins, sapogenins, flavonoiks and triterpenes, novel pregnan3 glycosides like desinine, indicine and hemidine, hemine-1 and hemidine-2. Air dried roots are used as tonic, diaphoretic, blood purifier, in rheumatism and in skin infections. Malathy and Pal (1998) succeeded in mass propagation through nodal segments. Root segments excised from aseptically grown 60 day old seedlings produced micro shoots upon horizontal placement on MS medium supplemented with 3.0mg/l BAP and 0.5mg/l NAA. Micro shoots elongated upon transfer of the responding root segments to hormone free half strength MS medium. Roots were formed from the vassal region of the shoots which later developed into healthy roots. The regenerated shoots with well-developed roots after hardening were transplanted plants showed 85 per cent survival (Ramulu *et al.*, 2003).

Leptadenia reticulata, commonly called as jiwanti, dodi is an important twining medicinal shrub possessing abundant alkaloids. The plant is a stimulant and a restorative and is used for curing nose and ear problems. Leaves and roots are used for the treatment of skin infections and wounds. It is regarded a good cure for tuberculosis and eye diseases. Hariharan *et al.* (2002) achieved mass micropropagation from leaves through somatic embryogenic callus. Subculture of these embryonic calli on half/full strength hormone free MS medium or medium supplemented with 1.0 mg/l BAP or kinetin develops embryoids. MS medium fortified with 1.0 mg/l kinetin very effective for the germination of these embryoids. These germinated plantlets could be successfully transferred to the field with 50 per cent survival.

Boraginaceae

Rotula aquatica, a rare rheophyte with virgate branches is highly valued for its diuretic roots and is also used in cardiac swellings, uterine disorders, piles and stone in the bladder and venereal diseases. A rapid clonal propagation has been achieved by Sebastain *et al.*(2002) through mature nodal explants on woody pant medium (WPM. Lloyd and Mc Cown 1981) supplemented with 6.0 mg/l BAP. *In vitro* developed micro shoots elongated on transfer on WPM same medium fortified with 0.5 mg/l kin and 600mg/l activated charcoal. The shoots so obtained rooted on half strength WPM supplemented with 0.5 mg/l IAA and rooted shoots were successfully transferred to the field.

Campanulaceae

Lobelia nicotianifolia, known by various names like Indian tobacco, wild tobacco, pulse weed, vomiroot, etc. is important due to presence of a myriad of alkaloids particularly lobelin which is widely used in the preparation of herbal medicines. In India, it is used in the treatment of asthma, liver diseases, ulcers and nervous disorders. Export of this species is banned and now protected under the Convention on International Trade: Endangered species of wild fauna and flora. Jabeen *et al*. (2002) succeeded in obtaining high frequency callus initiated from seeds inoculated on MS medium supplemented with 2.0 mg/l 2,4-D. sub-culturing of the callus on medium adjunct with 2.0 mg/l and 1.0 g/l activated charcoal produced multiple shoots (163). The shoots rooted on half strength MS with 1.0 mg/l IBA upon excision. *In vitro* developed plantlets were transferred to the soil with 90 per cent success.

Capparidaceae

Crataeva magna, known as varuna or three leaved caper, is widely used in vitiated conditions of *vats* and *kapha*, dyspepsia, colic, flatulence, helminthiasis, strangury, renal and vesicle calculi, cough, asthma, bronchitis, pruritus, skin diseases, inflammations and hepatopathy. Its bark and leaves are astringent, bitter, acrid, diuretic, demulcent depurative, astiperiodic and tonic. Benniamin *et al*. (2004) employed nodal segments taken from mature plants on MS medium supplemented 8.8 µM BAP which produced multiple shoots (4.4 per explants). Rooting of the excised shoots could be obtained on half strength MS medium supplemented with 9.84 µM of IBA and 0.54 µM of NAA. Rooted shoots after attaining a height of 2.0-2.5 cm were hardened successfully and transferred to soil with 68 per cent success.

Cucurbitaceae

Momordica tuberose, commonly called as kakrol, is a wild edible medicinal member. Plant is characterized by relatively high crude fibric content that delays food digestion and conversion of starch to sugars, a feature desirable for diabetic patients (Gapalan *et al*., 1993). Micropropagation protocol of this species has been established using nodal segments and shoot apices from field grown mature plants. Nodal segments showed highest regeneration efficiency on MS medium supplemented with 4.40 µM BAP and 4.60 µM Kn (9±0.49 shoots/explant). Mean length of these shoots was 10±0.47 cm. MS medium supplemented with 13.30µM BAP induced shoot regeneration from shoot apex with mean shoot length of 7.8±0.66 cm. micro shoots could easily be rooted on MS medium adjuncted with 4.9 µM IBA and established in the greenhouse with 90 per cent survival rate (Aileni *et al*., 2009).

Fabaceae

Pterocarpus marsupium, the Indian kino tree is of immense importance as wood infusion or decoction is used to control diabetes. The bark is used for the treatment of stomachache, cholera, dysentery, urinary complaints, tongue diseases and toothache. Nodal segments obtained from 35-40 day old seedling produced maximum number of shoots (3 per segment) on growth regulator free MS medium. *In vitro* developed plantlets developed roots on the same medium which were acclimatized in the green house and finally in the field. The success of establishment in the field was 68 per cent and showed the high homogeneity.

Guttiferae

Callophyllum inophyllum, the Indian laurel, is known to have cancer chemopreventive and has coumarins and xanthones with antimicrobial activity. It is also gaining importance as a source of anti-HIV medicine. Thengane *et al*. (2006a) successfully employed decapitated shoots of germinated seedlings for shoot proliferation on WPM with 0.91 µM TDZ. Elongation of the stunted shoots could be achieved on half strength WPM without any growth regulators and these shoots rooted on WPM supplemented with 2.46 µM IBA. The micro propagated plants could be acclimatized successfully and 72 per cent plants showed good growth and development.

Gracinia indica, commonly known as kokun, a polygamodioecious tree is known for being a rich source of garcinic acid and hydroxycitric acid in its fruit rind. The fruits have remarkable properties *i.e.* anthelmintic, cardio-tonic, and useful in piles, dysentery, tumours and heart complaints. Direct somatic embryogenesis (48.53 per seed explants) without an intervening callus phase could be induced from immature seeds on woody plant medium (WPM) supplemented with 22.19µM BAP. These embryos matures after 12 weeks of culture on a medium containing 16.08 µM BAP, 5.71 µM IAA and 4.66 µM Kn. The regenerated somatic embryos germinated upon transfer on half strength 15 roots per explants. The well-developed plantlets were transferred to half strength MS medium supplemented with 0.89 µM BAP for further elongation and growth, and successfully acclimatized (92 per cent) in the greenhouse (Thengane *et al.*, 2006b).

Hypoxidaceae

Curculigo archioides, commonly known as 'kali musali' is a perennial herb with tuberous roots. It is reported to have hypoglycemic, spasmolytic and anti-cancer principles. Rhizome is prescribed for the treatment of piles, jaundice, and asthma, diarrhoea (Kirtikar and Basu 1950) and pimples (Bhamare 1998). Meristem tip culture on MS medium supplemented with 2.21 µM BAP induced multiple shoots. These shoots could be rooted either on MS medium or on medium supplemented with 0.53 µM NAA and acclimatized in pots containing mixture of vermiculite and soil (1:1) for a period of 2-3 weeks (Wala and Jasrae, 2003) on the other hand found that leaf explants inoculated in MS medium incorporated with 2.0 mg/12, 4-D formed multiple shoots along with profuse rooting after a period of 60 days, without intervening callus formation. The regenerated plantlets were excised then proper hardening transferred to the field condition exhibited 82.57 per cent survival rate.

Lamiaceae

Mentha piperita, peppermint is an aromatic, stimulant, stomachic, carminative and used for allaying nausea, flatulence, and headache and vomiting. Ghanti *et al.* (2004) developed a high frequency and rapid regeneration protocol from shoot tip and nodal explants. Multiple shoot buds (45-50 per explant) emerged from shoot tip and nodal explants on MS medium supplemented with 1.0 mg/l BAP. Regeneration of dwarf shoots from these buds could be achieved on MS medium supplemented with 1.0 mg/l GA_3. Rooting of these shoots excised from shoot clumps was evident on medium containing 1.0 mg/l NAA. The rooted plantlets were hardened on MS basal liquid medium and subsequently in poly cups. Plantlets, thus obtained were successfully transferred in the soil with 90 per cent survival rate.

Plectrathus vetiveroides, known as valak possesses diuretic. trichogenous and antipyretic properties. The roots, stems and leaves are used to treat hyperdipsia, vitiated conditions of biliousness, burning sensation, strangury, leprosy, skin diseases, leucoderma, fever, vomiting, diarrhoea and ulcers. Culturing of nodal explants from adult plant on MS medium supplemented with 4.44 µM BAP stimulated multiple shoot formation (12 per explants). Multiple shoots rooted on half strength MS medium fortified with 7.38 µMM IBA after hardening in the green house for 3-4 weeks have been successfully transferred to the field and 82 per cent plants become established without morphological variation (Sivasubrananian *et al.*, 2002).

Menispermaceae

Tinospora cordifolia, the Giloe, is known to possess tonic like vitalizing properties and used as a remedy for diabetes, arthritis and metabolic disorders. The aqueous and alcoholic extract of the plant causes reduction in the fasting blood sugar. The stem contains a glycoside, giloin, a non-glycoside, bitter gilenin and a gilo-sterol. Kumar *et al.* (2003) cultured nodal explants from mature plants on MS medium supplemented with 2.0 mg/l BA and 0.06 mg/l NAA which induced shoot formation in 96 per cent cultures along with root formation. The plantlets were excised and transplanted in pots after sequential hardening.

Mimosaceae

Entada pursaetha, an endemic woody liana, is potential source of drug for various ailments such as cancer, liver disorders, dropsy, eye diseases, cuts, wounds, snakebites respiratory problems, debility, tuberculosis and anasarca. Efforts have been made by Vidya *et al.*(2005) to conserve this species through *in vitro* micropropagation using cotyledonary node as explants. Cotyledonary nodes excised from *in vitro* frown seedlings produced adventitious shoots (9.8 per explants) on MS medium supplemented with 5.0 mg/l BAP and 0.5 mg/l NAA. Micro shoots rooted on MS half strength medium supplemented with 2.0mg/l IBA. Seventy per cent of hardened rooted plants could be established in the field.

Salvadoraceae

Salvadora oleoides, commonly called as Khabbar, Pilu, is a small tree. Its root bark is used in painful rheumatic infections and fruits are used for the treatment of piles, bronchitis and diseases of spleen. The seeds contain quercitin, rutin and an alkaloid. Among different explants (leaf, nodal segments, internodes and cotyledons) only nodal segments showed best induction and subsequent growth (6-18 buds/node) on MA+BAP (2 mg/l) + 0.1 mg/l NAA. Culturing of these nodes with shoots after 45 days on the same medium increased in number further of proliferated shoots. Rooting of these shoots could be achieved on MS + 3.0 mg/l IBA which were successfully transferred to the pots (Singh and Goyal, 2007).

Solanaceae

Withania somnifera, the indian ginseng is endowed with medicinal properties ranging from ant-inflammatory, anti-arthritic, and curative for nervous and gynecological disorders, immunomodulatory to antitumour. It has found its application in chemotherapy and radiotherapy against cancer. Its medicinal value is due to alkaloids like somniferine and withasomnine and in particular withaferine, which possess anti-tumor and antibacterial properties. Nodal and internodal segments from seedlings have been successfully used for micropropagation (Kulkarni *et al.*, 2000). regeneration of plantlets from somatic callus (Rani and Grover, 1999) and from leaf segments taken from field grown plants is reported (Sivanesan and Munigesan 2005). Kannan *et al.* (2005) developed a large scale production method through deployment of nodal buds. Nodal buds explants excised from aseptically germinated 15 days old seedlings produced multiple (6.7 per explants) shoots upon inoculation on MA medium supplemented with 10.0 µM BA and 0.1 µM NAA. Elongation of these micro shoots could be achieved on medium containing 10.0 µM GA_3 in combination with 0.1 µM BA. Rhizogenesis in the regenerated shoots could be obtained on MS medium supplemented with 0.1 µM NAA. The rooted plants were transferred to hormone free MS medium for further growth and development. The hardened plantlets could be successfully transferred to the field with 90 per cent success.

Verbenaceae

Vitex negundo is a large woody medicinal and aromatic shrub/small tree. Unfortunately, its natural regeneration and conventional propagation through vegetative cuttings is slow and a large number of cuttings do not survive during transport and plantation. Sahoo and Chand (1998) succeeded in bud break of nodal segments from mature tree on MS medium supplemented with 2.0 mg/l BAP. Addition of 0.4 mg/l GA_3 led to an enhanced frequency of shoot development influenced by the explanting season. Prolific shoot cultures free from proximal callusing and high frequency multiplication rates could be achieved by repeated sub culturing of nodal segments harvested from *in vitro* formed auxenic shoots of MS medium + 1.0 mg/l BAP + 0.4 mg/l GA_3. The percentage shoot multiplication as well as the number of shoots/node was highest (6-8) during the first three culture passages which gradually declined thereafter. These shoots rooted best on 1/2 MS medium augmented with 1 mg/l each of IAA and IBA.

Secondary Metabolites

Higher plants in addition to their normal metabolic pathways produce a variety of compound which are termed as secondary metabolites like terpenoids, phenols, alkaloids, etc. functionally, some secondary metabolites are as viral to plants as primary ones. For instance, both ABA and GAs belong to class of terpenoids but these phytohormones are involved in controlling the development, differentiation and dormancy in plants. Phytosterols, an essential component of cellular membranes, are indeed classified as terpenoids. The most widely distributed sterols in higher plants are stigmasterol and β-sitosterol with 29 C atoms. On the other hand, there are secondary metabolites like alkaloids whose precise functional significance to their producer plant is still uncertain. It is difficult to comprehend, why plants should produce such substances on a large scale when those find no functional significance in their biochemistry economy. This contrasts with their significance in their biochemistry economy. This contrasts with their significance to animals and man who use them as powerful drugs. The *Cinchona* alkaloid, quinine, is the famous anti-malarial drug. Resperine from *Rauwolfia* plants is very effective in the treatment of hypertension. Higher plants till today are an important source of widely used pharmaceuticals. This is so because these pharmacologically active compounds have a rather complex structure and thus can't be synthesized chemically at a competitive price. As a result morphine and codeine are still isolated from the latex of unripe seed-pods of *Papaver somniferum* or quinine and quinnidine are to be extracted from the cortex of *Cinchona* trees. The tropane alkaloids are still produced by extraction from *Atropa*, *Hyoscyamus* or Datura.

Until the mid1970's, diosgenin was the exclusive source for corticosteroids and contraceptive production but due to the shortage of this material, the search for altenative sources lead to identification of solanosodine, which bears close structural similarity with diosgenin. This occurs in relatively high concentrations in *Solanum aviculare* and *S. laciniatum* (Beatson and Rohravach 1985). It is usually found that a series of sugar residues are attached to the oxygen at C 3 position and the most common forms are triglycosides, solamarhine. On the other hand, solasodine (SD) occurs in free form in *S. cyananthum* (Rizk and Abou-Zied 1970). The higher glycosides starting from the triglycoside are more polar, more water soluble than lower homologues. The isolation and degradation of SD is easier than the degradation of diosgenin but its content is much lower than that of diosgenin in Mexican *Dioscorea composite* (Crabbe' 1970). Extension of these studies by Schreiber (1968) has revealed that out of 267 *Solanum* species, 52 contain solasodine aglycone; some of these accumulate it in the whole plant while tropical species concentrated the alkaloids in the fruits only (Telek 1979). Ripperger and Schreiber (1981) observed solamargine and SD to be predominant alkaloids in 60 per cent of the tested species.

Since these pharmacologically active compounds have a rather complex structure, these can't be synthesized chemically at a competitive price. Therefore, it would be of particular interest if factory type production of these pharmaceuticals were possible. PTC technique offers to play a significant role to achieve this objective. Heble *et al*. (1968) isolated diosgenin and β-sitosterol from the tissue cultures of *Solanum xanthocarpum*. Kaul and Staba (1968) isolated diosgenin from *D. deltoidea* callus and suspension cells. The diosgenin content was highest (1.02 per cent) in 3-4 week old suspension cells whereas the differentiated cells showed only trace amounts of diosgenin. Addition of labeled cholesterol on the 10th day of culturing has been reported to stimulate diosgenin production by seedling callus while intitial feeding is reported to be inhibitory. Tomita *et al.* (1970) observed that cultures of seedling explants origin in *D. tokora* Makino retained the ability to synthesize tokorogenin, dosgenin, yonogenin, stigmasterol, β-sitosterol, campesterol and cholesterol even after being cultured on LS medium±10^{-3} M 2,4-D + 0.2 ppm kinetin after 1 year. Bhatt *et al.* (1983) demonstrated stimulation of SD content by IAA and sucrose in *Solanum nigrum* cultures. BAP was stimulatory in shoots and inhibited production in callus when applied in combination with IAA. The differentiated tissue produced more SD than callus when incubated in darkness. However, reverse was the trend when frown under a 16th photoperiod. They further reported that total 4-desmothylsterol content was higher *in vitro* differentiated shoots than in the young leaves of intact plants or in callus (Bhatt *et al.*, 1984). Sterol production was higher in shoot producing cultures than in root producing ones (Bhatt and Bhatt, 1984). In an attempt to correlate texture of callus with alkaloid content, Kokate and Radwan (1979) found higher levels of alkaloids in compact cell aggregates than in friable callus of *S. nigrum* and *S. khasianum*. Uddin and Chaturvedi (1971) observed variations in SD content depending upon explants of the same plant in *S. khasiana*. Generally the SD content increased initially up to 6-8 weeks and then gradually decreased. Heble *et al.* (1971) succeeded in isolating lupeol (triterpene) from callus of *S. xanthocarpum*. They succeeded in improving the diosgenin content by substituting 2,4-D by IAA or IBA however the cultures lost SD forming ability. Ehmke and Eilert (1986) found 0.2 mg/g DM alkaloids (Soladulcidine, SD, tigogenine and diosgenin) in green callus of *S. dulcamara* while the content was 0.1 mg/g DM in green suspension cultures. However, no alkaloid could be detected in heterotrophic cell line. Interestingly, alkaloid production in green cultures was not restricted to a particular growth period.

Bhatt *et al.* (1986) compared sterols and alkaloids in plants regenerated from cell suspension cultures of *S. dulcamara* with intact plants. Cholesterol was the predominant sterol (42 per cent of total) in the leaves but its level was reduced to 1-7 per cent in cell suspension cultures. Chandler and Dodds (1983a) observed an enhancement in SD content by induction of organogenesis in both primary leaf explants and callus in *S. laciniatum*. Leaf explants which produced roots and callus contained significantly higher concentrations of SD than those producing callus only. Addition of 0.04 mg/l ABA increased SD yield in dark grown callus. SD concentration is reported to increase further when medium phosphate or nitrogen are reduced or when sucrose on concentration is increased (Chandler and Dodds 1983b). Rerabck and Gorelova (1986) found that excess N or N-NO_3 form enhance steroid production and biomass yield but decreases SD content in hypocotyls derived suspension cultures of *S. laciniatum*. Medium enriched with ammonical N is reported to improve growth and alkaloid production.

Among different explants tested, root is reported to yield highest concentration of steroids followed by stem and leaf (0.31, 0.27 and 0.22 mg/g Dm, respectively) in *S.*

ariculare on a medium supplemented with 2,4-D and Kinetin. Galancs *et al*. (1984) identified steroids like squalene, lanosterol, stigmosterol, cholesterol, campesterol, β-sitaserol and culoaretivol in callus / suspension cultures of *S. ariculare* raised on MS medium +1mg/l 2,4D in dark from roots and hypocotyls. However, diosgenin was not detectable.

Establishment of Productive Cell Culture Systems

First question related to establishment of cultures for secondary metabolite production is whether special considerations or additional measures should be taken into account? The answer is simple that such cultures are establishment in the same manner as those used for other purposes. Secondly how important is the plant material and the choice of culture initiation medium for the later appearance of culture? If one wants to establish a culture producing a compound which is only synthesized in roots, is it then necessary to start culture initiation from roots only? The clear answer is 'no' since the plant cells are totipotent which differentiate into complete plantlets. Thus, biosynthesis of a compound may be repressed in leaf cells but as soon as these cells enter into another stage of differentiation, biosynthesis may resume. Conversely, the root derived callus may or may not synthesize the compound of the culture medium. Physiological state of the cell(s) from which callus is derived and its genetic makeup are important determinants for the initial development of cultures. Kinnerslley and Dougall (1980) reported a strong correlation between the nicotine content of callus cultures and the plants derived from them. Kurz *et al*. (1985) formulated that one of the cultivars of *Catharanthus roseus* produced four times more catharantine than others thereby indicating variability of secondary metabolite production among cultivars of the same species.

Influence of Culture Conditions on Productivity

Another question needing attention is whether the composition of the callus induction and growth medium is same as that of production medium? Secondary metabolite production is influenced by factors like medium composition and physical, biotic and abiotic elicitors. Medicinal plants in general, are very often found in secondary growth or perturbed habitats, highly associated with modified landscapes. Plant defense theory support this contention as perturbed habitats are sources of biologically active secondary compounds (Coley and Barone 1996). As long as the cells are not damaged by a given treatment, these changes are reversible and represent different physiological states. In general, the best growth medium is seldom a best production medium. The most desirable change in the composition of growth medium include (i) reduction or depletion of 2,4-D or other phytohormones with or without replacement by others (ii) reduction of phosphate levels (iii) high sucrose and alteration of C/N ratios (Lindsey and Yeoman 1983). The effect of most production media is through imposition of stress on the cells which inhibits early growth by nutritional deficiencies (Diicosmo and Tower 1984). Indeed growth is greatly reduced in production media or continues for only one growth cycle. Growth inhibition in cultures is often followed by cytodifferentiation and the induction of enzymes of secondary metabolism. It is indicative of positive effect of growth limiting culture conditions on specific production rates. Knobloch *et al*. (1981) observed that young cells loaded with phosphate showed smaller increases in cinnamoyl putrescine in the production medium than the older phosphate depleted cells. The effect of 2,4-D may be seen during the second passage when the carried over or stored 2,4-D has been sufficiently diluted in the cells. The initial ratio of fresh mass (cell number) to the total amount of constituents is also very important for further responses. Some of cell cultures rapidly absorb all phosphate from the medium and store in their vacuoles from where it is distributed within the growing cell population. Such cells may enter a different physiological state or may be delayed in moving to another state. Since cell cultures exist as heterogeneous population of cells in different physiological states and each may react differently to an external event such as medium variation. This supports the occurrence of many different secondary metabolites accumulating cells. High sucrose mediated increase in dry mass is possibly due to an enhanced storage or incorporation of the C source but not due to cell division and growth. Therefore, a good production medium must be carefully balanced between growth and best specific production rate for optimal results. In case where production of compound occurs in the late stationary phase, when one or more media constituents have been depleted, transfer of seemingly non-producing cell to a production medium may yield optimal results as more healthy cells are induced for product formation under controlled conditions. This has been substantiated by Kaul and Staba (1968) that addition of labeled cholesterol on the tenth day of culture stimulated diosgenin production by seedling callus while initial feeding was inhibitory. Heble and Staba (1980) reported that the metabolic events leading to the synthesis of steroids are more active during stationary phase cells under appropriate nutritional and hormonal conditions. They recorded lower diosgenin content of cells grown on the medium lacking sucrose while 2,4-D had a beneficial RT medium supplemented with 0.5 mgl^{-1} BAP has been reported to be optimal for multiple shoot bud formation and diosgenin synthesis. The diosgenin content of the cultures was 72 per cent of the value obtained from *in vivo* shoots. Tal and Goldberg (1982) found that 6 per cent sucrose resulted in higher diosgenin production. They further reported that ammonium and nitrate nitrogen is required for growth and steroid production. It has been shown that cell culturs in some plants produce higher product level than the intact plant thereby indicating that in some cases expression of a pathway is evidently better in cultured cells. However, in most cases their expression is not maximal in cultured cells as higher levels are reported in untact plants. Linsey and Yeoman (1983) and Yoshikawa and Furaya (1985) reported distinct increase in product level upon morphological differentiation. Low level of important compounds demonstrates inability to manipulate expression of secondary pathways. Some pathways such as those for protoberberines are spontaneously expressed in many different culture systems and levels upon 15 per cent are reported. Whether the production of berberine via PTC has a viable future is yet to be seen. Ulbrich *et al.* (1985) reported a production rate of 0.9 g/l/d rosmarinic acid. Since it possesses antiphlogistic activity, this seemingly unimportant compound may become an interesting candidate for production in cell culture. Shikonins, on the other hand, seem to be the only compounds which are produced commercially via cell cultures (Fujita *et al.*, 1982). Kaneda *et al.* (1987) observed that unorganized cell suspension cultures and root cultures of *Yucca schidigera* contained insignificant concentration of glucosides and sapogenins. Shoot organ culture also contained the same major sapogenins as the plant shoot but in different concentrations and had glycoside pattern resembling mature rhizome. Unexpectedly, none of growth regulators improved sapogenin concentration in shoot cultures.

Feeding of precursors has been employed with variable success to increase *de novo* synthesis of metabolites wherever, the productivity of cell line traced to the enzyme level, it has evinced that low or high productivity is related to the corresponding enzyme activities (Berlin *et al.*, 1982). The more productive a cell line is, the better are chances of increasing product level by precursor feeding (Berlin and Witte 1982). Therefore, it is questionable whether the precursor feeding to low producing cell lines is worthwhile expenditure. Since in nature a large number of compounds are synthesized in response to chemical or microbial attack, it is worthwhile to find whether eliciting compounds could be useful in secondary metabolism in cultured cells. A 26 fold increase of sanguanarine amounting to 2.9 per cent of dry mass has been reported upon addition of *Botrytis* sp. preparations to cultures of *Papaver somniferum* (Eilert *et al.*, 1985). Berlin *et al.* (1985) observed spontaneous production of sanguinarine and other benzophenathridines upon 6 per cent in cell cultures of Papaveraceae. The formation of ajamalicine and catharanthine by *Catharanthus roseus* cell cultures is known to be stimulated by fungal cultures (Dicosmo *et al., 1987*) it thus evinces that addition of fungal preparation represents an additional and superior method of re-inducing these alkaloids in cultures. Similarly, co-cultivation of *S. laciniatum* cell suspension with cyanobacterium enhances production of SD (Gorelova *et al.*, 1985). Therefore search for suitable and best elicitors of a distinct secondary pathway should continue, as additional tools with different mode of actions especially for those which don't respond to medium variations. The worthwhile question is whether such elicitors exist that are able to stimulate morphinan formation in *P. somniferum* to levels that are required for commercial exploitation. Synthesis of flavonoids (Hahlbrock and Griesebach 1979), cardenolides (Ohlsson *et al.*, 1983) and betacyanins (Berlin *et al.*, 1986) is influenced by light and the response depended upon the physiological state of the cells. From the technological point of view, however, it is more convenient if light is not required for achieving relevant product levels and could be replaced by other effectors.

The ultimate aim of PTC groups worldwide had been to establish productive cell culture system to obtain compounds for pharmaceutical industry. Despite some exciting improvements in yield of secondary metabolites by media variations, some experiments have yielded disappointing results. The relevant question that arises is what are the limits of media variations for product enhancement? First, the cells have to be competent *i.e.* they must be able to enter the production state. β- carboline production in *Peganum* cells (0.1-2 per cent alkaloid) increased only in cultures showing expression of its pathway at least to a low extent. It is also reported that total levels of low producing cell lines are often difficult to improve by media variations. The methods of controlling the depression of metabolic genes, to direct the precursor flow from primary to secondary metabolism (Berlin *et al.*, 1985) are likely to provide the main thrust of future work. In addition to transformation and immobilization techniques, some positive results can also be expected from exploitation of artificially elicited stress metabolism. This contention is supported by the fact that plants of *Solanum aviçulare* enhance their production of antifungal metabolites including SD in conditions of stress (Rowen *et al.*, 1983).

References

Aileni M, Reddy Kotta S, Kokkirala V R, Umate P and Abbagani S 2009. Efficient *in vitro* regeneration and micropropagation of medicinal plant *Momordica tuberose*. *J Herbs Species Med Pl* **15**: 141-148.

Arora R and Bhojwani S S 1989. *In vitro* propagation and low temperature storage of *Saussurea lappa* CB Clarke- An endangered medicinal plant. *Pl Cell Rep* **8**: 44-47.

Aziz Z A, Davey M R, Power J B, Anthony P, Smith R M and Lowe K C 2007. Production of asiaticoside and madecosside in *Centella asiatica in vitro* and *in vivo*. *Biol Pl* **51**: 34-42.

Banerjee N and Sharma A K 1989. Chromosome constitution and alkaloid content in *Rauwolfia* L. (Apocynaceae). *Cytologia* **54**: 723-728.

Beatson R A and Rohrabach J F 1985. Genotype-environment interaction in poroporo (*Solanum aviculare*). *NZ J Agric Res* **28:** 125-128.

Benniamin A, Manickam V S, Johnson M and Joseph L H 2004. Micropropagation of *Crataeva magna* (Lour.) DC- A medicinal plant. *Ind J Biotech* **3**: 136-138.

Berlin J 1986. Secondary products from plant cell culture, **In:** Rehm H J, Reed G (eds), Biotechnology VCH Weinheim 630-658.

Berlin J and Witte L 1982. Metabolism of phenylalanine and cinnamic acid in tobacco cell line with high and low yields of *Cinnamoyl putrescines*. *J Nat Prod* **45:** 88-93.

Berlin J, Knolbach K H, Hoffe G and Witte L 1982. Biochemical characterization of tobacco cell line with different levels of *Cinnamoyl putrescines*. *J Nat Prod* 45:83-87.

Berlin J, Beier H, Forche E, Noc W, Sasse F, Schielgo J and Wray V 1985. Conventional and new approaches to increase the alkaloid production of plant cell culture. *In*: Neumann K H, Barz W and Reinhard E (eds) Primary and Secondary Metabolites of Plant Cell Culture. Springer, Berlin 272-280.

Berlin J, Sieg S, Strack D and Bokern M 1986. Production of betalains by cell suspension cultures on *Chenopodium rubrum*. *Pl Cell Tiss Org Cult* **5:** 163-167.

Bhamare P B 1998. Traditional knowledge of plants for skin ailments of Dhule and Nandurbar, Maharashtra. *J Phytolo Res* **11:** 196-198.

Bhatt P N and Bhatt D P 1984. Changes in sterol content during leaf ageing and *in vitro* differentiation in *Solanym nigrum*. *J Natl Prod* 426-432.

Bhatt P N, Bhatta D P and Sussex I 1983. Studies on some factors affecting solasodine contents in *Solanum nigrum*. *Physiol Plant* **57:** 159-162.

Bhatt P N, Bhatt D P and Mehta A R 1984. Strategies to increase steroidal production in cell cultures of *Solanum* Sp., Proc. 7th Intl Biotechnol Symp, New Delhi 116-117.

Bhatt D P, Bhatt P N and Mehta A R 1986. Steroid analysis and plant regeneration from suspension cultures of *Solanum dulacamara* L. *Beitr Biol Pflanzen* **61:** 202-213.

Biahayee A and Chatterjee M 1994. Hypolipidaemic and antihyper-cholesteromic effects of oral *Gymnema sylvestre* R. Br. Leaf extract in albino rats fed on high fat diet. *Phytothera Res* **8:** 118-120.

Chandler S F and Dodds J H 1983a. Solasodine production in rapidly proliferating tissue culture of *Solanum laciniatum* Ait. *Plant Cell Rep* **2:** 69-72.

Chandler S F and Dodds J H 1983b. The effects of phosphate, nitrogen and sucrose on the production of phenolics and solasodine in callus cultures of *Solanum laciniatum*. *Plant Cell Rep* **2:** 205-208.

Chaturvedi H C 1968. *In vitro* Growth and Controlled Morphogenesis in Callus Tissue of *Rauwolfia serpentina* Benth. *Ph D Thesis*, Agra University, Agra.

Chaturvedi H C 1975. Propagation of *Dioscorea floribunda* from *in vitro* culture of single node segment. *Curr Sci* **41:** 839.

Chaturvedi H C 1979. Tissue culture of economic plants in progress in Plant Research, Vol. I Khoshoo TN and Nair PKK (eds.) Today and Tomorrow's Printers and Publishers, New Delhi. 265.

Chaturvedi H C and Sinha M 1975. Propagation of *Dioscorea floribunda* from *in vitro* culure of single-node stem segments. *Curr Sci* **41:** 839.

Chaturvedi H C and Sinha M 1979. Mass Propagation of *Dioscorea floribunda* by Tissue Culture. EBIS National Botanical Research Institutem Lucknow 12.

Coley P D and Barone J A 1996. Derbivory and plant defenses in tropical forests. *Ann Rev Ecol* Syst **27:** 305-355.

Crabbe P 1979. Some aspects of research based on natural products from plant origin. *Bull Soc Chim Belg* **88:** 345-388.

Deka P and Kalita M C 2005. *In vitro* clonal propagation and organogenesis in *Spilanthes acmella* (L) Murray: A herbal pesticidal plant of North-East India. *J Plant Biochem Biotech* **14:** 69-71.

Dicosmo F and Towers G H N 1984. Stress and metabolism in cultured cells. In: Timmermann B A, Steelink C, Loewus F A (eds.) *Phytochemical Adaptation to Stress*, Plenum, New York, 92-175.

Dicosmo F, Quesnel A, Misawa M and Tallevi S G 1987. Increased synthesis of ajamalicine and catharanthine by cell suspension cultures of *Catharanthus roseus* in response to fungal culture filtrates. *Appl Biochem Biotechnol* **14:** 101-106.

Ehmke A and Eilert U 1986. Steroidal alkaloids in tissue culture and regeneration of *Solanum dulcamara*. *Pl Cell Rep* **5:** 31-34.

Eilert U, Kurz W G W and Censtabel F 1985. Stimulation of sanguinarine accumulation in *Papaver somniferum* cell cultures by fungal elicitors. *J Pl Physiol* **119:** 65-76.

Emmanuel S, Iganacimuthu S and Kathiravan K 2000. Micropropagation of *Wedelia calendulacea* Less. *Phytomorphology* **50:** 195-200.

Fujita Y, Tabata M, Nishi A and Yamada Y 1982. New medium and production of secondary compounds with the two stage culture method. *In*: Fujiwara A (ed). Plant Tissue Culture. Maruzen, Tokyo 399-400.

Galancs I T, Webb D T and Rosaro O 1984. Steroid production by callus and cell suspension cultures of *Solanum aviculare*. *J Nat Prod* **47:** 373-376.

Ghanti K C and Banerjee N 2003. Influence of plant growth regulators on *in vitro* micropropagation of *Rauwolfia tetraphylla* L. *Phytomorphology* **53:** 11-19.

Gopalan C, Rama Sastri B V and Balasubramanian S C 1993. Nutritive value of Indian Foods, 2nd Edition, Natl Instt Nutri, ICMR, Hyderabad.

Gorelova O A, Rerabek J, Korzhenevskaya T G, Butenko R G and Gusev M V 1985. Influence of cyanobacterium *Chlorogloeopsis fritschii* on growth and biosynthetic activity of *Solanum laciniatum* cells in mixed cultures. *Fiziol Rast* **32:** 1158-1165.

Gorman C 1992. The Power of Potions. *Time* 52-53.

Gudhate P P, Lokhande D P and Dhumal K N 2009. Role of growth regulators for improving andrographolide in *Andrographis paniculata*. *Pharmac Mag* **5:** 249-253.

Guens J M C 1978. Steroid hormones and plant growth and development. *Phytochem* **17:** 1-14.

Hariharan M Sebastian D, Benjamin S and Prashy P 2002. Somatic embryogenesis in *Leptadenia reticulate* Wright and Arn. – A medicinal plant. *Phytomorphology* **52:** 155-160.

Hahlbrock K and Griesebach H 1979. Enzymatic controls in the biosynthesis of lignins and flavonoids. *Ann Rev Pl Physiol* **30:** 105-130.

Heble M R and Staba E J 1980. Diosgenin synthesis in shoot cultures of *Dioscorea composita*. *Planta Med Supp* 124-128.

Heble M R, Narayanwamy S and Chadha M S 1968. Diosgenin and Sitosterol: Isolation from *Solanum santhocarpum* tissue culture. *Science* **161:** 1145.

Heble M R, Narayanwamy S and Chadha M S 1971. Lupeol in tissue culture of *Solanum xanthocarpum*. *Phytochemistry* **10:** 910-11.

Jabeen M, Ananthan R, Arvinthan K M and Narmatha Bai V 2002. High frequency plant regeneration from embryo-derived callus of *Lobelia nicotianifolia* Roth Ex Roem and Schult- A medicinal plant. *Phytomorphology* **52:** 121-127.

Jasrai Y T and Wala B B 2000. *Curculigo orchioides* Gaertn (kali musli): An endangered medicinal herb. *In*: Khan I A and Khanna A (eds.) *Role of Biotechnology in Medicinal and Aromatic Plants*. Ukaaz Publ, Hyderabad 89-95.

Johnson M, Vallinayagam S, Manickam V S and Seeni S 2002. Micropropagation of *Rhinacanthyus nasutus* (L.) Kurz.- A medicinally important plant. *Phytomorphology* **52:** 331-336.

Kaneda N, Nakanishi H and John Stab E 1987. Steroidal constituents of *Yucca shidigera*. *Pl Tiss Cult* **26:** 1425-1429.

Kannan P, Ebenezer G, Dayanandan P, Abraham G C and Ignacimuthu S 2005. Large scale production of *Withania somnifera* (L.) Dunal using *in vitro* technique. *Phytomorphology* **55:** 259-266.

Kaul B and Staba E J 1968. *Dioscorea* tissue cultures 1. Biosynthesis and isolation of diosgenin from *Dioscorea deltoidea* callus and suspension cells. *Llyodia* **31:** 171-179.

Kinnersley A M and Dougall D K 1980. Correlation between nicotine content of tobacco plants and callus cultures. *Planta* 205-206.

Kirtikar K R and Basu B D 1935. Indian medicinal plants. Blatter E, Caim JF and Mhaekar K S (eds.). Bishen Singh, Mahendra Pal Singh Dehradun 2469-2470.

Kokate C K and Radwan S S 1979. Enrichment of *Solanum khasianum*. Callus generating rootlets with steroidal glycoalkaloids. *Z Naturefors* **34:** 634-636.

Komalavalli N and Rao M V 2000. *In vitro* propagation of *Gymnema sylvestre*- A multipurpose medicinal plant. *Pl Cell Tiss Org Cult* **61:** 97-105.

Kurihara Y 1992. Characteristics of antisweet substances, sweet proteins and sweetness inducing proteins. *Crit Rev Food Sci Nutr* **32:** 231-252.

Kurz W G W, Chatson K B and Constabel F 1985. Biosynthesis and accumulation of indole alkaloids in cell suspension cultures of *Catharanthys roseus* cultivars, **In:** Neumenn K K, Barz W, Reinhard E (eds.) Primary and secondary metabolism of plant cell culture. Springer Berlin 143-153.

Kulkarni A A, Thangane S R and Krishmurthy K V 2000. Direct shoot regeneration from node, internode, hypocotyls and embryo explants of *Withania somnifera*. *Pl Cell Tiss Org Cult* **62:** 203.

Kumar S, Narula A, Sharma M P and Srivastava P S 2003. Effect of copper and zinc on growth, secondary metabolite content and microporpagation of *Tinospora cordifolia*: a medicinal plant. *Phytomorphology* **53**: 79-91.

Lal N, Ahuja P S, Kukreja A K and Pandey B 1988. Clonal propagation of *Picorrhiza kurroa* Royule ex Benth by shoot tip culture. *Pl Cell Rep* **7:** 202.

Lemma A 1991. The potentials and challenges of Endod, the Ethopian soapberry plant for control of schistosomiasis. *In*: Science in Africa: Achievements and Prospects, American Association for Advancement of Science, Washington, USA.

Llyod C and Mc Cown B 1980. Commercially feasible micropropagation of mountain laurel, *Kalmia latifolia* by use of shoot tip culture. *Int Pl Propag Soc Proc* **30:** 421-427.

Malathy S and Pai R S 1998. *In vitro* propagation of *Hemidesmus indicus*. *Fitoterap* **69:** 533-536.

Mitra G C and Chaturvedi H C 1970. Fruiting plants from *in vitro* grown leaf tissue of *Rauwolfia serpentina* Benth. *Curr Sci* **39:** 128.

Nath S and Buragohain A K 2003. *In vitro* method for propagation of *Centella asiatica* (L) Urban by shoot tip culture. *J Pl Biochem Biotech* **12:** 167-169.

Nath S and Buragohain A K 2005. Micropropagation of *Adhatoda vasica* Nees-A Woody medicinal plant by shoot tip culture. *Ind J Biotech* **4:** 396-399.

Ohlsson A B, Bjark L and Gatenbeh 1983. Effect of light on cardenolide production in *Digitalis lantana* tissue cultures. *Phytochemistry* **22:** 2247-2250.

Payne G, Bringi V, Prince C and Shuler M 1992. The quest for commercial production of chemicals from plant cell culture, in plant cell tissue culture in liquid systems. Hanser Publ, Munich.

Prajapati H A, Patel D H, Mehta S R and Subramanian R B 2003. Direct *in vitro* regeneration of *Curculigo orchioides* Gaertn. – An endangered anticarcinogenic herb. *Curr Sci* **84:** 747-749.

Prakash E, Khan S V, Meru E and Rao K R 2001. Somatic embryogenesis in *Pimpinella tirupatiensis* Bal. and Subr. – An endangered medicinal plant of Tirumala hills. *Curr Sci* **81:** 1239-1242.

Ramulu D R, Murthy K S R and Pullaiah T 2003. Regeneration of plants from root segments derived from aseptic seedlings of *Hemidesmus indicus* R. Br. *Phytomorphology* **53:** 293-298.

Rana A C and Avdhoot Y 1992. Experimental evaluation of hepato-protective activity of *Gymnsema sylvestre* and *Curcuma zedoaria*. *Fitoterap* **63:** 60-62.

Rani G and Grover I S 1999. *in vitro* callus induction and regeneration studies in *Withania sonmifera*. *Pl Cell Tiss Org Cult* **57:** 23.

Ripperger I T and Schreiber K 1981. *Solanum* steroid alkaloids. *In*: Rodrigo R G A (ed.). The alkaloids: Chemistry and Physiology. Academic Press New York 81-192.

Rizk A M and Abou-Zied E N 1970. The steroidal constituents of *Solanum cyananthum*. *Planta Med* **18:** 347-349.

Rowen D D, MacDonald P E and Skip R A 1983. Antifungal stress metabolites from *Solanum auiculare*. *Phytochemistry* **21:** 2012-2014.

Sahoo Y and Chand P K 1998 Micropropagation of *Vitex negundo* L.- A woody aromatic medicinal shrub, through high frequency axillary shoots proliferation. *Pl Cell Rep* **18:** 301-307.

Sebastian D, Benjamin S and Hariharan M 2002. Micropropagation of *Rotula aquatica* Lour.- An important woody medicinal plant. *Phytomorphology* **52:** 137-144.

Seetharam Y N, Rajanna L N, Jyothishwaran G, Aravind B, Sharanabasappa G and Ballikkharjun P B 2007. *in vitro* shoot regeneration from leaf and explants of *Vernonia cinerea* (L.) Less. *Ind J Bitotech* **6:** 418-420.

Selvanayagan Z M, Gnanavendhan S G, Chandrasekhran P, Balakrishna K and Rao R B 1995. Plants with antisense venom activity- a review on pharmacological and clinical studies. *Fitoterap* **65:** 99-111.

Sharma N and Devi V 2006. *In vitro* multiplication and conservation of *Peristrophe bicalyculata* Nees.- A lesser known medicinal plant with potential value. *Phytomorphology* **56:** 121-125.

Sharma R K, Wakhlu A K and Boleria M 2004. Micropropagation of *Anethum fraveolens* L. through axillary shoots proliferation. *J Pl Biochem Biotech* **13:** 157-159.

Singh R 2005. Morphogenetic and biochemical studies in callus cultures on *Chlorophytum borivilianum* Sant. Et Ternand. *Ph D Thesis* CCS Haryana Agricultural University, Hisar.

Singh R and Goyal S C 2007. Micropropagation of *Sallvodora oleoides* Decne from nodal explants of mature tree. *J Pl Biol* **34:** 155-159.

Sivanesan I and Murugesan K 2005. *in vitro* adventitious shoot formation from leaf explants of *Withania somnifera* Dunal. *Pl Cell Biotech Mol Biol* **6:** 163.

Sivasubramanian S, Vallinayagam S, Patric R D and Manicram V S 2002. Micropropagation of *Plecttantus vetiveroides* (Jacob) Singh and Sharma- A medicinal plant. *Phytomorphology* **52:** 55-59.

Sudha C G, Krishna P N, Seeni S and Pushpangadan P 2000. Regeneration of plants from *in vitro* root segments of *Holostemma annulare* (Roxb.) K. Schum.- A rare medicinal plant. *Curr. Sci* **78:** 503-506.

Tal B and Godberg I 1982. Growth and diosgenin production by *Dioscorea deltoidea* cells in batch and continuous cultures. *Pl Med* **44:** 107-110.

Telek L 1979. Preparation of solasodine from fruits of *Solanum* species. *Planta Med* **37:** 92-94.

Thengane S R, Bhosle S V, Deodhar S R, Pawar K D and Kulkarni D K 2006. Micropropagation of Indian laurel (*Calophyllum inophyllum*), a source of anti-HIV compounds. *Curr Sci* **90:** 1393-1397.

Thengane S R, Deodhar S R, Bhonsle S V and Rawal S K 2006. Direct somatic embryogenesis and plant regeneration in *Garinia indica* Choiss. *Curr Sci* **91:** 1074-1078.

Tiwari S, Shah P and Singh K 2004. *in vitro* propagation of *Pterocarpus marsupium* Roxb. : An endangered medicinal tree. *Indian J Biotechn* **3:** 422-425.

Tomita Y, Uomori A and Minato H 1970. Steroidal sapogenins and sterols in tissue culture of *Dioscorea tokaro*. *Phytochemistry* **9:** 111-114.

Uddin A and Chaturvedi H C 1979. Soasodine in somatic tissue cultures of *Solanum khasianum*. *Pl Med* **37:** 90-92.

Ulbrich B, Weismer W and Arens H 1985. Large scale production of rosmarinic acid from plant cultures of coleus blumei. **In:** Neumann K H, Barz W, Reinhard W (eds.) Primary and secondary metabolism of plant cell cultures. Springer, New York 293-303.

Uma Devi P 1996. *Withania somnifera* Duanl (Ashwagandha) potential plant source of a promising drug of cancer chemotheraphy and radiosensitization. *Ind J Exp Biol* **34:** 115-118.

UNESCO 1996. Culture and Health. Orientation Texts- World decade for cultural development (1988-1997), Paris, France.

UNESCO 1998. Termianal Report: Promotion of Ethnobotany and the sustainable use of plant resources in Africa, Paris.

Uniyal M R 1993. Some popular and traditional ayurvedic herbs useful in family planning. *Sachitra Ayurved* **45:** 665-668.

Vidya S, Krishna V, Manjuntha B K and Shankarmurthy K 2005. Micropropagation of Entada pursaetha DC- An endangered medicinal plant of Western Ghats. *Ind J Biotech* **4:** 561-564.

Wala B B and Jasrai Y T 2003. Micropropagation of an endangered medicinal plant: *Curculigo Orchioides* Gaertn. *Pl Tiss Cult* **13:** 13-19.

Wickens G E 2001. Economic Botany- Principles and Practices, Kluwer Academic Publishers Dordrecht/Boston 459-473.

Yoshikawa T and Furuya T 1985. Morphinan alkaloid production by tissues differentiated from cultured cells of *Papaver somniferum*. *Planta Med* **12:** 110-113.

Zenktler M 1971. Development of new plants from leaves and root of *Atropa belladonna* L. in the *in vitro* culture. *Acta Soc Bot* **40:** 305.

Medicinal Plants: Aspects and Prospects (2014) ***Pages* 302–310**
***Editors:* Mukesh Kumar, Anjali Khare and C.P. Shukla**
ISBN: 978-81-7622-309-6
***Published by:* BIOTECH BOOKS, NEW DELHI**

Chapter 20

Neem (*Azadirachta indica* A. Juss.) A Versatile Tree Belonging to Indian Sub-Continent

Swati Dhyani and Sadhna Tripathi

Wood Preservation Discipline, Forest Products Division, Forest Research Institute, Dehradun – 248 001

ABSTRACT

Neem is a native of Indian sub-continent finding numerous applications in various fields including medicinal, pesticidal, insecticidal, cosmetics etc. Its medicinal property has been thoroughly evaluated and well documented. Neem is considered to be a store-house of various biologically active compounds, which have been found to be effective against a number of pathogens. Its mode of action is either by disrupting the feeding habit of the pathogen or by bringing a change in the metamorphosis of the insect. Although, neem extracts and seed oil are found to be very effective against the targeted pathogens, it does not have any negative effect on other species. Research has been carried out in evaluating the antifungal and anti-termicidal properties of neem leaves extracts and seed oil in the field of wood preservation. The results obtained were very encouraging. Thus, neem can be best referred as a versatile tree possessing remarkable activity.

Keywords: Insecticidal, Medicinal, Neem, Pesticidal, Wood preservation.

Introduction

Neem is the versatile tree belonging to the order Rutales and the family Meliaceae of the plant kingdom. Neem tree is indigenous to India from where it has spread to many Asian and African countries. For centuries the tree has been held in esteem by Indian folk because of its medicinal and insecticidal values. It is a native of dry forest areas of India, Pakistan, Srilanka, Malaysia, Indonesia, Thailand and Myanmar. In India, the tree is found wild on Shiwalik hills, Uttar Pradesh, and in dry forests of Andhra Pradesh, Karnataka and Tamil Nadu. In India it occurs in Tropical dry deciduous and thorn forests and in tropical dry evergreen forests (Champion and Seth 1968).

It has been extensively evaluated for its medicinal, pesticidal and insecticidal activities. It is considered to be storehouse of various biologically active compounds and its activities against pathogens, have been thoroughly evaluated. People have recognized that the leaves, bark, wood and fruits of neem tree either repel or otherwise discourage insect pests, and they incorporate these plant parts into traditional soil preparation, grain storage and animal husbandary practices. Through more recent the chemical analysis, of active compounds in neem tissues have been identified (Jotwani and Srivastava 1983). Neem plant contains several thousands of chemical constituents, containing hundreds of terpenoids isolated from its different parts. Medicinally, all parts of neem have been used including fruits, seeds, oil (extracted from the seeds), leaves, roots and bark. Alkaloids are major class of natural products that have physiological effect in organisms. The bark contains the bitter alkaloid margosine and the fruit contains the alkaloid azaridine, both having anti-inflammatory and anti-fungal properties.

Antifeedant Property of Neem

Schmuttere and Rambold (1980) carried out laboratory studies in the German Federal Republic. Some fractions of neem seeds from West Africa were purified, and 4 pure fractions had strong to very strong growth disruption effects on *Epliachna varivestis* Muls. The most promising antifeedant compounds were extracted from neem. Pereira and Wohlgemuth (1982) tested the effectiveness of ground neem seeds and leaves obtained from Togo as a protectant of stored maize against attack by pests in the laboratory. Neem seed was found to be toxic to adults of *Sitophilus oryzae* L., *S. zeamais* Motsch., *Cryptolestes ferrugineus* Steph. and *Rhyzopertha dominica* F. but not to *Tribolium castaneum* Hbst. Neem seeds and leaves effectively reduced the progeny production of all the pests and screened the adult emergence of *S. cerealella* and *Ephestia cautella* Wlk. The ground leaves and seeds acted by disrupting larval development or by reducing adult fecundity. Neem seeds have been found to be effective in protecting stored maize against *S. zeamais* and *Rhyzopertha dominica*.

Luca (1982) reported that oil obtained from *Azadirachta indica* acted as physical barriers to the pests. The oil did not prevent seed germination and were harmless to vertebrate consumers. Sowunmi and Akinnusi (1983) added powdered kernel of neem seed to stored seeds of cowpea in the laboratory in Nigeria which were then kept in open glass jars at ambient temperatures for 8 months. Malik and Mujtaba (1984) tested 7 plant species for their repellent activity against *T. castaneum* and antifeedant activity against *R. dominica*. The best repellent activity was found in the rhizomes of *Saussurea lappa* and antifeedant activity in the leaves of *Chenopodium ambrosioides* and in azadirachtin isolated from neem. Singh and Singh (1985) conducted studies to explore the possibility of utilizing deoiled neem kernel powder against Khapra beetle. Powdered neem seed kernel was extracted seven times with hexane to remove oil contents and then used for experiments against 1st, 2nd and 3rd instar larvae. The growth retardation effect and death of the larvae was due to the biologically active compound, azadirachtin, present in deoiled kernel.

Pandey *et al*. (1986) tested five plant extracts *viz.* leaves and twigs of *Azadirachta indica* leaves, flowers of *Lantana camara*, flowers and buds of *Ageratum conyzoides*, leaves twigs and young branches of *Theretia merifolia* and leaves of *Ipomoea carnea* Jacq. against C. *chinensis* Linn. Three concentrations *viz.* 0.5, 1.0 and 1.5 per cent were taken for each of the extracts. The plant extracts had been compared with treatments of absolute control and benzene. It is evident that all the plant extracts at the three doses repelled the beetles from 98.83 to 100 per cent. Benzene proved to have a potent repellent effect keeping 99 per cent beetles away from sacklets. It was also observed that the plant extracts at a higher dose (1.5parts/100parts of seed byw/w) and benzene alone proved very effective in reducing the egg laying. The observations further revealed that *A. indica, L. camera* and *A. conyzoides* gave better performance in terms of weight loss.

Neem in Pest Management

Discovery of the presence of biologically active principles in neem, in recent decades, and their effectiveness against numerous harmful insect species have focused worldwide attention on this wonder tree (Tewari 1992). The neem seed kernel and cake powder are now in use to control several pests including stem borer pests. The seed oil is very effective against a large number of sucking insects and others. Siddiqui *et al*. (1987) applied neem extracts, ground fruits and leaves soaked in water at 1kg/ 40 litres, to control foliage pests, including the aleyrodid, *Bemisia tabaci*, the cicadellid, *Jacobiasca lybica* and *Aphis gossypii*, and increased the yield by 0.5 tonnes/ha.

Utility of neem extracts have been discussed by Ketkar (1976), Mitra (1979), Attri and Prasad (1980a), Attri (1982) and Singh (1983). They highlighted the utility and pesticidal potential of neem seed extracts, which are rich in azadirachtin, a biologically active compound of high value. These compounds are potential insect gustatory phagodeterrent and have been employed for controlling foliage, feeding insects *i.e.*, against *Schistocerca gregaria* Forsk. Bhandari *et al*. (1988) have studied the effect of neem seed extracts on poplar defoliator, *Clostera cupreata* Butler in laboratory. They tested the methanolic extracts and nine chromatographic fractions of neem seed against the larvae of *Clostera cuperata*. A potential antifeedant activity was noticed in the methanolic extracts and chromatographic fraction.

Kumar *et al*. (1990) observed that 1 per cent kerosene in combination with 4 per cent neem extract was effective in the integrated pest management of *Coccus viridis* on coffee. This combination causes about 80-90 per cent mortality in 13-19 days. Pascual *et al*. (1990) observed that azadirachtin induced imaginal moult deficiencies in *Tenebrio molitor* L. It induced a delay of the development and an inhbition of apolysis and acdysis. Its potential for integrated control of stored food pests is emphasized. Lavie *et al*. (1967) isolated a triterpenoid alcohol, meliantriol from the fruits of dharek (*Melia azadarach*) and neem (*Azadirachta indica*) oil. The compound inhibited feeding in desert locust, *Schistocerca gregaria* at 8mg/cm^2 of filter paper (Singh 1983).

Neem seed cake mixed with soil has been found appreciably toxic to termites, *Microtermes* spp. (Dutta 1974). In addition to antifeedant and growth regulatory properties, neem is also known to possess some insecticidal actions. Simple macerated juice of leaves is reported to have 25 per cent mortality of leucerne weevil, *Hypera postica* (Singh 1983). Though neem is primarily considered to be antifeedant, affecting only the taste organs, but it also acts as repellent (Jacobson *et al.*, 1978). In a four year field trial, neem oil, neem seed kernel extract (NSKE), neem emulsion (NE) and neem wettable powder (NWP) along with insect growth regulators (IGR) and synthetic insecticides were evaluated against *H. armigera* infesting chicpea seed crop. All these treatments gave significantly higher yield of seed in both years (Attri and Prasad 1980b, Srivastava 2001). Das and Singh (1998) conducted field experiments for the integrated management of jute pests by integrating cultural practices and minimum use of pesticides.

Neem in Wood Preservation

The alcoholic extracts of neem leaves and neem seed oil are found to be very effective in controlling/checking the activity of wood destroying agencies such as the fungi and termites. The methanolic extract of neem leaves and neem seed oil when impregnated into the wood samples of hardwood and softwood and subjected against wood decaying fungi, a white rot (*Trametes versicolor*) and brown rot (*Oligoporus placentus*), provided maximum protection to the test samples even at a very low concentration. Similar, results were observed when the test samples of both the wood species treated with low concentration of methanolic extract and seed oil were exposed to wood destroying termites in laboratory (*Microcerotermes beesoni*) and in field (*Odentotermes obesus*) (Dhyani and Tripathi 2007).

Medicinal Properties of Neem

Neem is widely recognized throughout the world for its marked medicinal activity. The tender leaves in combination with *Piper nigrum* are found to be effective in intestinal helminthiasis. Leaves paste is useful in alceration of cowpox and the fresh mature leaves along with seeds of *Psoralea corylifolia* are used to prepare a very effective medicine for leucoderma. The aqueous extract (10 per cent) of tender leaves is reported to possess anti-viral properties against vaccinia, variola, foulpox and New Cattle disease viruses. It is also reported that the leaf extract fractions considerably delay the clotting time of blood. A crude extract of the leaves was studied for its effects on the cardiovascular system of anaesthetized guinea pigs and rabbits. It also exhibited a weak anti-arrhymthmic activity in rabbit against ouabain induced dysrhythmia. Limonoids isolated from neem leaves showed *in vitro*, cytotoxic activity against human tomor cell lines (Kigodi *et al.*, 1989). The water-soluble fraction separated from the crude leaf extract lowered the hyperglycemia in streptozotocin diabetes (Chakraborty *et al.*, 1989). The Khasi and Jaintia tribes of Meghalaya (India) use the leaves for diarrhoea, dysentery, tuberculosis and heart diseases (Kharkongor *et al.*, 1981).

The alcoholic extract of fresh stem bark yields a number of bitter principles; nimbin (0.04 per cent), nimbinin (0.002 per cent) and nimbidin (0.4 per cent) (Anon

1985). These compounds showed antiprotozoal property due to nimbidin and sodium nimbidinate. The oil is reported to have anti-fertility properties (Koul *et al.*, 1990). It also possesses antiseptic and anti-fungal activity and is found to be active against gram-positive and gram-negative organisms. An antitumor polysaccharide N9GI (Terumo Corporation Patent 1985) and antiinflammatory polysaccharide (Terumo Corporation Patent 1983) were isolated from the bark. Two new water-soluble polysaccharides named as GIII DO$'_2$Ia and GIII DO$'_2$Iia were also reported. Besides polysaccharides, 18 diterpenoids were also isolated from the stem bark; nimbisonol, dimethylnimbionol, margosone, margosolone (Ara *et al.*1989), nimbonone, nimbonolone (Ara *et al.*, 1989), nimbionone, nimbionol (Siddiqui *et al.*, 1988) and 6 from root bark; margocin, margocinin, margocilin (Ara *et al.*, 1990), nimbilicin, nimbocidin (Ara *et al.*, 1990), nimbidiol (Majumdar *et al.*, 1987).

Active compounds of Neem

Fresh matured leaves, on steam distillation gave an odourous viscous essential oil, the GC-MS analysis of this oil indicated it to be a mixture of tetrasulphides of C_3, C_5, C_6 and C_9 units (Pant *et al.*, 1986). White crystalline flakes m.p. 64°-66°C was obtained from the petroleum ether extract by column chromatography which were found to be a mixture of C-14, C-24, C-31and C-34 alkanes (Chavan and Nikam 1988). Systematic fractionation of water-soluble extract of neem leaves led to the isolation of quercetin-3-O-β-D-glucoside, myricetin-3-O-rutinoside, quercetin-3-O- rutinoside, kaempferol-3-O- β-D-glucoside and quercetin-3-O-rhamnoside (Chakraborrty *et al.*, 1989). An isoprenylated flavanone, nimbaflavone ($C_{26}H_{30}O_5$) was isolated from the chloroform soluble fraction of the ethanolic extract of neem leaves by column chromatography (Garg and Bhakuni 1984).

A number of limonoids have been isolated from neem leaves out of which nimbin ($C_{30}H_{36}O_9$) was the first bitter principle to be reported (Narayanan *et al.*, 1964). Nimbinene ($C_{24}H_{34}O_7$), 6-deacetylnimbinene ($C_{26}H_{32}O_6$) and nimbandiol ($C_{26}H_{32}O_7$) were isolated from the ether extract of the air-dried neem leaves. Their structures were elucidated from IR, high resolution mass, ^{1}H- and ^{13}C-NMR spectral data (Devakumar and Mukerjee 1986). A number of terpenoids, Nimocinol ($C_{28}H_{36}O_5$), a tetranortriterpenoid, was isolated from the ethanolic extract, nimbocinone ($C_{30}H_{46}O_4$), isolated from the ethylacetate extract, nimocinolide ($C_{28}H_{36}O_7$), from chloroform extract, nimbocinolide ($C_{32}H_{42}O_{10}$), from chloroform extract, nimbolide ($C_{27}H_{30}O_7$) have also been isolated from neem leaves and their structures have been determined.

A number of active compounds have been isolated from neem seeds possessing marked antifeedant and anti-microbial property, among them azadirachtin ($C_{35}H_{44}O_{16}$) is of great value (Kraus *et al.*, 1985). The other compounds isolated from the seeds are 22,23-Dihydro-23β-methoxyazadirachtin (C36$H_{48}O_{17}$); 3-desacetyl-11-deoxyazadirachtin ($C_{33}H_{42}O_{15}$); 1,3-diacetylvilasinin ($C_{30}H_{40}O_7$). The compounds isolated from the whole fruit are nimocin ($C_{33}H_{38}O_4$), nimbocinol ($C_{26}H_{32}O_4$), meldenin ($C_{28}H_{38}O_5$), azadirachtol ($C_{32}H_{46}O_6$), nimbin ($C_{30}H_{36}O_9$), vepinin ($C_{28}H_{36}O_5$), gedunin ($C_{28}H_{34}O_7$), nimbandiol ($C_{26}H_{32}O_7$), nimbanal ($C_{29}H_{34}O_8$) and salannolide ($C_{34}H_{44}O_{11}$).

Other Research on Neem

Neem oil has long been produced in Asia on an industrial scale in soaps, cosmetics and pharmaceuticals. During the 1980's companies began commercial production and distribution of pest control formulations that use azadirachtin as the principle active ingredient. Interest in biological pest control (BPC) agents has developed along with the environmental and consumer rights movements, and the recognition that integral pest management (IPM) strategies are needed to sustain agricultural production. New markets for organically grown products and natural products also spurred the development of the BPC industry, and azadirachtin was among the first to be commercialized. In United State neem –based BPC's were first approved for use on non-food crops in 1985. After subsequent testing, the Environmental Protection Agency (EPA) regulated the use of Dihydro azadirachtin (DAZA), a reduced derivative of azadirachtin, for use on food crops. In 1996 the EPA exempted raw agricultural commodities from meeting DAZA residue requirements, as long as the chemical is applied as an insect growth regulator or antifeedant at not more than 20gm/acre with a maximum of seven applications per growing season (EPA 1997).

Neem extracts can be made from leaves and other tissues, but the seeds contain the highest concentrations of azadirachtin. Industrial scale extraction processes use solvents such as alcohol, ether and hydrocarbons instead of water. Some sources claim that the water-based extracts work nearly as well. In Pakistan a process freeze-drying the water-based neem extract produces a crystalline powder called 'bitters' that is water-soluble (Read and French 1993). Researchers believe that even modest efforts at genetic improvement could result in higher seed yields, higher levels of azadirachtin, and other useful compounds in the seed. In toxicological studies carried out in USA and Germany, different neem products have been found neither mutagenous nor carcinogenic, and they did not produce any skin irritations or organic alternations to mice even at high concentrations. In another study at Canada, Neem was found to be harmless to aquatic invertebrates and other non-target species. Neem pesticide, being natural product is absolutely non-toxic, 100 per cent biodegradable and environmentally friendly.

References

Anon 1985a. Terminal Report (1967-85) of ICAR Schemes. All India Co-ordinated Research Project for Investigation on Agricultural Byproducts and Industrial Waste material for Evolving Economic Rations for Livestock, Anand Center, Gujarat Agricultural University, Anand Campus, Anand, India.

Anon 1985b. The Wealth of India, Raw Materials Publication and Information Directorate, C.S.I.R., New Delhi.

Ara I, Siddiqui B S, Faizi S and Siddiqui S 1989. Diterpenoids from the root bark of *Azadirachta indica. Z Natur Foresch B Chem Sci* **44:** 1279-1282.

Ara I, Siddiqui B S, Faizi S and Siddiqui S 1989. Isolation of meliacin cinnamates from root bark of *Azadirachta indica. Hetrocycles* **29:** 729-735.

Ara I, Siddiqui B S, Faizi S and Siddiqui S 1990. Tricyclic diterpenes from the stem bark of *Azadirachta indica. Planta Medica* **56:** 84-86.

Ara I, Siddiqui B S, Faizi S and Siddiqui S 1990. Tricyclic diterpenoids from root bark of *Azadirachta indica. Phytochemistry* **29:** 911-914.

Ara I, Siddiqui B S, Faizi S and Siddiqui S1990c. Three new diterpenoids from the stem bark of *Azadirachta indica. J Nat Prod* **53:** 816-820.

Attri B S and Ravi Prasad G 1980a. Neem oil extractive- an effective mosquito larvicide. *Ind J Entomol* **42:** 371-374.

Attri B S 1982. Neem (*Azadirachta indica* A. Juss.) as a source of Pest Control Material, Cultivation and Utilization of Medicinal Plants. Regional Research Laboratory, Jammu, India. 680-689.

Attri B S and Ravi Prasad G 1980b. Studies on the pesticidal value of Neem oil byproducts. *Pestology* **4:** 16-20.

Bhandari R S, Lal J, Ayyar K S and Singh P 1988. Effect of neem seed extract on poplar defoliator abd *Pygaera cupreata* Butler (Lepidoptera: Notodontidae) in laboratory. *Indian Forester* **114:** 790-795.

Chakarbortty T, Verotta L and Poddar G 1989. Evaluation of *Azadirachta indica* leaf extract for hypoglycemic activity in rats. *Phytotherapy Res* **3:** 30-32.

Champion H G and Seth S K 1968 Revised Survey of the Forest types of India. The Manager of Publications, Government Press, Delhi-6.

Chavan S R and Nikam S T 1988. Investigations of alkanes from neem leaves and their mosquito larvicidal activity. *Pesticides* **22:** 32-33.

Das L K and Singh B 1998. Integrated management of jute pests. *Environ and Ecol* **16:** 218-219.

Devakumar C and Mukerjee S K 1986. Partial synthesis of nimbinene and 6-desacetylnimbinene- the new pentanortriterpenoids. 73rd Session, Indian Science Congress, Delhi. **157:** 91.

Dhyani S and Tripathi S 2008. Quercetin- a potential compound from *Azadirachta indica*. A. Juss. (Neem) leaves exhibiting potential activity against wood decaying fungi and termites. Paper presented in the 39th Annual Conference of International Research Group on Wood Protection, Turkey.

Dutta N 1974. Report on the performance of Neem oil cake powder against *Microtermes* sp. Department of Agriculture Ent. University of Kalyani, West Bengal.

Garg H S and Bhakuni D S 1984. An isoprenylated flavanone from leaves of *Azadirachta indica*. *Phytochemistry* **23:** 2115-2118.

Jacobson M, Reed D K, Crystal M M, Moreno D S and Soderstrom E L 1978. Chemistry of Biological activity of insect feeding deterrents from certain weeds and crop plants. *Entomologia Exp Appl* **24:** 248-257.

Jotwani M G and Srivastava K P 1983. A review of neem research in India in relation to insects. Proceeding 2nd International Neem Conference, Rouischholzhausen, Germany: 43-56.

Ketkar C M 1976. Final Technical Report- Utilization of Neem, (*Azadirachta indica* A.Juss) and its byproducts. Nana Dengle Sadhna Press, Pune (India). **3:** 47-56.

Kharkongor P and Joseph J 1981. Folkore medicobotany of rural khasi and Jantia. **In:** Jain S K (Ed) Glimpses Indian Ethnobot. Oxford and IBH Publications, New Delhi: 127.

Kigodi P G K, Blasko G, Thibiaranonth Y, Pizzoto J M and Cordell G A 1989. Spectroscopic and biological investigation of nimbolide and 28-Deoxonimbolide from *Azadirachta indica*. *J Nat Prod* **52:** 1246-1251.

Koul O, Isman M B and Ketkar C M 1990. Properties and Uses of Neem (*Azadirachta indica*). *Canadian J Bot* **68:** 1-11.

Kraus W, Bokel M, Klenk A and Pohnl H 1985. The structure of azadirachtin and 22,23-dihydro-23β-methoxyazadirachtin. *Tetrahedron Letters* **26:** 6435-6436.

Kumar M, Gokuldas, Krishnamoorthy P, Bhat and Ramaiah P K 1990. Potential role of kerosene and neem derivatives in Integrated management of mealybugs on coffee. *J Coffee Res* **19:** 17-29.

Lavie D, Jain M K and Shpan-Gabrielith S R 1967. A locust phagorepellent from two *Melia* species. *J Chem Soc Chem Commun* 910-911.

Luca Yde 1982 Products of vegetable origin that can be used against bruchids (Col.) (Attractants, antifeedants, deterrents, repellents, lethal). *Frustula Entomologica* **2:** 19-29.

Malik M M and Naqvi S H, Mujtaba 1984. Screening of some indigenous plants as repellents or antifeedants for stored grain insects. *J Stored Products Res* **20:** 41-44.

Mitra C R 1979. Waste fruits and seeds: Chemistry, Technology and Utilization. *Plant Res* **2:** 165-183.

Mojumdar P L, Maiti D C, Kraus W and Bokel M 1987. Nimbidiol, a modified diterpenoid of the root bark of *Azadirachta indica*. *Phytochemistry* **26:** 3021-3023.

Narayanan C R, Pachapurkar R V, Pradhan S K, Shah V R and Narasimhan N S 1964. Structure of nimbin. *Ind J Chem* **2:** 108-113.

Pascual N, Marco M P and Belles X 1990. Azadirachtin induced imaginal moult deficiencies in *Tenebrio molitor* L. (Coleoptera: Tenebrionidae). *J Stored Products Res* **26:** 53-57.

Pandey N D, Mathur K K, Pandey S and Tripathi R A 1986. Effects of some plant extracts against pulse beetle, *Callosobruchus chinensis* Linnaeus. *Ind J Entomol* **48:** 85-90.

Pant N, Garg H S, Madhusudanan K P and Bhakuni D S 1986. Sulfurous compounds from *Azadirachta indica* leaves. *Fitoterapia* **57:** 302-304.

Pereira J and Wohlgemuth R 1982. Neem (*Azadirachta indica* A.Juss) of West African origin as a protectant of stored maize. Zeitschrift fur Angewandte Entomologie **94:** 208-214.

Read M D and French J H 1993. Genetic Improvement of Neem : Strategies for the future. Proceedings of the International Consultation on Neem Improvement held at Kasetsart University, Bangkok, Thailand, 18-22

Schmutterer H and Rambold H 1980. Effects of some pure fractions from seeds of *Azadirachta indica* on feeding activity and metamorphosis of *Epilachna varivestis* (Col. Coccinellidae). *Zeitschrift fur Angewandte Entomologie.* **89:** 179-180.

Siddq S A, Schmutterer H and Ascher K R S 1987. A proposed pest management program including neem treatments for combating potato pests in the Sudan. Proceedings of the 3rd International Neem Conference, Nairobi, Kenya. 449-459.

Siddiqui S, Ara I, Faizi S, Mahamood T and Siddiqui B S 1988. Phenolic tricyclic diterpenoids from the bark of *Azadirachta indica. Phytochemistry* **27:** 3903-3907.

Singh, R P 1983. Neem in Insect Pest Managemant. *Proc. Principles and Concepts of Integrated Pest Managemant*: 106-113.

Singh R P and Singh S 1985. Evaluation of deoiled neem (*Azadirachta indica* A.Juss) seed kernel against *Trogoderma granarium* Eversts. *Curr Sci* **54:** 950-951.

Sowunmi O E and Akinnusi O A 1983. Studies on the use of neem kernel in the control of stored cowpea beetle (*Callosobruchus maculatus* F.). *Tropical Grain Legume Bulletin* **27:** 28-31.

Srivastava R P 2001. *Neem and Pest Management*. Published by International Book Distributing Co. Charbagh, Lucknow.

Terumo Corporation Japan. Kokai Tokkyo Koho. Antiinflammatory polysaccharide from *Melia azadirachta*, JP 58, 225, 021 [83, 225, 021] (CI. A61 K35/78), 27 Dec. 1983. Appl. 82/105, 532; 21 June, 1982: 9.

Terumo Corporation Japan. Kokai Tokkyo Koho. Antitumor polysaccharide (N9GI) from *Melia azadirachta*, J P 60, 42, 331[85, 92, 331] (CI. A61 K35/78), 06 March. 1985. Appl. 83/1505. 341; 19 Aug., 1983: 8.

Tewari D N 1992. *Monograph on Neem*. International Book Distributor,Dehra Dun.

Medicinal Plants: Aspects and Prospects (2014) ***Pages* 311–336**
***Editors:* Mukesh Kumar, Anjali Khare and C.P. Shukla**
ISBN: 978-81-7622-309-6
***Published by:* BIOTECH BOOKS, NEW DELHI**

Chapter 21

Medicinal Flora of Billawar Forest Division, Kathua District, Jammu Division, Jammu and Kashmir

***Bharat Bhushan*[1], *Mukesh Kumar*[1] *and A.K. Sharma*[2]**

*[1]Department of Botany,
Sahu Jain P.G. College, Najibabad – 246 763, U.P.
[2]Principal Chief Conservator of Forest (Research & Training),
2A, Forest Campus, 17-Rana Pratap Marg, Lucknow – 226 001, U.P.*

ABSTRACT

Ethnomedicinal plants are widely used in district Kathua. The knowledge of medicinal plants varies from region to region. The Billawar Forest Division is rich in ethnic and biological diversity. Due to varied altitude and topography, the study area bears a diverse flora ranging from the sub-tropical in the South to the alpine meadows on the higher peaks in the North. The information regarding the medicinal plants has been collected from the local inhabitants of the study area.

Anthropogenic activities like industrialization, deforestation, habitat destruction, urbanisation, etc. are increasing day by day, which pose a serious threat to many species. It is, therefore, necessary to document the useful ethnobotanical flora and conserve it for future generations. Hence the step towards the conservation of the species appropriate measures involving the precipitation of the local people has been adopted in the present research work. In the present study, a total of 90 medicinal plants belonging to 80 genera and 46 families have been investigated along with the documentation of significant information regarding their scientific names, common names, families and plant parts used for different purposes.

Keywords: Ethnic diversity, Ethnobotanical flora, Jammu Division.

Introduction

The word Ethnobotany literally means the study of botany of the primitive human race. Ethnobotany is the total relationship between man and vegetation which meant more than even the scope of economic botany. Ethnobotanist use to visit interior villages and localities and interact with the Folk to record plant usages among primitive people. Most of the ethnobotanists link themselves to confine studies of plants and human relations in urban setting for convenience. The culture and the habitat of ethnic groups are both changing to urban patterns at quite fast rate. Hence their knowledge, as also the environment where that knowledge took birth, have been diminishing. During the last 100 years the attention of ethnobotanists has been more on primitive people, many such societies have been studied and elements of their knowledge systems documented. Indian systems of medicines and Indian thought on health care can be traced back to sometimes between 600-100 B.C. Even today tribal and local people adjoining forest are not only engaged in collection, processing and marketing of medicinal plants to boost their income but also use them to cure a number of diseases in interior villages where standard medical facilities are not available.

Plant wealth of the Indian Himalayan region is known for its uniqueness, naturalness and socio-economic values. The inhabitants of the area have been dependent on plants resources for medicine, fuel, food, fodder, fibre, timber and various other purposes. The traditional use of plant as medicines is well known among the native communities of the Himalayas. All folk medicines used by villagers of Kathua are collected from the forest. The persons who recognize medicinal plants and know their names do not disclose easily. They tell that, according to need, the spirit under man's control comes in a dream and tell the names of the medicines and its whereabouts. Therefore, during the study of indigenous medicinal use of plants, the elder people were visited several times and developed confidence in them to reveal the necessary information. Contacts were made with several older and experienced Dogras. Tribal doctors were often taken to the forest as guides and informants. These people also assisted us in the collection of plant specimens and ornaments. The prehistoric man must have used plants for various purposes such as food, fodder and medicinal purposes. The history of medicine and surgery dates back perhaps to the origin of the human race. But, as no mode of recording events existed in prehistoric times, there are no data on the methods of the treatment practiced in that period. In those days, the subject of human suffering and its alleviations was intimately associated with religion, myth and magic. Whenever the curiosity of the present day man probes into the past and bring to light even fragmentary information on the indigenous methods of our ancestors, it makes a fascinating study.

In India, the references to the curative properties of some herbs in the Rig Veda seem to be the earliest records of use of plants in medicine. The period of Rig Veda is estimated to be between 3500 and 1800 B.C. But the references to plants in the Rig Veda are very brief. Some curative properties of herbs are also mentioned in the Atharva Veda (1500 B.C.), Upanishads (1000-600 B.C.), Mahabharata and Puranas

(700-400 B.C.), etc. These include uses of plants in worship as medicine, food, fuel and for tools of Agriculture (Jain, 2007). That time appeared the two most important works on Indian system of medicine, the works of Charak and Susruta, namely the Charak-Samhita and Sasruta-Samhita. The latter deals with about 700 drugs, some of these are not indigenous to India. With the passing of time, more and more plants found entry into natives medicine taking the number of Indian medicinal herbs to about 1500.

Numerous large and small books dealing with medicinal plants of India now exist; some of them run into several volumes of hundreds of pages. The useful properties attributed to one and the same plant in different parts of India sometimes greatly vary; this has resulted in vast literature in regional languages too. It has also been reported that hundreds of drugs including plant species, *i.e.*, opium poppy, pomegranate, Aloe, Onion were commonly used by the Egyptians about 6000 years ago. Zarathustra (600 B.C.) has mentioned about 1200 drugs of plant origin in his writings. Aristotle (384-322 B.C.) listed more than 500 plants. During the last few centuries particularly in the 19th and 20th century, this knowledge has been threatened and is gradually becoming extinct due to the destruction of many areas of forest. It is perhaps this pressure and also the renewal of interest shown by people in natural foods and drugs which has made ethnobotany an organized science within a very short period.

In India, organized ethnobotanical research started about five decades ago. Within last forty years, organized field work among the tribal of different areas has been undertaken and continued by more than 20 Indian Institutions. Several institutions are also engaged in studies of allied interdisciplinary aspects like ethnopharmacology and ethnopalaeobotany, etc. Botanists of Botanical Survey of India undertook ethnobotanical studies in 14 states of India covering more than 30 different tribes and have recorded data on about 2000 plants used in medicine, food, fodder, fibre for house building, musical instruments, fuel, oil seeds, narcotics, beverages in material culture, for medico-religious purposes and for veterinary medicines. Over 9500 wild plant species used by tribal for meeting their varied requirement have been recorded so far under All India Co-ordinated Research Project on Ethnobiology involving BSI, National Botanical Research Institute (Lucknow), Birbal Sahani Institute of Palaeobotany (Lucknow), Central Drug Research Institute (Lucknow), several regional research laboratories and department of various institutions and universities.

Traditionally important medicinal plants of Bhaderwah hills Jammu Province were enumerated by Kapur (1996).). The traditional knowledge is inherited and transmitted from one generation to another. They also emphasized to conserve traditional knowledge and bio-resources, documentation of this empirical knowledge is must. The traditional knowledge is very much relevant to the conservations of biodiversity and the sustainable use of plant resources. However, there is paucity of data on ethnobotanical aspects in Billawar Forest Division.

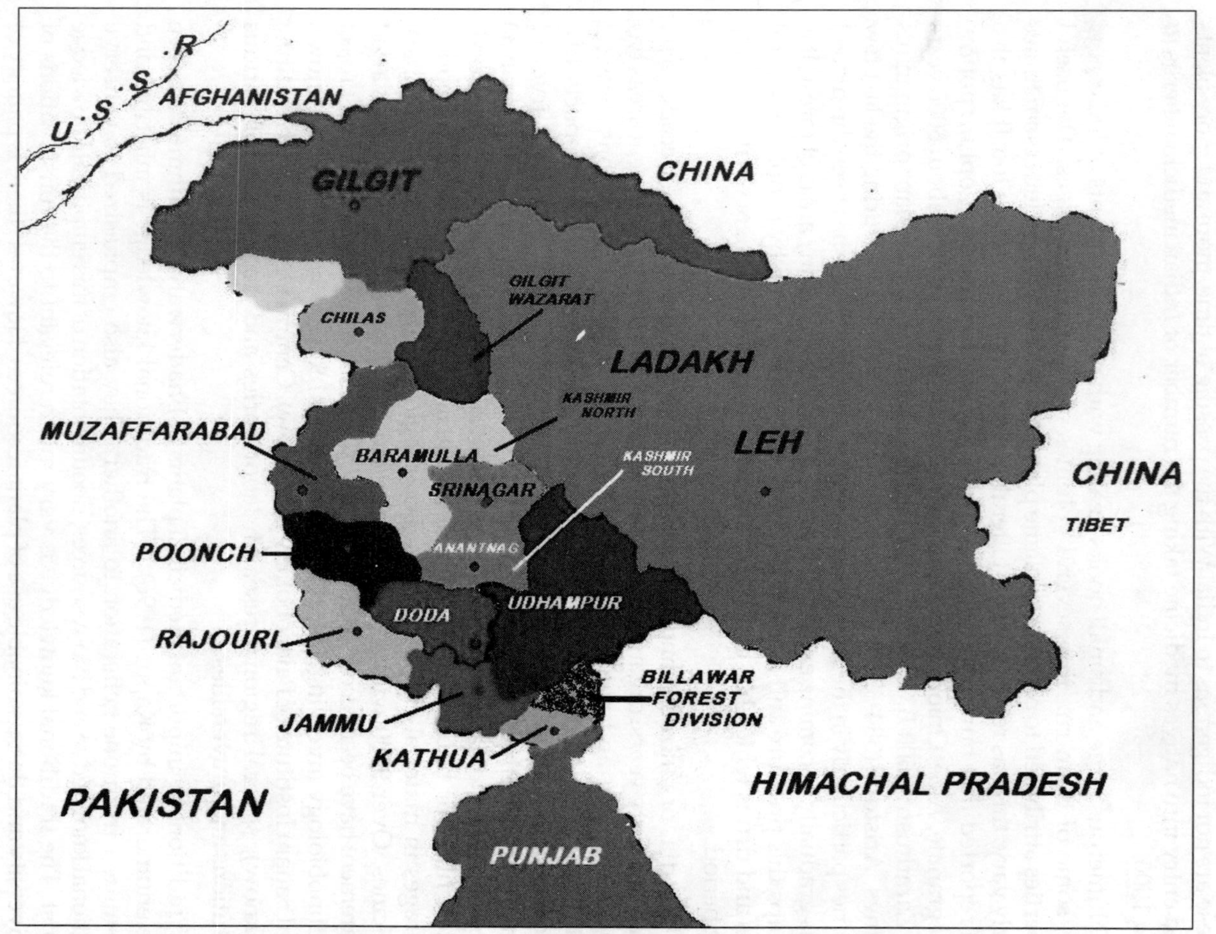

Figure 21.1: Jammu and Kashmir map showing the location of Billawar Forest Division in Kathua district

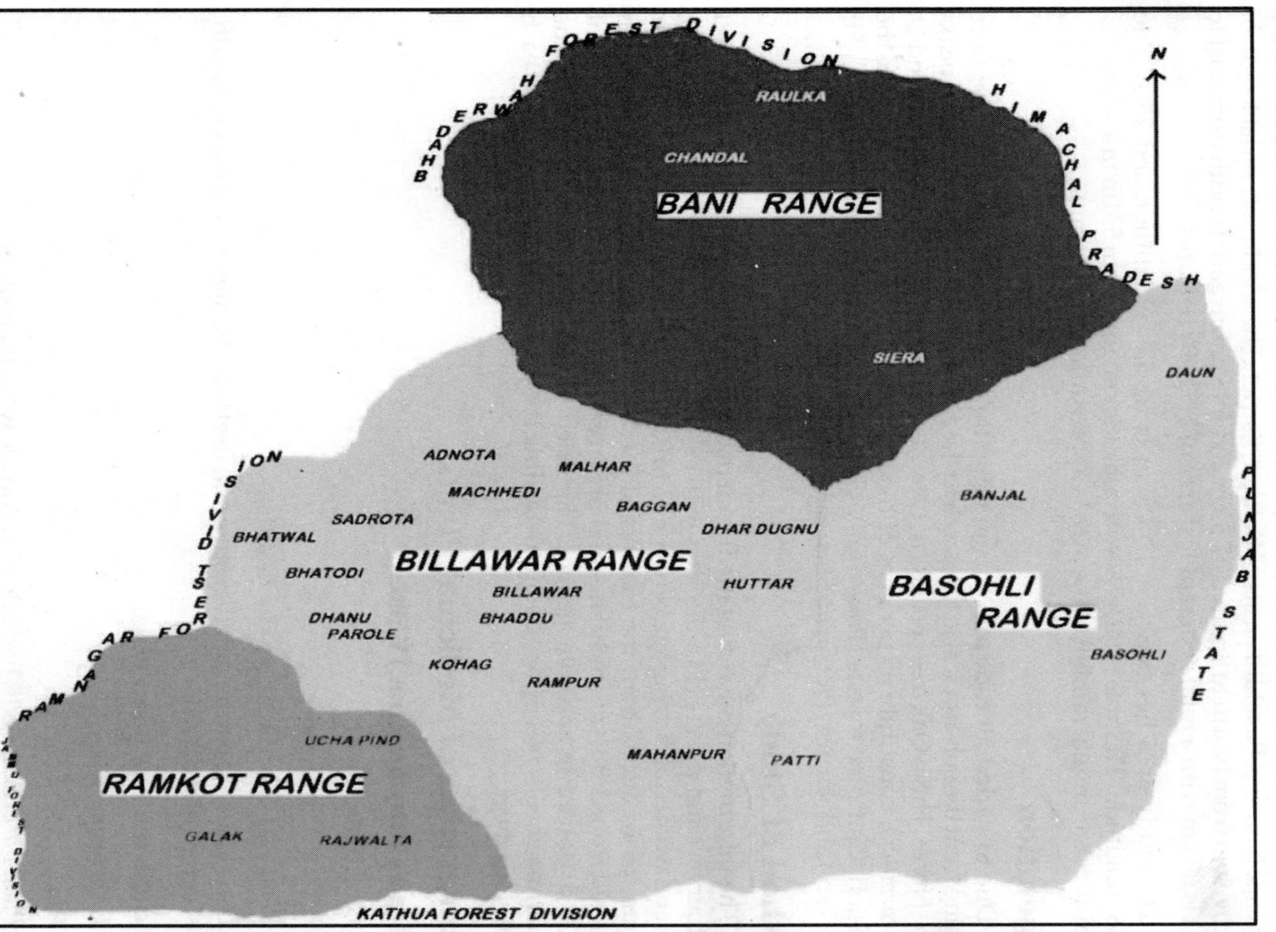

Figure 21.2: Billawar Forest Division showing Bani, Basohli, Billawar and Ramkot ranges

Study Area

The present study of medicinal plants was carried out in Billawar Forest Division, District- Kathua, Jammu Division, J&K. The Billawar Forest Division is located in the Billawar, Bani and Basohli Tehsils of Kathua district. The maximum aerial length of the division from North to South is 47.5 km and maximum aerial width from East to West is 60 km with a total geographical area of 1,52,300 hectares.

The study area lies between 32° 28' N to 32°53' N latitude and 75°17'E to 75°56' longitude. The area is all hilly with altitude ranging from 531m to 4341ma SL. Total average annual rainfall of the area is approximately 1798 mm.

Vegetation

Due to varied altitude and topography, the study area bears a diverse flora ranging from the sub-tropical in the South to the alpine meadows on the higher peaks in the North. Three Oak tree species *i.e.*, *Quercus leucotrichophora*, *Q. floribunda* and *Q. semicarpifolia*, constitute 65 per cent of the total broad leaved plant species in the Division. Chir pine forests (*Pinus roxburghii*), Deodar forests (*Cedrus deodara*), Fir (*Abies* sp.) and Spruce (*Picea smithiana*) forests make up conifers of the area.

Method of Study

The information regarding medicinal plants was gathered by means of direct field surveys and direct field interviews with native people. About 35 people including men and women all having age between 45 to 95 years were interviewed to collect or to gain information about the traditional medicinal plants. The plants had been collected from different study sites and preserved after a treatment with 2 per cent $HgCl_2$ dissolved in absolute alcohol. The dried plants were then stitched on the herbarium sheets. Necessary details regarding the plants had been written on each herbarium sheet.

Enumeration of Medicinal Plants

1. *Acacia catechu* (Linn.) Willd.

Local Name: Khair

Family: Mimosaceae

Plant Part Used: Stem (product Kattha)

Uses: Kattha is astringent, digestive and useful in the ailments of throat, mouth, gums, cough and diarrhoea.

2. *Acacia nilotica* (Linn.) Del.

Local Name: Kikar

Family: Mimosaceae

Plant Part Used: Pods, bark, flower, gum, leaves and roots.

Uses: (a) Pods are effective in urinogenital disorders.

(b) Gum is used along with *Calotropis procera* latex to cure asthma, stop bleeding and urinary and vaginal discharges.

(c) It is also useful in diabetes, skin diseases and bleeding piles.

(d) Bark is used in asthma.

(e) Flowers are used as tonic in diarrhoea and dysentery.

(f) Paste of burnt leaves is effective ointment in itch.

(g) Roots and trunk paste is used to heal wounds

3. *Achyranthes aspera* Linn.

Local Name: Parkanda

Family: Amaranthaceae

Plant Part Used: Leaves and Seeds

Uses: (a) The roasted seed powder mixed with honey is given during cough and throat irritations.

(b) Leaf juice is given to cure diarrhoea.

4. *Acorus calamus* Linn.

Local Name: Baryan

Family: Araceae

Plant Part Used: Rhizomes

Uses: The rhizome is boiled in water for half an hour, that water is effective against cough and throat pain.

5. *Adhatoda vasica* Nees.

Local Name: Brenkar

Family: Acanthaceae

Plant Part Used: Flower and Leaves

Uses: (a) Flower ash with honey is given to cure whooping cough.

(b) The smoke from the burning leaves is inhaled for the cure of asthma and cough.

6. *Adiantum capillus* Linn.

Family: Pteridaceae

Plant Part Used: Leaves

Uses: (a) It breaks up stone in the bladder and kidneys.

(b) Paste of leaves is useful in headache and chest pains.

(c) Leaves are also used in respiratory problems

7. *Aegle marmelos* Corr.

Local Name: Bel

Family: Rutaceae

Plant Part Used: Leaf, Fruit and Root.

Uses: (a) The unripe or half-ripe fruits improve appetite and digestion.

(b) The antibiotic activity of the leaf, fruit and root has been confirmed.

(c) The tribal take an infusion of root bark in fever.

8. *Ageratum conzoides* Linn.

Family: Asteraceae

Plant Part Used: Leaves and Flower buds.

Uses: (a) Leaves are useful in piles, boils, wounds and sores.

(b) Flower buds heal cancerous growth.

9. *Ajuga parviflora* Benth.

Family: Lamiaceae

Plant Part Used: Whole plant

Uses: (a) The paste of the plant is applied on the wounds.

10. *Albizia lebbeck* (Linn.) Willd

Local Name: Sarih

Family: Mimosaceae

Plant Part Used: Leaves, Bark and Flowers.

Uses: (a) The mild heated juice of the leaves of sarih and mango dropped into the ear to remove the pain.

(b) The leaves are also effective against cough.

(c) Bark strengthens gums.

(d) Flowers used in chronic cough and asthma.

11. *Aloe vera* Mill.

Local Name: Kuargandal

Family: Asphodelaceae (Liliaceae)

Plant Part Used: Leaves

Uses: (a) Leaf juice is used as a remedy for intestinal worm in children.

(b) Juice of roasted leaf is given with honey for cold and cough.

(c) The paste of the leaves is effectively used against skin diseases *i.e.,* sun burn.

12. *Asparagus racemosus* Willd.

Local Name: Sanspod

Family: Liliaceae

Plant Part Used: Roots

Uses: (a) The fresh juice of roots along with equal amount of til oil is applied on the head to remove pain.

(b) The dried powder of roots taken along with milk results in the improvement of milk in females.

13. *Azadirachta indica* A. Juss.

Local Name: Neem

Family: Meliaceae

Plant Part Used: Stem, Bark, Leaves and Root's bark.

Uses: (a) The bark is bitter tonic, it is astringent and antiperiodic *i.e.*, it is useful in fevers; it breaks the periodic sequence of fevers (like malaria) and is useful in skin diseases.

(b) The leaves are applied on skin diseases and boils.

(c) A decoction of leaves is also taken internally to expel out the germs and worms present in the intestine.

14. *Barleria prionitis* Linn.

Local Name: Kali Brenkar

Family: Acanthaceae

Plant Part Used: Bark of stem

Uses: The dried bark is powdered and with honey is given to cure whooping cough.

15. *Bauhinia variegata* Linn.

Local Name: Karal

Family: Caesalpiniaceae

Plant Part Used: Stem bark, Root bark, flower and flower buds.

Uses: (a) Bark is tonic, blood purifier and anthelmintic.

(b) Bark of roots is useful in flatulence.

(c) Flowers are laxative.

(d) Flower buds are used against piles and cough.

16. *Berberis aristata* DC.

Local Name: Kaemblu

Family: Berberidaceae

Plant Part Used: Root bark, Roots and Lower stems

Uses: (a) Root bark, roots and lower stems are boiled in water, strained and evaporated till a semi-solid mass is obtained; this is called Rasaut, soluble in water.

(b) Rasaut mixed with butter and alum is applied externally on eye lids to cure ophthalmia and other eye diseases.

(c) The use of rasaut is also helpful in curing ulcers and in early detection of malarial fever.

17. *Bombax ceiba* Linn.

Local Name: Simbal

Family: Bombacaceae

Plant Part Used: Root, bark and young fruits

Uses: (a) Roots are used in the treatment of diarrhoea.

(b) Bark is mucilaginous, and used for healing wounds and also to stop bleeding.

(c) Young fruits are useful in ulceration of bladder and kidney.

18. *Bryophyllum pinnatum* Kurz.

Family: Crassulaceae

Plant Part Used: Leaf

Uses: (a) The slightly roasted leaves are applied on wounds and cuts to heal rapidly.

(b) Leaves also reduce swellings.

19. *Butea monosperma* (Lamak.) Taubert.

Local Name: Pala

Family: Fabaceae

Plant Part Used: Gum, seeds and root bark

Uses: (a) The gum is valuable for treatment of diarrhoea.

(b) Seeds are useful against ringworms, roundworms and tapeworms.

(c) The root bark has been found to have some action on blood pressure.

20. *Calotropis gigantea* R. BR.

Local Name: Daryai aak

Family: Asclepiadaceae

Plant Part Used: Leaves and Roots

Uses: (a) The smoke from the burning leaves is inhaled for the cure of asthma and cough.

(b) A paste of the charcoal prepared from roots mixed with some bland oil is applied over skin diseases.

21. *Cannabis sativa* Linn.

Local Name: Bhang

Family: Cannabinaceae

Plant Part Used: Leaves

Uses: (a) The main use of hemp is for easing pain and inducing sleep.

(b) The tincture helps parturition and all painful urinary infections.

22. *Cassia occidentalis* Linn.

Local Name: Aarua

Family: Caesalpiniaceae

Plant Part Used: Whole plant

Uses: Extract of the plant is given to cure urinary disorders.

23. *Cassia tora* Linn.

Local Name: Chhoti aarua

Family: Caesalpiniaceae

Plant Part Used: Whole plant

Uses: (a) The whole plant is used to expel ringworms.

(b) The paste of the leaves cure skin diseases.

24. *Catharanthus roseus* (L.) G. Don.

Local Name: Sadabahar

Family: Apocynaceae

Plant Part Used: Leaves and flowers

Uses: (a) The leaves and flowers are good for the diabetic patients.

(b) Juice of leaves is used for wasp stings.

25. *Cedrus deodara* Loud.

Local Name: Deodar

Family: Pinaceae

Plant Part Used: Stem

Uses: Decoction of the wood is used in the treatment of urinary disorders, piles, kidney stones and diabetes.

26. *Centella asiatica* (L.) Urban.

Local Name: Brahmi

Family: Apiaceae

Plant Part Used: Leaves

Uses: The leaves powder is given with milk in small doses in mental weakness and to improve memory.

27. *Cinnamomum tamala* (Ham.) Nees and Ebrm.

Local Name: Tejpatta

Family: Lauraceae

Plant Part Used: Leaves

Uses: (a) The dried power of leaves along with honey is taken against cough.

(b) The leaves are also recommended against jaundice.

28. *Cinnamomum zeylanicum* Bl.

Local Name: Dalchini

Family: Lauraceae

Plant Part Used: Bark

Uses: (a) Bark has the property of destroying certain germs and fungi.

(b) The branches are lopped and their bark is removed, which is used in diarrhoea, nausea and vomiting.

(c) It is commonly used as condiment

(d) It is also useful for improving the blood and nervous system.

29. *Cissampelos pareira* Linn.

Local Name: Dordu

Family: Menispermaceae

Plant Part Used: Leaves

Uses: The leaves are effective against the diarrhoea in children

30. *Citrus medica* Linn.

Local Name: Jambhiri

Family: Rutaceae

Plant Part Used: Seeds and Fruit juice

Uses: (a) The powder of dried seeds is effective to expel out the intestinal worms.

(b) Fruit juice is very effective against scurvy.

31. *Colebrookia oppositifolia* J.E.Smith.

Local Name: Dusa

Family: Lamiaceae

Plant Part Used: Leaves

Uses: The leaves are wrapped over the painful portion of the body to remove pain.

32. *Cordia dichotoma* G. Forst.

Local Name: Lusade

Family: Boraginaceae

Plant Part Used: Fruits

Uses: The fruits are used against cholera, dropsy and dysentery.

33. *Curcuma aromatica* Salisb.

Local Name: Ban haldi

Family: Zingiberaceae

Plant Part Used: Rhizome

Uses: The rhizome powder is very effective to stop bleeding from the wounds.

34. *Cuscuta reflexa* Roxb.

Local Name: Amarbel

Family: Convolvulaccae

Plant Part Used: Whole plant

Uses: (a) The seeds are anthelmintic and are used in the treatment of bilious disorders.

(b) The stems are also used in the treatment of bilious disorders.

(c) The whole plant is used externally in the treatment of itchy skin.

35. *Cynodon dactylon* (Linn.) Pers.

Local Name: Khabbal

Family: Poaceae

Plant Part Used: Roots

Uses: (a) Decoction of roots, diuretic in dropsy.

(b) Infusion of roots for stopping bleeding from piles.

36. *Cyperus rotundus* Linn.

Local Name: Dila

Family: Cyperaceae

Plant Part Used: Rhizome

Uses: The decoction of the roots is a good antidote to all poisons.

37. *Datura metel* Linn.

Local Name: Datura

Family: Solanaceae

Plant Part Used: Leaf, Twigs and Fruits

Uses: (a) The dried leaves and twigs are smoked for cure of asthma and whooping cough.

(b) The juice of the fruits is useful to check dandruff and falling of the hair.

38. *Dalbergia sisoo* Roxb. Ex DC.

Local Name: Talli

Family: Fabaceae

Plant Part Used: Leaves

Uses: The fresh juice of leaves mixed with honey dropped into the eyes for the improvement of eyesight.

39. *Dendrocalamus* sp.

Local Name: Banj

Family: Poaceae

Plant Part Used: Flowers

Uses: (a) Flowers are used against diarrhoea.

(b) The decoction of flowers is used against cancer.

40. *Eclipta prostrata* Linn.

Local Name: Bhringraj

Family: Asteraceae

Plant Part Used: Entire plant

Uses: (a) Fresh aerial parts are used to treat snake bite.

(b) Entire plant is used for tuberculosis.

(c) The juice of the plant is good for hair and skin.

41. *Phyllanthus officinalis* Gaertn.

Local Name: Amla

Family: Euphorbiaceae

Plant Part Used: Fruit

Uses: (a) The fruits are very effective against jaundice.

(b) Dried fruit are good blood purifier.

(c) It is also used in vomiting and habitual constipation

42. *Eucalyptus citriodora* Hook.

Local Name: Safeda

Family: Myrtaceae

Plant Part Used: Leaves

Uses: The leaves are used in the treatment of asthma and bronchitis.

43. *Eupatorium purpureum* Linn.

Family: Asteraceae

Plant Part Used: Leaves

Uses: (a) The leaves are diuretic

(b) The paste of leaves is applied over foul ulcers, wounds and burns.

44. *Euphorbia heliscopia* Linn.

Local Name: Kalbatra

Family: Euphorbiaceae

Plant Part Used: Whole plant

Uses: The paste of the plant is applied on the wounds to heal the injured part of the body.

45. *Euphorbia hirta* Linn.

Local Name: Dudli

Family: Euphorbiaceae

Plant Part Used: Entire plant

Uses: (a) It is useful in removing worms in children, in bowel complaints, asthma and cough.

(b) The plant promotes formation and flow of milk in women.

(c) Roots of the plant stop vomiting.

(d) Large doses of the drug, however, cause irritation in stomach and result in vomiting and nausea.

46. *Euphorbia nerifolia* Linn.

Local Name: Shuu

Family: Euphorbiaceae

Plant Part Used: Milky juice

Uses: (a) The white milky juice of such plants is acrid and poisonous.

(b) Careless handling of the juice of these plants can cause skin and other diseases.

(c) Only in restricted doses, they are used as purgatives, etc.

(d) The juice is applied externally to kill maggets in wound and on skin diseases and warts

47. *Ficus benghalensis* Linn.

Local Name: Bado

Family: Moraceae

Plant Part Used: Latex

Uses: Its latex is used to expel out the thorns which are broken down inside the body.

48. *Ficus palmata* Linn.

Local Name: Fukada

Family: Moraceae

Plant Part Used: Latex

Uses: The latex is used to expel out the broken thorns from the skin

49. *Ficus racemosa* Linn.

Local Name: Rumbal

Family: Moraceae

Plant Part Used: Fruits and Latex

Uses: (a) Fruits are useful against kidney diseases.

(b) Fruits are used in treatment of dry cough and loss of voice.

(c) Latex applied externally on infected wounds to promote the healing.

50. *Ficus religiosa* Linn.

Local Name: Peepal

Family: Moraceae

Plant Part Used: Leaves, Roots and Fruits

Uses: (a) The roots are chewed to prevent gum diseases.

(b) Ripe fruits are antidote against poison or venom.

(c) The powder of fruits is used for asthma.

(d) The leaves are used to treat constipation.

51. *Juglans regia* Linn.

Local Name: Ban akhrot

Family: Juglandaceae

Plant Part Used: Bark

Uses: Bark is used as toothache and provides strength to the gums.

52. *Lantana camara* Linn.

Local Name: Panjfuli jadi

Family: Verbenaceae

Plant Part Used: Leaves

Uses: (a) The leaves are used as an antiseptic for wounds and externally for scabies.

(b) The leaf juice is used for the treatment of skin itches.

53. *Lawsonia inermis* Linn.

Local Name: Mehndi

Family: Lythraceae

Plant Part Used: Leaves

Uses: (a) Leaves are applied on boils, burns and skin diseases.

(b) Paste of leaves is used in headache, burning sensation in feet, etc.

54. *Mallotus philippensis* Muell.-Arg.

Local Name: Kamla

Family: Euphorbiaceae

Plant Part Used: Powder of the seeds

Uses: The powder of the fruits is highly beneficial for expelling out intestinal worms in old and children.

55. *Mangifera indica* Linn.

Local Name: Aam

Family: Anacardiaceae

Plant Part Used: Leaves and Fruits

Uses: (a) The leaves, which are dried in shade, are more effective against diabetes.

(b) Fruits make the nervous system strong.

56. *Melia azadirach* Linn.

Local Name: Drak

Family: Meliaceae

Plant Part Used: Leaves and Bark

Uses: (a) With honey, the juice is prescribed for jaundice and skin diseases.

(b) The fresh juice of the leaves is given for intestinal worms.

(c) The paste of leaves is applied over skin diseases especially for boils.

(d) An ointment to destroy lice, for scalp and other skin diseases.

(e) The bark and leaves are effective against the tapeworms, so it is anthelmintic.

57. *Mentha arvensis* Linn.

Local Name: Jangli putna

Family: Lamiaceae

Plant Part Used: Whole plant

Uses: (a) It is good cardiac tonic.

(b) It is also used in swollen gums, mouth wash and toothache.

(c) It also acts as a good blood cleanser.

58. *Mimosa pudica* Linn.

Local Name: Chui-mui

Family: Fabaceae

Plant Part Used: Leaves

Uses: (a) Decoction of leaves is used for diabetes.

(b) Paste of leaves arrests bleeding and fasten the wound healing process.

59. *Morus alba* Linn.

Local Name: Toot

Family: Moraceae

Plant Part Used: Leaves

Uses: (a) Leaves inhibit the premature graying of hair and can promote hair growth if used regularly.

(b) It also stimulates appetite.

60. *Nelumbo nucifera* Gaertn.

Local Name: Kamal

Family: Nymphaeaceae

Plant Part Used: Rhizome and Leaves

Uses: The rhizome or leaves are used with other herbs to treat vomiting of blood.

61. *Nymphaea lotus* Linn.

Local Name: Safed Kamal

Family: Nymphaeaceae

Plant Part Used: Rhizome and Flower

Uses: (a) Rhizome is useful in diarrhoea.

(b) Flowers are cardiotonic

62. *Opuntia vulgaris* Mill.

Family: Cactaceae

Plant Part Used: Fruits

Uses: The ripe fruits are good remedy in whooping cough and asthma.

63. *Origanum vulgare* Linn.

Local Name: Najbo

Family: Lamiaceae

Plant Part Used: Leaves

Uses: The juice of leaves is very effective against eyes and ear pain.

64. *Oroxylum indicum* (Linn.) Vent.

Local Name: Tantu

Family: Beignoniaceae

Plant Part Used: Stem bark, Leaf and Fruit

Uses: (a) Stem bark paste is used for the cure of scabies.

(b) Leaf decoction is given in stomachache.

(c) Mature fruits are used in treating cough, piles and cardiac disorders.

65. *Oxalis corniculata* Linn.

Local Name: Khatti butti

Family: Oxalidaceae

Plant Part Used: Leaf

Uses: (a) Leaves are used in the treatment of urinary tract infections.

(b) Leaf juice is applied to insect bites, burns and skin eruptions.

(c) Leaves are rich in Vitamin C and used for the treatment of scurvy.

66. *Plantago major* Linn.

Local Name: Isabgol

Family: Plantaginaceae

Plant Part Used: Whole plant

Uses: It is used in the treatment of dysentery, constipation and disorders of digestive system.

67. *Plumbago zeylanica* Linn.

Local Name: Chitrak

Family: Plumbaginaceae

Plant Part Used: Roots

Uses: (a) It promotes appetite and helps in digestion.

(b) Roots have been used as abortificant in some indigenous practices *i.e.* internally to the uterus.

68. *Punica granatum* L.

Local Name: Daduni

Family: Punicaceae

Plant Part Used: Bark, Roots, Seeds and Leaves

Uses: (a) The bark of stem and root is very effective against tapeworms.

(b) The fruit is very useful against the cough and jaundice.

(c) Leaves, seeds, roots and bark are effective in anthelmintic activity.

69. *Rauvolfia serpentina* Benth.

Local Name: Sarpgandha

Family: Apocynaceae

Plant Part Used: Roots

Uses: Roots are very effective against the snake bite *i.e.*, poison of snake.

70. *Ricinus communis* Linn.

Local Name: Arandi

Family: Euphorbiaceae

Plant Part Used: Leaves and Oil

Uses: (a) Castor leaves are used externally by nursing mother to increase the flow of milk

(b) Castor oil is used externally against itching.

(c) Leaves are applied to head to relieve headache.

71. *Sapindus mukorossi* Gaertn.

Local Name: Retha

Family: Sapindaceae

Plant Part Used: Fruits

Uses: Fruits are used for washing hair and acts as hair tonic.

72. *Saraca asoca* Linn.

Local Name: Ashok

Family: Caesalpiniaceae

Plant Part Used: Bark and Flower

Uses: (a) The decoction of bark is effective remedy for piles.

(b) The dried flowers are given in diabetes.

73. *Solanum nigrum* Linn.

Family: Solanaceae

Plant Part Used: Fruits

Uses: (a) Ripe fruits are very useful in respiratory problems.

(b) Ripe fruits are given in fever and their decoction cures cough and bronchial irritations effectively.

74. *Solanum xanthocarpum* Schrad and Wendl.

Local Name: Kateli

Family: Solanaceae

Plant Part Used: Whole plant

Uses: (a) The leaves are used as anthelmintic.

(b) Its decoction is given in chest pain and heart diseases.

75. *Spilanthes acmella* (Linn.) Murr.

Local Name: Piprament

Family: Compositae

Plant Part Used: Leaves and Flower

Uses: It is used to treat toothache and gum infections.

76. *Syzygium cumini* (Linn.) Skeels.

Local Name: Jamunu

Family: Myrtaceae

Plant Part Used: Leaves and Seeds

Uses: (a) The leaves give strength to tooth and gums.

(b) The seeds of the fruit are used for treatment of diabetes.

77. *Tamarindus indica* Linn.

Local Name: Imli

Family: Caesalpiniaceae

Plant Part Used: Leaves, Flower and Fruit.

Uses: (a) Dried or boiled tamarind leaves and flowers are used as poultices for swollen joints.

(b) Ripe fruits are prescribed for jaundice treatment.

78. *Terminalia arjuna* (Rox. Ex DC) Wight and Arn.

Local Name: Arjun

Family: Combretaceae

Plant Part Used: Bark

Uses: Decoction with milk is given every morning on an empty stomach which is used for cleaning sores and ulcers.

79. *Terminalia bellirica* Roxb.

Local Name: Bahera

Family: Combretaceae

Plant Part Used: Fruits

Uses: (a) The fruits are useful in digestion and diarrhoea.

(b) It is also useful in piles and, leprosy, dropsy and fever.

(c) It is the constituent of triphala, in which amla and harad are included.

80. *Terminalia chebula* Roxb.

Local Name: Harad

Family: Combretaceae

Plant Part Used: Fruit

Uses: (a) The powder or of the fruit is used as dentifrice for the strength of gums.

(b) The fruit is very effective against cough.

81. *Tinospora cordifolia* (Willd.) Miers.

Local Name: Gudo

Family: Meninspermaceae

Plant Part Used: Stem

Uses: (a) It is used in urinary disorders.

(b) It is used in the treatment of jaundice.

82. *Toona haxandra* (Wall Ex. Roxb.)

Local Name: Tooni

Family: Meliaceae

Plant Part Used: Leaves

Uses: (a) Leaves are tonic, useful in chronic dysentery.

(b) Flowers used in menstrual disorders

83. *Vanda cristata* R.Br.

Family: Orchidaceae

Plant Part Used: Leaf

Uses: The leaves of vanda play an important role in bone joints and wounds.

84. *Viola odorata* Linn.

Local Name: Banafsha

Family: Violaceae

Plant Part Used: Flower

Uses: The flowers are given in lung troubles/diseases and kidney diseases.

85. *Vitex negundo* Linn.

Local Name: Bana

Family: Verbenaceae

Plant Part Used: Flower and Leaves

Uses: (a) The extract of the leaves is used to expel out worms in children.

(b) Fresh flowers extract cures diarrhoea.

(c) Leaves are chewed in cough and associated colds.

86. *Withania somnifera* Dunal.

Local Name: Ashwgandha

Family: Solanaceae

Plant Part Used: Leaves and Roots

Uses: (a) Leaves are used to promote the healing processes of ulcers, boils and swellings.

(b) It also reduces the pus formation.

(c) Leaves are used for the treatment of inflammation of joints.

(d) Roots are used in cough and female disorders.

87. *Woodfordia fruticosa* Kurz.

Local Name: Dhai

Family: Lythraceae

Plant Part Used: Flowers and Leaves

Uses: (a) The flowers are useful against burning sensation, skin diseases, diarrhoea, fever, headache, ulcers and wounds.

(b) The leaves juice is effective against gall bladder problems.

(c) The juice of fresh flowers applied on the forehead, reduces the headache.

88. *Xeromphis spinosa* (Thunb.) Keay.

Local Name: Rade

Family: Rubiaceae

Plant Part Used: Fruits

Uses: (a) The fruits are used against cough.

(b) The fruits are used to wash the hair infected with lice.

89. *Zanthoxylum alatum* Roxb.

Local Name: Tirmiru

Family: Rutaceae

Plant Part Used: Stem

Uses: (a) The bark is used against toothache and stomachache.

(b) The seeds are used as tonic.

90. *Ziziphus jujuba* Lamk.

Local Name: Ber

Family: Rhamnaceae

Plant Part Used: Leaves, Root and Fruit

Uses: (a) The leaves promote hair growth.

(b) The root is made into powder and applied to old wounds and ulcers.

(c) The dried fruits are used against cancer.

Results and Discussion

A total of 90 medicinal plants belonging to 80 genera and 46 families have been investigated from Billawar Forest Division, District Kathua, Jammu Division, J&K state of India. The study aimed to record the indigenous knowledge of senior villagers regarding medicinal uses of the neighbouring/local plants. Moreover, it is an undesirable fact that the knowledge of indigenous people is invaluable in the present day context of biological diversity conservation and its sustainable utilization.

The tribal people *i.e.*, Gujjars and Bakarwals live in complete harmony with nature and their daily needs are met by the natural surroundings. In the absence of modern facilities, they usually depend upon herbs and other materials for the treatment of disease and ailments. 80 per cent of the world's population relies on traditional medicine to maintain their health. More emphasis should be made to utilize this practice in scientific manner. Non-scientific approach should be discouraged for the use of the plants being made for various medicinal purposes.

It is well known that due to the impact of other systems of medicine, particularly, lately by the rapid progress and spread of modern medicine and surgery, faith in and popularity of the herbal medicines has been gradually declining. But, there is much to say in favour of the use of native medicinal herbs or the indigenous system of medicine. This is not the place for any detailed analysis of this subject. About 80 per cent of the Indian population resides in small and often remote villages, where per capita income is very meager, as such the people find it very difficult to procure expensive and complicated prescriptions of allopaths. Moreover, qualified allopathic practitioners are not available in most of the hilly regions of the study area. The traditional knowledge of the medicinal plants is, therefore, very necessary for younger generations.

References

Babu C R 1997. *Herbaceous Flora of Dehradun*. Council of Scientific and Industrial Research, New Delhi, India.

Jain S K and Mudgal V 1999. *A Handbook of Ethnobotany*. Bishen Singh,Mahendra Pal Singh. 23-A, New Cannaught Place, Dehradun.

Jain S K 2006. *Medicinal Plants*. National Book Trust, New Delhi, India.

Kant S and Dutt H C 2004. Plant species causing dermatitis from Bhaderwah, J&K, India. *Nat J Life Sci* **1**: 449-452.

Kapoor S K and Srivastava T N 1996. Traditionally important medicinal flora of Udhampur district (Jammu Province). *J Econ Taxon Bot* 82-88.

Kapur S K 1996. Traditional important medicinal plants of Bhaderwah Hills – Jammu Province – *IV J Econ Taxon Bot* **12:** 70-74.

Kaul M K 1997. *Medicinal Plants of Kashmir and Ladakh*. Indus Publ Co, New Delhi.

Kaul M K, Sharma P K and Singh V 1989. Ethnobotanical studies in North-West and Trans Himalaya. Contribution to the Ethnobotany of Basohli-Bani region, J&K, India. *Bull Bot Surv Ind.* **31:** 89-94.

Pandey B P 2006. *Economic Botany*. S. Chand and Company Ltd. New Delhi.

Shah A, Abass G and Sharma M P 2012. Ethnobotanical study of some medicinal plants from Tehsil Budhal, District Rajouri, J&K. *Int Multi Res J* **2:** 05-06.

Sharma B M and Kachroo P 1983. *Illustrations to the flora of Jammu and plants of neighbourhood.* Bishen Singh Mahendra Pal Singh, Dehradun.

Singh J B and Kachroo P 1994. *Forest flora of Pir Panjal range (North-West Himalaya)*, Bishen Singh Mahendra Pal Singh, Dehradun, India.

Singh N P, Singh D K and Uniyal B P 2000. *Flora of Jammu and Kashmir-Vol-I.* Botanical Survey of India, Kolkata.

Slathia P K and Paul N 2012. Traditional practices for sustainable livelihood in Kandi belt of Jammu. *Indian J Tradit Knowle* **11:** 548-552.

Swami A and Gupta B K 1996. A note on some commonly occurring medicinal weeds of Udhampur district J&K. *J Econ Taxon Bot* **12:** 89-91.

Swami A and Gupta B K 1998. *Flora of Udhampur district*. Bishen Singh Mahendra Pal Singh, Dehradun.

Medicinal Plants: Aspects and Prospects (2014) ***Pages* 336–346**
***Editors:* Mukesh Kumar, Anjali Khare and C.P. Shukla**
ISBN: 978-81-7622-309-6
***Published by:* BIOTECH BOOKS, NEW DELHI**

Chapter 22

Medicinal and Aromatic Flowering Plants of Uttarakhand: An Overview

***Rajesh Kumar*[1]*, Manish Kumar Singh*[2] *Prabhat Mudgal*[3] *and Kumar Avinash Bharati*[4]**

[1]*Department of Botany, Bareilly College, Bareilly – 243 005*
[2]*Arid Forest Research Institute, Jodhpur – 342 005*
[3]*Department of Botany, NREC College, Khurja – 203 131*
[4] *Raw Materials Herbarium and Museum Delhi (RHMD), CSIR-NISCAIR, New Delhi – 110 012*

ABSTRACT

The Himalaya is the one of the hot-spot of biodiversity. The state of Uttarakhand has diverse range of eco-climatic conditions due to variation in topography. It harbors plenty of medicinal and aromatic floras. In present study, published literatures were surveyed for the assessment of the diversity of medicinal and aromatic species found in Uttarakhand. Apart from latest published literature, the old literatures were also consulted: Thomas Hardwick (1796), Royle (1823), Atkinson (1882), Thomson (1855), Smythe, (1932- 38), Jacquemont (1934) and Duthie (1906). A total of 167 species of plants belongs to 69 families are enlisted with local names, useful parts, and uses. The maximum representation is made by family Fabaceae (20 sp.), it is followed by Lamiaceae (12 sp.), Apocynaceae (9 sp.), Asteraceae (08), Malavaceae (7 sp.), Solanaceae (7 sp.), Asparagaceae (5 sp.), Euphorbiaceae (5 sp.) and rest of the families are represented by 4 or less number of species. In the remedy preparation, maximum numbers of species are harvested for leaves (42), followed by rhizome/root (30), stem/stem bark (28), seeds (18 sp.), fruit (13 sp.), whole plant (9 sp.), Gum/latex (9 sp.), flowers (6 sp.), oil (6 sp.), etc. The plants enlisted in present communication have potential for being used for commercial purpose.

Keywords: ***Medicinal plants, Aromatic plants, Flowering plants, Traditional knowledge, Uttarakhand.***

Introduction

The Himalayas have a great wealth of medicinal plants and traditional medicinal knowledge. The Indian Himalayan region alone supports about 18,440 species of plants (Angiosperms: 8000 spp., Gymnosperm: 44 spp., Pteridophytes: 600 spp., Bryophytes: 1736 spp., Lichens: 1159 spp. and Fungi: 6900 spp.), it is believed that of which about 45 per cent are having medicinal properties (Negi *et al.*, 1985-87). Thomas Hardwick was the first person to collect the plants from Himalayas during his visit in Alaknanda valley in 1796. William Spencer Webb, collected plants from Jamnotri, Gangotri and Kali River on the boundary of Kumaon in 1802-03 and sent them to Dr. Wallich, who later employed his plant collectors Blinkworth and Kamrup with him to collect plants more extensively from Garhwal and Kumaon (Duthie, 1906). Royle (1823) as Superintendent of Botanic Garden collected several plants from North-West Himalaya and published, *Illustrations of Botany and other branches of natural history of the Himalayan mountains and of the flora of cashmere in 1833-1840.* William Moor Craft (1812), a veterinary officer of Bengal Government crossed over the Nitipass into the Mansarover area and collected several plants along Alaknanda river and from the Niti valley in the interior of Garhwal Himalaya. Extensive collections were made by two army officers Richard Strachy and J. E. Winterbottom during 1846-1849 from western Himalaya between Nepal in the East and Bhagirathi River in the west, and published a catalogue containing over 2000 species, popularly known as the Strachey Winterbottom herbarium. The original catalogue was reprinted and published in 1882 in *"Gazetteer of Himalayan District and Northwestern Provinces and Oudh"* by Atkinson. This work was revised by Duthie (1906) after incorporating his own collections and those of several other notable botanists like Wallich (1830-1832), Jacquemont (1934), Pennel (1943), Thomson (1852), Hooker and Thomson (1855). Comprehensive informations on Garhwal flora has been mentioned in *"Flora of British India"* by Hooker (1872-1897). Schlich and Brandish (1878-1880) have given an important contribution in botanical explorations in the Northwest Himalaya. Brandish in association with Stewart subsequently published a document *"Forest flora of North-West and Central India (1874)"*, which embodies frequent references to the plants of Dehradun (Babu, 1977).

A few historic reports of alpine vegetation were published by Holdorth and Smythe (1932) who were members of the Kamet expedition of 1931. They have listed plants of Bhuinder valley and published in *"Kamet Conquered (1932)"*. Smythe was so much impressed by the scenic beauty and wealth of flowers of this valley, that he revisited the valley in 1937 and collected seeds and bulbs of more than 250 species of beautiful plants for the Royal Botanic Garden Edinburgh and published his hill collection in the famous book *"The valley of flowers"* in 1938. Inspired by the Smythe and his writings, Joan Margaret Legge continued his work and lived in the valley for many months and unfortunately felt to her death here on 4th July, 1939. Her tombstone is placed near the center of the valley with the inscriptions *"I will lift up mine eyes unto the hills from whence Cometh my help."*

After independence, a regional centre of Botanical Survey of India had been established in Dehradun (1956), many botanists started survey work in Northern Zone, and a fresh impetus was given to the exploration of peaks and vallies of Western Himalaya among them are Gupta (1955, 1960, 1972), Ghildiyal (1957), Babu (1977), Rau (1963, a, b, 1964, 1968, 1975), Naithani (1969-85), Raizada and Saxena (1978), Semwal and Gaur (1986), Gaur (1987), Uniyal *et al.*(1989), etc.

The Himalaya is a source of indigenous drugs had always remained a point of attraction throughout the ancient ages. Being the main source of divine medicinal herbs, several ancient Rishi-Munies had tried to investigate the Jadi-buties from Himalayas and described them in our ancient treatises as Rig-Veda, Viriksha-Ayurveda, Yajur-veda, Agnipuran, Shushrut, Charak-samhita, etc. Many indigenous drugs used to cure the people has been described in this Vedic literature, but the systematic studies and Latin diagnosis of plants were initiated only after the publication of *"Species Plantarum"* written by Carolus Linnaeus in May, 1753. Atkinson (1882) was the first person who described the medicinal plant of the Kumaon and Garhwal region. The organized study in this field was initiated by Dr. E.K. Janaki Ammal (1954), Botanical survey of India. Some interesting ethnomedicinal studies in western Himalayas have been done by De (1962), Shah and Joshi (1971), Nautiyal (1981), Shah (1977), Negi and Gaur (1994), Aswal (1993). Gaur (1999) in his flora of the district Garhwal has described several plants with their ethnic uses. The work incorporated in ensuing chapter is the thorough study of published literatures on medicinal flora used as indigenous drugs by the people living in Uttarakhand.

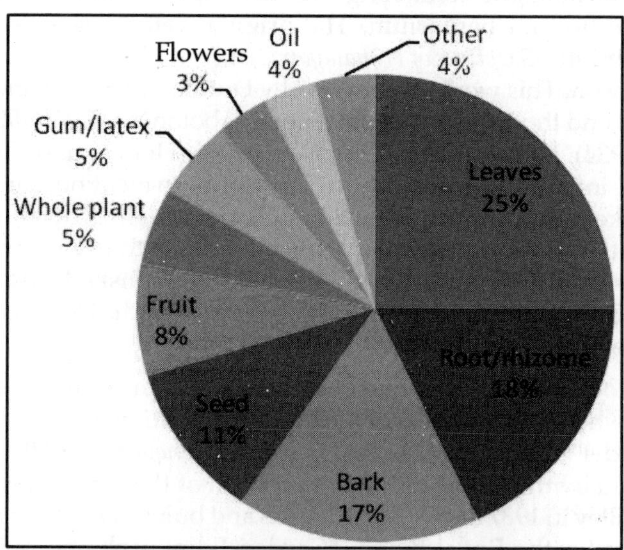

Figure 22.1: Statistics of plant parts used

Enumeration

A total of 167 species of plants belongs to 69 families are enlisted here and their respective uses are given (Table 22.1). The maximum representation is made by family Fabaceae (20 sp.), it is followed by Lamiaceae (12 sp.), Apocynaceae (9 sp.), Asteraceae (08), Malavaceae (7 sp.), Solanaceae (7 sp.), Asparagaceae (5 sp.), Euphorbiaceae (5 sp.) and rest of the families are represented by 4 or less number of species. Leaves are most frequently used plant parts (42 sp.), followed by root (30 sp.), bark (28 sp.), seeds (18 sp.), fruit (13 sp.), whole plant (9 sp.), Gum/latex (9 sp.), flowers (6 sp.), oil (6 sp.) and other parts (7 sp.; Figure 22.1).

Table 22.1: Enumeration of medicinal plants and aromatic plants having indigenous use(s)

Sl.No.	*Plant Name [Family]*	*Local Name*	*Part Used*	*Uses*
1.	*Abelmoschus moschatus* Medik. [Malvaceae]	Lata kasturi	Seeds	Carminative
2.	*Abroma augusta* (L.) L.f. [Malvaceae]	Ulat kambal	Root, bark	Irregular estrus cycle
3.	*Abrus precatorius* L. [Fabaceae]	Ratti	Root, Seeds	Asthma
4.	*Abutilon indicum* (L.) Sweet [Malvaceae]	Atibala/khangi	Root	Fever, piles
5.	*Acacia catechu* (L.f.) Willd. [Fabaceae]	Khair	Bark	Skin disease
6.	*Acacia concinna* (Willd.) DC. [Fabaceae]	Shikakai	Leaves, fruit	Hair loss
7.	*Acacia nilotica* (L.) Delile [Fabaceae]	Babule	Bark	Leprosy
8.	*Achillea millefolium* L. [Asteraceae]	Achilia	Leaves	Dysentery
9.	*Achyranthes aspera* L. [Amaranthaceae]	Apamarg/ chircthata	Root	Asthma
10.	*Acorus calamus* L. [Acoraceae]	Bach	Stem	Cough
11.	*Adenanthera microsperma* Teijsm. and Binn. [Fabaceae]	Raktachandan	Leaves	Aphrodisiac
12.	*Aegle marmelos* (L.) Corrêa [Rutaceae]	Bael	Leaves, root	Digestion
13.	*Agave americana* L. [Asparagaceae]	Ram bans	Leaves	Diuretic
14.	*Ageratum conyzoides* (L.) L. [Asteraceae]	Phulanda	Leaves	Wounds
15.	*Albizia lebbeck* (L.) Benth. [Fabaceae]	Kala sirish	Bark, Seeds	Asthma
16.	*Albizia procera* (Roxb.) Benth. [Fabaceae]	Safad sirish	Bark	Hemorrhages
17.	*Allium cepa* L. [Amaryllidaceae]	Pyaj	Bulb	Anti periodic
18.	*Allium sativum* L. [Amaryllidaceae]	Lehsun	Bulb	Ear ache
19.	*Aloe vera* (L.) Burm.f. [Asparagaceae]	Ghequr	Leaves	Jaundice
20.	*Alstonia scholaris* (L.) R. Br. [Apocynaceae]	Saptparni	Leaves	Liver tonic
21.	*Amaranthus spinosus* L. [Amaranthaceae]	Choulai	Leaves	Leucorrhoea
22.	*Andrographis paniculata* (Burm.f.) Nees [Acanthaceae]	Kalmegh	Whol plant	Antipyretic
23.	*Anethum graveolens* L. [Apiaceae]	Sowa	Seeds	Colic pain

Contd...

Table 22.1–*Contd...*

Sl.No.	*Plant Name [Family]*	*Local Name*	*Part Used*	*Uses*
24.	*Annona acuminata* Saff. [Annonaceae]	Ramphal	Fruit	Vermicide
25.	*Annona squamosa* L. [Annonaceae]	Sarifa	Fruit	Purgative
26.	*Argemone mexicana* L. [Papaveraceae]	Pilikatali	Seeds	Opthalmia
27.	*Argyreia nervosa* (Burm. f.) Bojer [Convolvulaceae]	Bidhara	Leaves	Ulcer
28.	*Artemisia annua* L. [Asteraceae]	Malaria booti	Whole plant	Fever
29.	*Asclepias curassavica* L. [Apocynaceae]	Kaknasa	Leaves	Piles
30.	*Asparagus officinalis* L. [Asparagaceae]	Soop satvar	Shoots	As vegetable
31.	*Asparagus racemosus* Willd. [Asparagaceae]	Sataver	Root	Health tonic
32.	*Azadirachta indica* A.Juss. [Meliaceae]	Neem	Leaves	Fever
33.	*Bacopa monnieri* (L.) Wettst. [Plantaginaceae]	Brahmi	Whole plant	Cooling
34.	*Baliospermum solanifolium* (Burm.) Suresh [Euphorbiaceae]	Danti	Seeds	Dropsy
35.	*Bambusa bambos* (L.) Voss [Poaceae]	Katila bans	Root	Leprosy
36.	*Barleria prionitis* L. [Acanthaceae]	Pilabansa	Leaves	Toothache
37.	*Bauhinia variegata* L. [Fabaceae]	Kachanar	Bark, Seeds	Piles
38.	*Berberis aristata* DC. [Berberidaceae]	Daruhaldi	Root	Skin disease
39.	*Bixa orellana* L. [Bixaceae]	Sinduri	Bark	Gonorrhoea
40.	*Boerhavia diffusa* L. [Nyctaginaceae]	Punarnarva	Whole plant	Jaundice
41.	*Bombax ceiba* L. [Malvaceae]	Semal	Flowers	Piles
42.	*Boswellia serrata* Roxb. ex Colebr. [Burseraceae]	Salai	Gum	Rheumatic
43.	*Brassica rapa* L. [Brassicaceae]	Sarsoon	Seeds oil	Dental disorder
44.	*Bryophyllum pinnatum* (Lam.) Oken [Crassulaceae]	Patherchta	Leaves extract	Dysentery
45.	*Butea monosperma* (Lam.) Taub. [Fabaceae]	Dhak	Bark	Anthelminitic
46.	*Calendula officinalis* L. [Asteraceae]	African marigold	Whole plant	Chronic ulcer
47.	*Calotropis gigantea* (L.) Dryand. [Apocynaceae]	Madar	Stem	Hysteria
48.	*Calotropis procera* (Aiton) Dryand. [Apocynaceae]	Safed madar	Leaves	Leprosy
49.	*Cannabis sativa* L. [Cannabaceae]	Bhang	Dried leaves	Cough
50.	*Capsicum frutescens* [Solanaceae]	Merch	Fruit	Stimulant
51.	*Capsicum annuum* L. [Solanaceae]	Merch	Fruit	Arthritis
52.	*Carica papaya* L. [Caricaceae]	Papaya	Latex	Induce abortion

Contd...

Table 22.1–*Contd...*

Sl.No.	Plant Name [Family]	Local Name	Part Used	Uses
53.	*Cascabela thevetia* [L.] Lippold syn. *Thevetia peruviana* [Pers.] K.Schum. [Apocynaceae]	Pili Kaner	Bark	Purgative
54.	*Cayaponia laciniosa* (L.) C.Jeffrey [Cucurbitaceae]	Shivling	Leaves	Boils
55.	*Chrysopogon zizanioides* [L.] Roberty syn. *Vetiveria zizanioides* [L.] Nash [Poaceae]	Khas	Root oil	Headache
56.	*Combretum indicum* [L.] DeFilipps syn. *Quisqualis indica* L. [Combretaceae]	Madhumalti	Leaves	Diarrhoea
57.	*Cullen corylifolium* [L.] Medik. syn. *Psoralea corylifolia* L. [Fabaceae]	Babchi	Seeds	Leprosy
58.	*Diploknema butyracea* (Roxb.) H.J.Lam Syn. *Madhuca butyracea* (Roxb.) J.F.Macbr. [Sapotaceae]	Mahua	Leaves	Joint pain
59.	*Eryngium foetidum* L. [Apiaceae]	Phari dhania	Rots	Stomachic
60.	*Eucalyptus globulus* Labill. [Myretaceae]	Safeda	Leaves	Mosquito repellant
61.	*Euphorbia hirta* L. [Euphorbiaceae]	Dudhi	Whole plant	Asthma
62.	*Euphorbia pulcherrima* Willd. ex Klotzsch [Euphorbiaceae]	Poinsettia	Flowers	Laxative
63.	*Fagopyrum esculentum* Moench [Polygonaceae]	Kuttu	Seeds	Laxative
64.	*Ficus racemosa* L. [Moraceae]	Gular	Latex	Dysentery
65.	*Ficus religiosa* L. [Moraceae]	Papal	Bark	Skin infection
66.	*Gardenia gummifera* L.f. [Rubiaceae]	Nari –Hing	Gum	Worms
67.	*Gardenia gummifera* L.f. [Rubiaceae]	Gandharaj	Flowers	Essential oil for perfume
68.	*Gloriosa superba* L. [Liliaceae]	Kalhari	Tuber	Abortifacient
69.	*Grevillea robusta* A.Cunn. ex R.Br. [Protaceae]	Silver oak	Root	Diuretic, Spasmolytic
70.	*Hedychium spicatum* Sm. [Zingiberaceae]	Kapoor kachri	Rhizome	Vomiting
71.	*Helianthus annuus* L. [Asteraceae]	Sunflowers	Roots	Bronchitis
72.	*Helicteres isora* L. [Sterculiaceae]	Maror phalli	Root	Scabies
73.	*Hibiscus rosa-sinensis* L. [Malvaceae]	Gurhal	Roots	Cough
74.	*Hyoscyamus niger* L. [Solanaceae]	Khurasani ajwain	Flowers	Asthma
75.	*Ixora coccinea* L. [Rubiaceae]	Rukmani	Root	Dysentery
76.	*Jasminum sambac* (L.) Sol. [Oleaceae]	Mogra	Root	Anti-inflammatory
77.	*Jatropha curcas* L. [Euphorbiaceae]	Tattan jot	Stem bark	Scurvy
78.	*Justicia adhatoda* L. [Acanthaceae]	Vasaka	Root	Asthma

Contd...

Table 22.1–*Contd...*

Sl.No.	Plant Name [Family]	Local Name	Part Used	Uses
79.	*Kigelia africana* (Lam.) Benth. [Bignoniaceae]	Balam khira	Fruits	Stomach disorder
80.	*Lantana camara* L. [Verbenaceae]	Lantana	Whole plant	Intestinal worms
81.	*Lavandula angustifolia* Mill. [Lamiaceae]	Lamiaceae	Leaves	Stimulant
82.	*Lawsonia inermis* L. [Lythraceae]	Mehndi	Leaves	Asthma, cough
83.	*Lepidium sativum* L. [Brassicaceae]	Clandrasur	Seeds	Skin disease
84.	*Leucas cephalotes* [Roth] Spreng. [Lamiaceae]	Gumma	Whole plant	Cough
85.	*Lippia alba* (Mill.) N.E.Br. ex Britton and P.Wilson [Verbenaceae]	Lipia	Leaves	Essential oil used in perfume
86.	*Madhuca longifolia* var. *latifolia* (Roxb.) A.Chev. [Sapotaceae]	Mahua	Stem bark	Bronchitis
87.	*Magnolia champaca* [L.] Baill. ex Pierre syn. *Michelia champaca* L. [Magnoliaceae]	Champa	Whole plant	Regulation of estrus cycle
88.	*Martynia annua* L. [Martyniaceae]	Baghnakha	Leaves	Sore throat
89.	*Matricaria chamomilla* L. [Asteraceae]	Baboon	Inflorescence	Anti-inflammatory
90.	*Melia azedarach* L. [Meliaceae]	Bakain	Stem bark	Antihelminthics
91.	*Mentha piperita* L. [Lamiaceae]	Bergamot mint	Whole plant	Flavoring agent
92.	*Mentha arvensis* L. [Lamiaceae]	Japanese mint	Whole plant	Indigestion
93.	*Mentha spicata* L. [Lamiaceae]	Spear mint	Whole plant	Digestive disorder
94.	*Mesua ferrea* L. [Calophyllaceae]	Nagkesar	Flowers	Bleeding piles
95.	*Mimosa pudica* L. [Fabaceae]	Lajwanti	Root	Piles
96.	*Mimusops elengi* L. [Sapotaceae]	Maulsari	Bark	Increase fertility in woman
97.	*Mirabilis jalapa* L. [Nyctaginaceae]	Gulabasa	Leaves	Blood purifier
98.	*Momordica charantia* L. [Cucurbitacecae]	Karela	Whole plant	High blood pressure
99.	*Moringa oleifera* Lam. [Moringaceae]	Sajan	Stem bark	Indigestion
100.	*Morus alba* L. [Moraceae]	Sahoot	Bark	Sore throat
101.	*Mucuna pruriens* [L.] DC. [Fabaceae]	Kaunch	Seeds	Health tonic
102.	*Murraya koenigii* [L.] Spreng. [Rutaceae]	Karipatta	Root, bark leaves	Astringent
103.	*Murraya paniculata* [L.] Jack [Rutaceae]	Kamini	Root bark	Astringent
104.	*Musa paradisiaca* L. [Musaceae]	Kela	Roots	Cuts
105.	*Myrica nagi* Thunb. [Myricaceae]	Kaphal	Fruit	Fever
106.	*Neolamarckia cadamba* (Roxb.) Bosser [Rubiaceae]	Kadamb	Bark	Febrifuge
107.	*Nerium oleander* L. [Apocynaceae]	Lal kaner	Leaves	Cardio tonic
108.	*Nicotiana tabacum* L. [Solanaceae]	Tambakhu	Leaves	Cough

Contd...

Table 22.1–*Contd...*

Sl.No.	*Plant Name [Family]*	*Local Name*	*Part Used*	*Uses*
109.	*Nyctanthes arbor-tristis* L. [Oleaceae]	Har-Shringar	Leaves	Eye disease
110.	*Ocimum americanum* L. [Lamiaceae]	Van tulsi	Leaves	Dysentery
111.	*Ocimum basilicum* L. [Lamiaceae]	Tulsi/basil	Seeds	Asthma
112.	*Ocimum gratissimum* L. [Lamiaceae]	Rama tulsi	Leaves	Fever
113.	*Ocimum kilimandscharicum* Gürke [Lamiaceae]	Kapoor tulsi	Leaves	Cough
114.	*Ocimum tenuiflorum* L. [Lamiaceae]	Pooja ke- tulsi	Seeds	Cold and cough
115.	*Origanum majorana* L. syn. *Majorana hortensis* Moench [Lamiaceae]	Marjoram Sweet	Leaves	Cold
116.	*Oroxylum indicum* [L.] Kurz [Bignoniaceae]	Syonak	Root	Arthritis
117.	*Papaver somniferum* L. [Papaveraceae]	Poppy	Flowers, fruit	Narcotic
118.	*Pelargonium graveolens* L'Hér. [Geraniaceae]	Geranium	Leaves oil	Cosmetic
119.	*Phyllanthus amarus* Schumach. and Thonn.[Phyllanthaceae]	Bhoomi aonla	Whole plant	Liver tonic
120.	*Pimenta dioica* [L.] Merr. [Myrtaceae]	All spice	Leaves fruit	Spice
121.	*Pinus palustris* Mill. syn. *Pinus longifolia* Salisb. [Pinaceae]	Cheer	Resin	Arthritis
122.	*Piper longum* L. [Piperaceae]	Pipali	Spike root	Bronchitis
123.	*Plantago orbignyana* Steinh. ex Decne. Syn. *Plantago ovata* Phil. [Plumbagiaceae]	Isabgol	Seeds husk	Laxative
124.	*Plumbago zeylanica* L. [Plumbaginaceae]	Chitrak	Root	Abortifacient
125.	*Plumeria rubra* L. [Apocynaceae]	Pagoda	Bark	Rheumatism
126.	*Pogostemon cablin* [Blanco] Benth. [Lamiaceae]	Patchouli	Leaves	Regulate estrus cycle
127.	*Pongamia pinnata* [L.] Pierre [Fabaceae]	Karanj	Bark	Antiseptic
128.	*Prosopis caldenia* Burkart [Fabaceae]	Sami braksha	Pods	Astringent
129.	*Pueraria montana* var. *lobata* [Willd.] Sanjappa and Pradeep syn. *Phaseolus trilobus* [L.] Aiton [Fabaceae]	Jangli moong	Whole plant	Loss of mental balance
130.	*Punica granatum* L. [Lythraceae]	Anar	Root bark	Bleeding from nose
131.	*Putranjiva roxburghii* Wall. [Putranjivaceae]	Putranjeeva	Leaves	Cold fever
132.	*Raphanus sativus* L. [Bressicaceae]	Mooli	Root	Diuretic
133.	*Rauvolfia serpentina* [L.] Benth. ex Kurz [Apocynaceae]	Sarpgandha	Root	Hyper tension
134.	*Rauvolfia tetraphylla* L. [Apocynaceae]	Sarpgandha	Root	Skin troubles
135.	*Ricinus communis* L. [Euphorbiaceae]	Arand	Root	Purgative
136.	*Rosa damascene* [Rosaceae]	Gulab	Flowers	Cardio -tonic
137.	*Rosmarinus officinalis* L. [Lamiaceae]	Rosemerry	Leaves	Indigestion

Contd...

Table 22.1–*Contd...*

Sl.No.	Plant Name [Family]	Local Name	Part Used	Uses
138.	*Ruta graveolens* L. [Rutaceae]	Sadab	Whole plant	Aphrodisiac
139.	*Sansevieria roxburghiana* Schult. and Schult.f. [Asparagaceae]	Sansiveria	Rhizome	Antidote
140.	*Santalum album* L. [Santalaceae]	Chandan	Heartwood	Diuretic
141.	*Sapindus saponaria* L. [Sapindaceae]	Reeta	Fruits	Expectorant
142.	*Saraca indica* L. [Fabaceae]	Sita ashok	Dried bark	Regulate estrus cycle
143.	*Sesamum indicum* L. [Pedaliaceae]	Till	Root	Hair tonic
144.	*Sesbania sesban* [L.] Merr. [Fabaceae]	Dhaincha	Root	Antihelminthics
145.	*Shorea robusta* Gaertn. [Dipterocarpaceae]	Sal	Bark	Diabetes
146.	*Sida cordifolia* L. [Malvaceae]	Bala	Whole plant	Fever
147.	*Sida rhombifolia* L. [Malvaceae]	Mahabala	Roots	Aphrodisiac
148.	*Solanum americanum* Mill. syn. *Solanum nigrum* L. [Solanaceae]	Makoi	Whole plant	Dysentery
149.	*Solanum capsicoides* All. syn. *Solanum khasianum* C.B. Clarke [Solanaceae]	Kanthari	Fruit	Contraceptive
150.	*Solanum surattense* Burm. f. syn. *Solanum virginianum* L. [Solanaceae]	Kanthari	Seeds coat	Antihelminthics
151.	*Spilanthes acmella* [L.] L. [Asteraceae]	Akarkara	Root	Tongue paralysis
152.	*Syzygium cumini* [L.] Skeels [Myrtaceae]	Jamun	Bark	Diuretic
153.	*Tagetes minuta* L. [Asteraceae]	Gaynda	leaves	Antiseptic
154.	*Tamarindus indica* L. [Fabaceae]	Imli	Seeds, Leaves	Diuretic
155.	*Tectona grandis* L.f. [Lamiaceae]	Sagon	Whole plant	Antihelminthics
156.	*Tephrosia purpurea* [L.] Pers. [Fabaceae]	Sarpunkha	Whole plant	Pimples
157.	*Terminalia arjuna* [Roxb. ex DC.] Wight and Arn. [Combretaceae]	Arjun	Stem bark	Tonic
158.	*Terminalia belerica* Roxb. [Comberetaceae]	Bahera	fruit	Indigestion
159.	*Terminalia chebula* Retz. [Comberetaceae]	Harar	Fruit	Anorexia
160.	*Thuja occidentalis* L. [Cupressaceae]	Morpankhi	Stem	Fever
161.	*Tinospora cordifolia* [Willd.] Miers [Menispermaceae]	Giloe	Whole plant	Arthritis
162.	*Toona ciliata* M.Roem. [Meliaceae]	Toon	Bark	Dysentery
163.	*Tribulus terrestris* L. [Zygophyllaceae]	Chota gokhru	Fruit	Tonic
164.	*Trigonella foenum-graecum* L. [Fabaceae]	Methi	Seeds	Diuretic
165.	*Tylophora indica* [Burm. f.] Merr. [Apocynaceae]	Antmool	Leaves bark	Asthma
166.	*Vernonia anthelmintica* [L.] Willd. [Asteraceae]	Verrnonia	Fruit	Skin disease
167.	*Vitex negundo* L. [Lamiaceae]	Nirgundi	Leaves	Vermifuge

Acknowledgement

The authors are highly grateful to the CSIR-NISCAIR, New Delhi, for the library support extended towards the review of literature. Dr. K. A. Bharati is thankful to the Council for Scientific and Industrial Research (CSIR), India for providing financial assistance. Dr. R. Kumar is grateful to the Principal, Bareilly College, Bareilly for support and encouragement.

References

Aswal B S 1993. *Rare or Threatened Medicinal Plants of Garhwal Himalaya and their Conservation. Garhwal Himalaya Ecology and Environment,* Ashish Publishing House, New Delhi.

Atkinson E T 1882. *The Himalayan Gazetteer* Vols 2 (Reprint 1973), Cosmo Publ., New Delhi.

Babu C R 1977. *Herbaceous Flora of Dehradun* C S I R, New Delhi.

De A C 1962. *Folklore of Medicinal Plants of Bhagirathi Valley* (*Himalaya*) Colombo Govt Press, Ceylon.

Duthie J F 1906. Catalogue of the Plants of Kumaon and of the Adjacent Portions of Garhwal and Tibet based on the collections made by Strachey and Winterbottom during the years 1846-1849, London.

Gaur R D 1987. Contribution to Flora of Srinagar Garhwal. *J Econ Taxo Bot* **9:** 31-63.

Gaur R D 1999. *Flora of the Dist Garhwal: North-West Himalaya* (*With Ethnobotanical Notes*). Transmedia, Srinagar (Garhwal).

Ghildiyal B N 1957. A Botanical Trip to Valley of the Flowers. *J Bom Nat Hist Soc* **54**: 365-386.

Gupta R K 1955. Botanical Explorations in the Bhillangana Valley of Erstwhile Tehri Garhwal state. *J Bom Nat Hist Soc* **53**: 581-594.

Gupta R K 1960. On a Botanical Trip to the Source of River Ganga in Tehri Garhwal Himalayas. *Ind J Fores* **86**: 547-552.

Gupta R K 1972. Boreal and Arcto-Alpine Elements in the Flora of Western Himalayas. *Veg* **24:**159-175.

Jacquemont 1934. *Letter from India: Describing a Journey in the British Dominions of India Tibet Lahore and Kashmir During the Years 1847-1848,* London.

Naithani B D 1969. Plant with the Kedarnath Parbat Expedition 1969. *Bull Bot Sur Ind* **11:** 224-233.

Naithani B D 1984-85. *Flora of Chamoli.* Vol: 1-2. *Bot Sur Ind,* Howrah.

Naithani H B and Tiwari S C 1982. Flowering Plant of Pauri and its Vicinity. *Ind J For* **5:**142-148.

Naithani H B and Tiwari S C 1983. Flowering Plant of Pauri and its Vicinity. *Ind J For* **6:** 70-74.

Negi K S and Gaur R D 1994. Principal Wild Food Plants of Western Himalaya Uttar Pradesh India *Higher Plants of Indian Subcontinent* **3:** 1-147

Negi K S, Tiwari J K and Gaur R D 1985. A Contribution to the Flora of Dodital A High Altitude Lake in the Himalaya (Uttarkashi). *J Bom Nat Hist Soc* **82**: 258-270.

Negi K S, Tiwari J K and Gaur R D 1987. A Contribution to the Flora of Khatling Glacier in the Garhwal Himalaya (Dist Tehri). *J Bom Nat Hist Soc* **84:** 585-598.

Raizada M B and Saxena H O 1978. *Flora of Mussoorie,* Vol I, Bishen Singh Mahendra Pal Singh, Dehradun.

Rau M A 1963. *Illustration of West Himalayan Flowering Plants. Bot Suv Ind* , Howrah.

Rau M A 1963. The Vegetation around Jamnotri in Tehri Garhwal, UP. *Bull Bot Suv Ind* **5:** 280-287.

Rau M A 1964. A Visit to the Valley of Flowers and Lake Hemkund in North Garhwal, U P. *Bull Bot Suv Ind* **6:** 169-171.

Rau M A 1968. Flora of the Upper Gangetic Plain and the Adjacent Siwalik and Sub Himalayan Tracts. *Bull Bot Suv Ind* **10:** 1-87.

Rau M A 1975. *High Altitude Flowering Plants of West Himalaya. Bot Suv Ind,* Howrah.

Royle J F 1833-40. *Illustrations of the Botany and other Branches of the Natural History of the Himalayan Mountains and of the Flora of Kashmir,* London.

Semwal J K and Gaur R D 1986. Addition to the Flora of Tungnath in Garhwal Himalaya. *J Bom Nat Hist Soc* **83:** 267-271.

Shah N C 1977. Ethnobotany of *Acorus calamus. Ind J Pharmacol* **39:** 8-11.

Shah N C and Joshi M C 1971. An Ethnomedicinal study of the Kumaon region of India. *Econ Bot* **25:** 414-24.

Smythe F S 1932. *Kamet Conquered,* Victor Gallery Press, London.

Smythe F S 1938. *The Valley of Flowers.* Victor Gallery Press, London.

Strachey R 1906. *Catalogue of the Plants of Kumaon and of the Adjacent Portions of Garhwal and Tibet,* London.

Thomson T 1855. Western Himalaya and Tibet: A Narrative of a Journey through the Mountains of Northern India During the years 1847-1848, London.

Uniyal M R 1989. *Medicinal Flora of Garhwal Himalayas.* Vaidyanath Ayurved Bhawan Pvt Ltd, Nagpur.

Medicinal Plants: Aspects and Prospects (2014) Pages 347–353
Editors: Mukesh Kumar, Anjali Khare and C.P. Shukla
ISBN: 978-81-7622-309-6
Published by: BIOTECH BOOKS, NEW DELHI

Chapter 23

Enumeration of Medicinal Flora of District Kullu, Himachal Pradesh, India

Rita[1], Anjali Khare[1] and Bharat Bhushan[2]

[1]Department of Botany,
Advance Institute of Science and Technology, Dehradun – 248 001
[2]Department of Botany, Sahu Jain P.G. College, Najibabad – 246 763

ABSTRACT

Medicinal plants are important components of our natural wealth. They serve as important therapeutic agents as well as valuable raw materials for manufacturing numerous traditional and modern homeopathic medicines. An ethnobotanical study was carried out in different areas of Kullu district of Himachal Pradesh. The information regarding medicinal plants was gathered by means of direct field surveys and direct field interviews with native people. About 35 people including men and women, all having age between 45 to 95 years has been interviewed to collect or to gain information about the medicinal plants.

A total of 40 plants belonging to 38 families are listed in this chapter. Details of medicinal plants are described alphabetically with their botanical names, families, local names, plant parts used and disease/ailments for which they are used.

Keywords: Himachal Pradesh, Kullu, Medicinal plants.

Introduction

Medicinal plants are the plants that have a recognized use to get rid of some ailments. They range from those used in the production of mainstream pharmaceutical products to plants used in herbal medicine preparations. Herbal medicine is one of the oldest forms of medical treatment in human history and could be considered one of the forerunners of the modern pharmaceutical trade. These plants are used as natural medicines with no side-effects. This practice of using medicinal plants has existed since prehistoric times. There are three ways in which plants have been found useful in medicines.

1. First they may be used directly as tea or in other extracted forms for their natural chemical constituents.
2. Second, they may be used as an agent in the synthesis of drugs.
3. Finally, the organic molecules found in the plants may be used as models for synthetic drugs.

India has a rich wealth of medicinal plants, a large number of species and types occur in different parts of the country. A rapid extension of the allopathic system of medical treatment, which has taken place in India during the last few decades, has been responsible for creating a great demand for pharmaceutical drugs and their products. Attempts have, therefore, been made to undertake studies connected with the introduction of many drug plants in India and also their commercial cultivation.

The Himalayas have a great wealth of medicinal plants and traditional medicinal knowledge. Himachal Pradesh, one of the pioneer Himalayan States is a rich repository of medicinal flora. Because of its geographical position, hazardous means of transport communication; the traditions, myths, legends and folklores of the ancients are carefully preserved temples and historical places in existence at various regions of this state. Kullu District of Himachal Pradesh is well known medicinal plant hot spot in the Western Himalayas that has a rich diversity of flora and fauna.

North-Western Himalayan region with its wide range of altitudes, topography and climatic conditions, is a rich repository of medicinal wealth, which occupied an important place in Vedic treatise. More than 800 valuable medicinal species found in this part of India is extensively used by the locals since time immemorial for curing various diseases of humankind. It is now a well known fact that medicinal plant sector possesses great potential to uplift the economy of this part of India.

Eversince ancient times, in search for rescue for their disease, the people looked for drugs in nature. The beginning of the medicinal plants use was instinctive, as is the care with animals (Stojanoski 1999). Singh and Rawat (1998) have listed more than 250 species as ethnomedicinal and about 50 of them are commercially exploited. These species have been listed in the Red Data Book of IUCN (Singh and Rawat, 1998 and 2000). Dhaliwal and Sharma (1999) recorded more than 800 species of angiosperms from Kullu Valley. A brief ethnobotanical account of 109 plant species belong to 4l families and 86 genera of Kullu District in North Western Himalaya have been recorded by Singh (1999). Singh (2004) has enumerated about 58 locally used medicinal plants from Great Himalayan National Park, Kullu Valley. Negi and Subramani (2006) recorded 38 plants belonging to 23 families and 17 Red listed species in the Naggar area of Kullu Valley Himachal Pradesh. First hand information on 35 plant species was recorded from Malana, located in Parvati Valley of Kullu District by Sharma (2005).

The importance of recording the usage of plants in this region is especially imperative because of rapid loss of forest wealth and traditional wisdom due to increase in tourism and modernization. Thus an attempt has been made to enumerate the medicinal flora of Kullu so as they may be utilized by the people for treatment of various diseases.

Description of Study Site

Himachal Pradesh is located in the lap of Northern India. It spreads over an area of 21,495 sq mile (55,670 km^2) and is bordered by Jammu and Kashmir in the north, Punjab in the West and South-West, Haryana and Uttarakhand in the South-East, and also by the Tibet, an autonomous region at the East. The state is famous for its abundant natural beauty and its uniqueness. The geographical location of Himachal Pradesh is 30' 22' 40" North to 33' 12' 40" North latitude and its longitudinal extent is 75' 45' 55" East to 79' 04' 20" East.

Kullu is a district in Himachal Pradesh, India. The district stretches from the Village of Rampur in the South to the Rohtang Pass in the North. Kullu District is located at 31°20´25´´ to 32 °25 ´ N latitude and 76°56´30´´ to 77°52´20´´ E longitude. It is bounded by Shimla district on the South. Further, on the South-West and West it is bounded by Mandi district and, on the North-West by Kangra District.

Methodology

The study was carried out in different areas of Kullu district of Himachal Pradesh. The information regarding medicinal plants was gathered by means of direct field surveys and direct field interviews with native people. About 35 people including men and women all having age between 45 to 95 years were interviewed to collect or to gain information about the medicinal value of the local plants.

The plants were collected from different study sites, preserved and treated with 20 per cent $HgCl_2$ in absolute alcohol and then the dried plants were stitched on herbarium sheets. The plants were identified with the help of pertinent literature in the field.

Observation

A total of 40 species have been recorded and enumerated with its family names, local names, parts used, disease/ailments and uses (Table 23.1).

Result and Discussion

A total of 40 species were recorded from the study site. It includes about 40 medicinal plants which are being used for various ailments. The study aimed to record the indigenous knowledge from traditional medical practitioners regarding their medicinal uses. Moreover, it is an undesirable fact that the knowledge of indigenous people is invaluable in the present concept of biological diversity conservation and its sustainable utilization.

The tribal people of Gaddhi live in complete harmony with nature and their daily needs are met by the natural surroundings. In the absence of modern facilities,

Table 23.1: Enumeration of Medicinal Plants

Sl.No.	Vernacular Names	Local Names	Families	Status	Plant Parts Used	Uses
1.	*Achyranthes aspera* Linn.	Apmarga	Amaranthaceae	H	Flower	Asthma, cough, pneumonia, bites of snakes, ulcers, warts
2.	*Acorus calamus* Linn.	Vacha	Araceae	H	Rhizome	Stomach ache, bronchitis, vermifuge, chronic diarrhoea, memory loss
3.	*Adiantum capillus* Veneris Linn	Hansraj	Adiantaceae	Fern H	Leaves and rhizome	Diuretic, calm coughs, gallstone, heat burn, child birth, Diabetes
4.	*Allium carolinianm* D.C.	Piyaz	Alliaceae	H	Bulbs and leaves	Digestion, reducing cholesterol levels, moth repellent
5.	*Angelica glauca* Linn.	Chora	Umbelliferae	H	Root and the entire aerial plant body	Stimulant, carminative, constipation
6.	*Argemone mexicana* Linn.	Pharangi datura	Papaveraceae	H	Entire Plant	Infestation, skin disease, itching, inflammations, poisoning, constipations, fever
7.	*Berberis aristata* Linn.	Daru haldi	Berberidaceae	S	Root and fruit	Eyes disorders malaria, diuretic, stomachic anti-inflammatory, kapha and pitta doshars
8.	*Bergenia ligulata* (Wall) Engl.	Pashan-bhed	Saxifragaceae	H	Root	Diuretic, anti- inflammatory, kidney stone, wounds.
9.	*Celtha palustris* Linn.	Marsh Merigold	Arecaceae	H	Rhizome	Stomachic, carminative, bronchitis, remittent fever, vermifuge, chronic diarrhoea, memory loss
10.	*Cannabis sativa* Linn.	Bhang	Cannabinaceae	S	Flower Leaves and Seeds	Inducing sleep, urinary infection, asthma, cystitis, diarrhoea, fevers, nausea and vomiting
11.	*Capsella bursa-pastoris*	Mother's heart	Brassicaceae	H	Aerial parts	Astringent, anti-inflammatory, diuretic, hemorrhaging, eczema.
12.	*Catharanthus roseus* (Linn.) G.Don	Sadabahar	Apocynaceae	H	Root, shoot and leaves	Diabetes, malaria wasp stings
13.	*Cedrus deodara* (Roxb.) G.Don	Deodar	Pinaceae	T	Stem	Urinary disorder, piles, kidney stone and diabetes

Contd...

Table 23.1–*Contd...*

Sl.No.	Vernacular Names	Local Names	Families	Status	Plant Parts Used	Uses
14.	*Cinnamomum tamala* (Buch. Ham.) T. Nees & C.H.	Tejpat	Lauraceae	T	Leaves and Bark	Diuretic, cardiac disorders, diarrhoea, nausea
15.	*Convolvulus pluricaulis* Choisy.	Shankh pushpi	Convolvulaceae	H	Leaves and entire plant	Brain tonic, stress ulcers, anxiety, hypertension
16.	*Dioscorea deltoides* Wall. ex. Kunth	Singli Mingli	Dioscoreaceae	H	Root, tubers and leaves	Birth control, allergies, round worm, constipation
17.	*Ficus palmata* Linn.	Fagda	Moraceae	T	Latex and fruit	Constipation, lung disease, bladder and wart treatment
18.	*Fragaria nubicola* (Hook. F) Lindl.	Strawberry	Rosaceae	H	Leaves and fruit	Diarrhoea, urinary infection, blemishes on the tongue
19.	*Hippohaea rhamnoides* Linn.	Chharma	Eleaguaceae	S	Leaves and roots	Tuberculosis, lung diseases, blood pressure, indigestion and blood purifier.
20.	*Junlans regia* Linn.	Khor	Juglandaceae	T	Flowers bark and leaves	Diarrhoea, tape worms, itching, astringent, antiseptic
21.	*Lathyrus latifolia* Linn.	Jungli matter	Paplionaceae	H	Seeds	Nutritious, helps in digestion
22.	*Oxalis corniculata* Linn.	Tinapattiya	Oxalidaceae	H	Entire plant, leaves	Anal tissue, hemorrhoids, appetite, dysentery, bloody diarrhoea
23.	*Picrorhiza kurroa* Royle ex Benth.	Kutki	Scrophulariaceae	H	Entire Plant	Jaundice, lactation, bronchial, asthma, urinary discharge, blood tubules,
24.	*Pinus wallichiana* A.B. Jacks.	Kail, Chil	Pinaceae	T	Leaves	Antiseptic, diuretic, skin irritation, wounds, burn, kidney and bladder treatments, coughs, cold
25.	*Pistacia integerrima* J.L. stewartex Brandis	Kackda	Anacardiaceae	T	Fruit	Vata-kapha suppressant, anti-inflammatory, digestion
26.	*Punica granatum* Linn.	Jungli anaar	Punicaceae	T	Bark, roots, seeds and leaves	Tapeworms, antihelminthic, digestive disorders, urinary infections

Contd...

Table 23.1–*Contd...*

Sl.No.	Vernacular Names	Local Names	Families	Status	Plant Parts Used	Uses
27.	*Plantago ovata* Forssk.	Isabgol	Plantaginaceae	H	Roots and stem	Constipation, to lower blood sugar levels, reducing cholesterol
28.	*Rhododendron arboreum* Sin.	Bras	Ericaceae	T	Flowers and leaves	Headache, diarrhoea, dysentery, stomach disorders
29.	*Rosa brunonii*	Jungli gulab	Rosaceae	S	Stem, bark fruit and Petals	Cold and influenza, skin treatment.
30.	*Rubia cordifolia* Linn.	Manjistha	Rubiaceae	H	Flower, leaves and entire plant	Blood disorders, arthritis, infection and urinary tract problem
31.	*Rumex hastatus* Linn.	Churki	Polygonaceae	H	Shoot and leaves	Bloody dysentery, aches in throat, cough, headache and fever
32.	*Sapindus mukorossi* Gaertn.	Reetha	Sapindaceae	T	Fruit	Migraine, eye disease, disease of teeth, epilepsy
33.	*Solanum nigrum* Linn.	Tamatar	Solanaceae	S	Leaf	Liver, aches and pains
34.	*Tagetes minuta* Linn.	Jungli genda	Asteraceae	H	Entire plant	To kill intestinal worms, skin infection, fungal infection, insect repellant antimicrobial
35.	*Taraxacum officinale* F.H. Wigg.	Milk witch	Asteraceae	H	Entire Plant and latex	Diuretic, appetite, improving digestion, mosquito repellent
36.	*Thymus serpyllum* Linn.	Van ajwayan	Lamiaceae	S	Leaves	Antiseptic, disinfectant, depressions
37.	*Trifolium repens* Linn.	Clover	Fabaceae	H	Entire plant	Cough, cold, fever, arthritis, irritation of eyes
38.	*Urtica dioica* Linn.	Bicchu Buti	Urticaceae	H	Root, seed, leaf and stem	Allergies, paralysis, diuretic, minor fracture, jaundice
39.	*Valeriana jatamansi* Jones	Mushak bala	Caprifoliaceae	H	Rhizome and root	Constipation, scorpion poisoning, nervous disorders, Typhoid
40.	*Zanthoxylum armatum* DC.	Taramira	Rutaceae	H	Seed, leaf, and bark	Fever, cholera, toothache, roundworms diuretic property, cough, digestion.

these people usually depend upon herbs and other materials for the treatment of disease and ailments.

The present scenario offers a unique opportunity for government representatives, scientists and NGOs to work in coordination, devise mechanism to promote traditional health care system and engagement of rural inhabitants in the conservation, cultivation, processing and marketing of raw materials. Despite of some floristic studies undertaken time to time in this region a large number of native wealth remains unexplored due to scattered and no documentation of potential uses of medicinal plants. Realizing their importance there is a need to study the useful and potentially important medicinal flora of this region.

In view of the importance of this region, this analysis highlights the useful types that can uplifting the data status of medicinal flora in Kullu Distt. and also the scope of cultivation of species having potential importance.

References

Dhaliwal D S and Sharma M 1999. *Flora of Kullu District*. BSMPS, Dehradun

Negi P S and Subramani S P 2006. Ethnobotanical study in the Naggar area of Kullu Valley, Himachal Pradesh. *J Econ Taxon Bot* **30:** 349-358.

Singh S K and Rawat G S 1998. Traditional verses commercial use of wild medicinal plants of Great Himalayan National Park. Proceedings International Mountain Meet, Rishikesh.

Singh G S 1999. Ethnobotanical study of useful plants of Kullu District in North-Western Himalaya, India. *J Econ Taxon Bot* **23:**185-198.

Singh S K and Rawat G S 2000. *Flora of Great Himalayan National Park*. Himachal Pradesh. Bishen Pal Mahendra Pal Singh, Dehradun.

Singh S K 2004. Ethno-medicinal plants of Kullu valley, Himachal Pradesh. *J Non-Timber For Prod* **11:** 74-79

Sharma P K 2005. Studies on plants associated indigenous knowledge among the Malanis of Kullu district, Himachal Pradesh. *Ind J Trad Knowledge*. **4:** 403-408.

Stojanoski N 1999 Development of Health Culture in Veles and its region from the past to the end of the 20th Century. *Veles Soc Sci Art* 13-34.

Medicinal Plants: Aspects and Prospects (2014) ***Pages* 354–361**
***Editors:* Mukesh Kumar, Anjali Khare and C.P. Shukla**
ISBN: 978-81-7622-309-6
***Published by:* BIOTECH BOOKS, NEW DELHI**

Chapter 24

Algae: An Emerging Panacea

Minhaj Akhtar Usmani[1], M.R. Suseela[1] and Sarita Sheikh[2]

[1]Algology, CSIR-National Botanical Research Institute, Lucknow – 226 001, U.P.
[2]Ethelind School of Home Science, SHIATS, Allahabad – 211 001, U.P.

ABSTRACT

The global markets for both pharmaceuticals and nutraceuticals are burgeoning quickly worldwide. A growing proportion of today's promising pharmaceutical and nutraceutical research focus on the production of promising compounds from algae. Now, the importance of marine algae as a sources of functional ingredients has been well recognized due to their valuable health beneficial effects. Therefore, isolation and investigation of novel bioactive ingredients with biological activities from marine algae have also attracted great attention. Thus, the untapped potential of algae in the fields of pharmaceuticals and nutraceuticals has to be still explored to grow and capitalize on tremendous global marketing opportunities.

Kewwords: Marine algae, Nutraceuticals, Pharmaceuticals.

Introduction

Ever since the ancient times, especially in Asian countries, algae have been used, as a remedy to cure or prevent various physical ailments. Extensive scientific research has established a connection between these nutrient-rich sea plants and the body's immune system response. It all started when intensive studies of marine life began in the 1970s to locate potential sources of pharmacologically active agents (Baba 1988). As outcome, algae has emerged as a rich and varied source of pharmacologically active natural products and nutraceuticals.

Although the nutraceutical and pharmaceutical contents in the baseline algae strain is very small, but current market values for these products are extremely high. The major products currently being commercialized or under consideration for commercial extraction include carotenoids, phycobilins, fatty acids, polysaccharides, vitamins, sterols, and biologically active molecules for human use and animal health. According to Baba (1988), algae have several beneficial properties:

1. It is a complete protein with essential amino acids (unlike most plant foods) that are involved in major metabolic processes such as energy and enzyme production.
2. It contains high amounts of simple and complex carbohydrates which provide the body with a source of additional fuel. In particular, the sulfated complex carbohydrates are thought to enhance the immune system's regulatory response.
3. It contains an extensive fatty acid profile, including Omega 3 and Omega 6 acids. These essential fatty acids also play a key role in the production of energy.
4. It has an abundance of vitamins, minerals, and trace elements in naturally-occurring synergistic design.

This exclusive profile brings researchers' attention to focus on use of algae as a potential source of pharmaceutical and nutraceutical ingredients.

Algae as a Pharmaceutical Source

The prevalence of allergic diseases and metabolic disorders has increased during the last few decades and contributed a great deal to morbidity and an extensive amount of mortality all around the world. Until now, very few novel efficacious drugs have been discovered to treat, control or even cure these diseases and disorders with a low adverse-effect profile (Vo *et al.* 2012). Therefore, researchers have came to a fact that it is an urgent need to discover and isolate novel anti-allergic agents and bioactive metabolites from natural resources. In these contexts algae has received an ever increasing interest, especially the marine algae and cyanobacteria (blue-green algae) as they are a valuable source of chemically diverse bioactive compounds with numerous health benefit effects.

The Table 24.1 lists various algae and their uses in medicines.

There are a range of pharmaceutical products derived from micro and macroalgae. Some of them include:

1. Antibiotics, Antimicrobials, Antivirals & Antifungals
2. Neuro-protective Products
3. Therapeutic proteins
4. Drugs

1. Antibiotics, Antimicrobials, Antivirals and Antifungals

The presence of antagonistic compounds in algae could make them useful as antibiotics for the killing of bacteria, fungi, and viruses (Hoppe 1979). The compounds

Table 24.1: Lists of Various Algae and their Uses in Medicines.

Phyllum	*Genus*	*Uses*
Chlorophyta	*Enteromorpha*	Can be used to treat hemorrhoids, parasitic disease, goiter, asthma, coughing and bronchitis; they reduce fever and ease pain
	Ulva	Can be used to treat goiter; reduce fever, ease pain, induce urination
	Codium	Can be used to treat urinary diseases, treat edema, expel ascarid; is an ecbolic
	Acetabularia	Can be used to treat urinary diseases and edema
Phaeophyta	*Ishige*	Can be used to treat cervical lymphadenitis, to diminish inflammation and to induce urination
	Laminaria	Can be used to treat goiter, urinary diseases; is an ecbolic; contains iodine and potassium
	Endarachne	Can be used to treat urinary diseases, edema, gastric diseases and hemorrhoids
	Sargassum	Can be used to treat cervical lymphadenitis, edema; diminishes inflammation; reduces fever; induces urination; contains iodine and potassium
Rhodophyta	*Porphyra*	Can be used to treat goiter, bronchitis, tonsillitis and cough
	Gelidium	Laxative; can be used to treat tracheitis, gastric diseases and hemorrhoids; can be used to extract agar
	Pterocladiella	Laxative; can be used to treat tracheitis, gastric disease and hemorrhoids; can be used as an agar extract; can be used to make coating for pills
	Eucheuma	Can be used to treat goiter, tonsillitis, bronchitis, asthma, cough, gastric diseases and hemorrhoids
	Corallina	Marl, parasiticide
	Gracilaria	Can be used to treat goiter, edema, urinary diseases, can prevent ulcer; can be used to as an agar extract and make coating for pills
	Hypnea	Can be used to treat bronchitis, gastric diseases and hemorrhoids; can be used to make carragenate
	Chondrus	Can be used to treat bronchitis, tonsillitis, gastric diseases, asthma and cough; is an adhesive, can be used to make carragenate
	Centroceras	Can be used to treat gastro-intestinal intolerance
	Chondria	Ascaricide
	Grateloupia	Ascaricide; lowers blood pressure
	Gloeopeltis	Can be used to treat goitre, tonsillitis and bronchitis; prevents high blood pressure and scurvy

Source: Compendium of Materia Medica.

(http://web.ntm.gov.tw/seaweeds/english/e/e3_01.asp#top).

include fatty acids, bromophenols, tannins, phloroglucinol, and terpenoids from a variety of different algae (Hashimoto 1979). Both microalgae and macroalgae exhibit antimicrobial activity which finds use in various pharmaceutical industries.

a) Microalgae

Microalgae, such as *Ochromonas sp.*, *Prymnesium* and a number of blue-green algae produce toxins that may have potential pharmaceutical applications. Various strains of cyanobacteria are known to produce intracellular and extracellular metabolites with diverse biological activities such as antibacterial, antifungal and antiviral activities. The biological activities of the algae may be attributed to the presence of volatile compounds, some phenols, free fatty acids and their oxidized derivatives.

b) Macroalgae

There are numerous reports of macroalgae derived compounds that have a broad range of biological activities, such as antibiotic, antiviral, anti-neoplastic, antifouling, anti-inflammatory, cytotoxic and antimitotic. In the past few decades, macroalgae have been widely recognized as producers of a broad range of bioactive metabolites. Such antimicrobial properties enable macroalgae to be used as natural preservatives in the cosmetic industry. The highest percentage of antimicrobial activity was found in Phaeophyceae (84 per cent), followed by Rhodophyceae (67 per cent) and Chlorophyceae (44 per cent). Red and brown macroalgae extracts show significant potential as anti-pathogenic agents for use in fish aquaculture.

2. Neuroprotective Products

Both microalgae and macroalgae contains neuroprotective agents and promote nerve cell survival.

a) Microalgae

Among the various microalgal species, *Spirulina* is most commonly referred as a neuroprotective agent. *Spirulina platensis* may be useful in the development of novel treatments for neurodegenerative disorders such as Alzheimer's or Parkinson diseases. *Spirulina maxima* is found to partially prevent MPTP (1-methyl-4-phenyl-1,2,3,6 tetrahydropyridine) neurotoxicity and oxidative stress, suggesting it could be a possible alternative in experimental therapy.

b) Macroalgae

Several macroalgae are found to possess therapeutic potential for combating neurodegenerative diseases associated with neuroinflammation. A marine macroalgae, *Ulva conglobata*, has neuroprotective effects in murine hippocampal and microglial cells.

3. Human Therapeutic Proteins

Pharmaceutical companies could substantially reduce the expense of costly treatments for cancer and other diseases produced from mammalian or bacterial cells by growing human therapeutic proteins in algae. Microalgae usually find extensive use as therapeutic and diagnostic proteins.

a) Microalgae

Expression of recombinant proteins in green algal chloroplast holds substantial promise as a platform for the production of human therapeutic proteins. The percentage of human proteins produced in their algal cultures is comparable to the fraction produced by mammalian cell cultures and much better than that produced by bacterial systems. A study confirmed that diverse human therapeutic proteins could be produced in *Chlamydomonas reinhardtii*, a green alga. Algae can be used to produce:

- ✰ VEGF (Vascular Endothelial Growth Factor) for treating emphysema
- ✰ HMGB1 (High Moblility Group Protein B1) which activates immune cells
- ✰ Domain 14 of human fibronectin
- ✰ Domain 10 of human fibronectin used to increase the accumulation of other proteins
- ✰ Human pro-insulin could be produced by algae, but only at lower levels.

b) Macroalgae

The red seaweed *Ancanthophora delilei* has been described as a source of carnosine (β-alanyl-L histidine), a histidyl peptide with an antioxidant activity as well as is associated the ability of chelate transition metals (Fleurence, 2004). A brown seaweed *Scytosiphon lomentaria* was shown strong ROS scavenging activities after being hydrolysed by proteases (Ahn *et al.* 2004).

One of the first pronounced antioxidative effects was revealed from water-soluble, protease enzymatic extracts of seven species of marine edible brown seaweeds, including *Ecklonia cava, Scytosiphon lomentaria, Ishige okamurae, Sargassum fullvelum, Sargasum horneri* and *Sargassum thunbergii* around Jeju-Do coasts in South Korea (Heo *et al.* 2003).

C-Phycocyanin, a major biliprotein isolated from *Spirulina platensis* with high purity (>95 per cent) has been tested on the growth and multiplication of human chronic myeloid leukemia cell line (K562) and positive results were achieved (Subhashini *et al.* 2004).

High anticoagulant activities were reported from *Codium fragile* and *Sargassum horneri* against activated partial thromboplastin time (APTT) and prothrombin time (PT) assays. The anticoagulant activity of marine green algae, *Codium pugniformis,* was assessed by studying the APTT, PT and thrombin time (TT) using normal human plasma, comparing with the commercial anticoagulant heparin protein (Matsubara, *et al.* 2000). In this study, the anticoagulant activity has been shown due to proteoglycan including sulfated polysaccharide and protein polysaccharide (Samarakoon and Jeon (2012).

4. Drugs

Algal chemistry has attaracted many researchers in order to develop new drugs, as algae include compounds with functional groups which are characteristic from this particular source.

a) Microalgae

Researchers have been able to produce an exciting class of anti-cancer drugs originally isolated from blue-green algae. A compound named cryptophycin-1 has been isolated from blue-green algae which hold significant promise as an anti-cancer drug. Microalgae produce incredibly potent alkaloidal neurotoxins such as saxitoxin and polyketide neurotoxins such as the brevetoxins for use as anticancer drugs.

b) Macroalgae

The alkaloids found in marine macroalgae present special interest because of their pharmacological activities. These alkaloids in macroalgae are largely focused on finding drugs for cancer treatment.

Algae as a Source of Nutraceuticals

The growing use of algal biomass for nutraceutical purposes is expected to provide an attractive revenue stream for algae producers. Physiologically-active nutraceuticals from algae include food supplements, dietary supplements, value-added processed foods as well as non-food supplements such as tablets, soft gels, capsules etc.

Some of the noteworthy components that can be derived from algae are:

- ☆ Omega 3 polyunsaturated fatty acids (PUFA)
- ☆ Carotenoids
- ☆ Astaxanthin, and
- ☆ β-Carotene

1. Omega 3 Polyunsaturated Fatty Scids (PUFA)

As algae could be relatively easily cultivated at different stress conditions, they offer the prospect of a good source of PUFA for the nutraceutical market. Recently, attention has focused on μ-3 PUFAs from algae, especially eicopentaenoic acid (EPA) and docosahexaenoic acid (DHA), due to their association with the prevention and treatment of several diseases (atherosclerosis, thrombosis, arthritis, cancers, etc.). Long-chain polyunsaturated fatty acids of the omega 3 and omega 6 series are straight chain carboxylic acids of 20 or more carbon atoms that contain 3 or more double bonds. The conventional source of EPA and DHA is marine fish oil, but higher amount of EPA and some DHA can be produced by the use of algae.

a) Microalgae

Microalgae have been an attractive source of PUFA (Benemann *et al.* 1987) due to their inherently high PUFA content. Cyanobacterium *Spirulina* is rich in ã-linolenic acid (GLA) (and poor in the á-isomer) and thus is a good source for the purification of this PUFA. *Chlorella minutissima* is a eukaryotic species with a fast growth rate and high PUFA content and could be another important source of a PUFA-rich nutraceutical supplement. The red algae, *Porphyridium* contain more than 30 per cent of total fatty acids as AA (Arachidonic Acid) and equally high concentrations of EPA.

b) Macroalgae

Fatty acids (FA) from marine macroalgae are generally richer in polyunsaturated fatty acids (PUFA) and a higher degree of total unsaturation. Both Phaeophyta and Rhodophyta species of macroalgae are rich in arachadonic acid (AA) and eicosopentaenoic acid (EPA), and Ulvales in docosahexaenoic acid (DHA) contents.

2. Carotenoids

Carotenoids are a class of natural fat-soluble pigments found principally in algae where they play a critical role in the photosynthetic process. In human beings, carotenoids can serve several important functions. The most widely studied and well-understood nutritional role for carotenoids is their provitamin A activity. Carotenoids are found to be powerful anti-oxidants.

a) Microalgae

β-Carotene and astaxanthin are attractive microalgal products as they command a high market price and are present in high concentrations in some algal cells

β-Carotene- β-Carotene is usually derived from *Dunaliella salina*. β-Carotene is a natural pigment derived from green algae, is used as a yellow-orange food coloring and may help prevent certain types of cancers.

Astaxanthin- It is obtained from *Haematococcus pluvialis* which is believed to contain the highest known natural levels of astaxanthin. Astaxanthin has a number of biological functions, like:

- ✰ Protection against oxidation of essential polyunsaturated fatty acids
- ✰ Protection against UV light effects
- ✰ Protects against cancer, and
- ✰ Modifying immune function

Now, the importance of marine algae as sources of functional ingredients has been well recognized due to their valuable health beneficial effects. Thus, the untapped potential of algae in the field of pharmaceuticals and nutraceuticals has to be still explored to grow and capitalize on tremendous global marketing opportunities.

References

Ahn, C B, Jeon Y J, Kang D S, Shin, T S and Jung B M 2004. Free radical scavenging activity of enzymatic extracts from a brown seaweed *Scytosiphon lomentaria* by electron spin resonance spectrometry. *Food Res Int* **37**: 253–258.

Baba 1988. Mechanism of inhibitory effect of dextran sulfate and heparin in replication of human immunodeficiency virus *in vitro*. Proc Natl Acad Sci **85**: 6132-6136.

Fleurence J 2004. Seaweed proteins. In: R Y Yada (ed.), *Proteins in Food Processing* (pp. 197–213). Woodhead publishing limited, Cambridge, UK.

Hashimoto Y 1979. Marine Toxins and Other, *Bioactive Marine Metabolites*. Japan Scientific Societies, Press, Tokyo, 369 pp.

Heo S J, Lee K W, Song C B, and Jeon Y J 2003. Antioxidant activity of enzymatic extracts from brown seaweeds. Algae **18**: 71–81.

Hoppe H A, Levring T and Tanaka Y (eds.) 1979. Marine Algae in Pharmaceutical Science. Walter de, Gruyter, Berlin, 807 pp.

Importance and economic uses of algae -cited from- http://web.ntm.gov.tw/seaweeds/english/e/e3_01.asp#top

Matsubara K, Matsuura Y, Hori K, and Miyazawa K 2000. An anticoagulant proteoglycan from the marine green alga, *Codium pugniformis*. Journal of Applied Phycology **12:** 9–14.

Medical uses of algae-cited from-http://www.megaessays.com/viewpaper/2417.html

Oilgae-Wednesday, July 10, 2013 accessed from- http://www.oilgae.com/blog/2010/09/algae-as-a-source-of-pharmaceuticals-nutraceuticals.html.

Samarakoon K and Jeon Y J 2012. Bio-functionalities of proteins derived from marine algae: A review Food Research International **48**: 948–960.

Subhashini J, Mahipal S V K, Reddy M C, Reddy M M, Rachamallu A and Reddanna P 2004. Molecular mechanisms in C-Phycocyanin induce apoptosis in human chronic myeloid leukemia cell line-K562. *Biochem Pharmacol* **68**: 453–462.

Vo T S, Ngo D H and Kim S K 2012. Marine algae as a potential pharmaceutical source for anti-allergic therapeutics. *Proc Biochem* 386–394.

Medicinal Plants: Aspects and Prospects (2014) ***Pages* 362–378**
***Editors:* Mukesh Kumar, Anjali Khare and C.P. Shukla**
ISBN: 978-81-7622-309-6
***Published by:* BIOTECH BOOKS, NEW DELHI**

Chapter 25

Ashwagandha: Highly Potential Medicinal Plant

***Nishesh Sharma*[1], *Eapen P. Koshy*[1] *and Manjul Dhiman*[2]**

[1]Department of Tissue Engineering, Jacob School of Biotechnology, SHIATS, Allahabad – 211 001, U.P.
[2]Department of Botany, K.L. DAV P.G. College, Roorkee – 247 667, Uttrakhand

ABSTRACT

Withania somnifera (Ashwagandha), the Indian Ginseng is one of the most commonly utilized medicinal plant for health management and treatment of the ailments, ever since the practice of traditional medicine to modern medicine. Twenty three known species of *Withania* are known to be distributed in drier parts of tropical and subtropical zones ranging from Canary Islands, Mediterranean region and northern Africa to South-West Asia. The crop is cultivated in an area of about 4000 hectares in India mainly in drier parts of Madhya Pradesh, Punjab and Rajasthan and also in south India. Ashwagandha prefers a subtropical climate. It exhibits many medicinal properties such as anti-inflammatory, anti-stress, anti-tumor, antioxidant, anti-neoplastic effects, rejuvenating tonic, immune-modulatory activity, cardio-protective activity and hypothyroid activity. The plant has been utilized in Ayurveda system of medicine. Ashwgandha has been a crucial component of Ayurvadic Rasayan. The biologically active chemical constituents are alkaloids (ashwagandhine, cuscohygrine, anahygrine, tropine etc), steroidal compounds including ergostane type steroidal lactones, withaferin- A, withanolides A-Y, withasomniferin-A, withasomidienone, withasomniferols A-C, withanone etc. Withanolides are the major chemical compounds and are mainly localized in the leaves, and their concentration normally ranges from 0.001 per cent to 0.5 per cent per gram dry weight. There is urgent need to conserve and intensify the crop production.

Introduction

Withania somnifera (L.) Dunal commonly known as Ashwagandha, Indian Ginseng or Winter Cherry is one of the esteemed medicinal plant used in Indian Ayurvedic system for over 3000 years. It is one of the most commonly utilized medicinal plant for health management and treatment of ailments ever since the practice of traditional medicine to modern medicine. The roots of *Withania somnifera* are specifically utilized in various medicinal formulations, although its seeds and leaves also have significant medicinal value. It is mostly collected from wild stands, but in India it has been well cultivated in Rajasthan and Madhya Pradesh. *Withania somnifera* occurs naturally or has been naturalized in many areas of tropics.

Withania is a branched, erect unarmed undershrub having a height of about 0.3-1.5 meter. Plant exhibits simple leaves, alternate or sub opposite pairs at node (Figures 25.1A and B). Flowers are small, greenish or lurid yellow, sessile and shortly pedicelled. Calyx of the plant is gamosapalous and campanulate around 5mm in length (Figure 25.1C). It increases up to 18mm becoming inflated and nearly globose and encloses fruit. Stamens are five in number and are inserted on the tube of corolla near its base. Corolla are gamopetalous and campanulate, greenish yellow in colour. Ovary of the plant is ovate and globose, surrounded at the base by thin glandular annular disc, bilocular or falsely four chambered. Fruits of *Withania* are small, smooth, globose which are pea green when unripe (Figure 25.1D) while brick red on ripening (Figures 25.1E and F) and are enclosed in large calyx. Seeds are lens shaped or kidney shaped and show epigeal germination. The root system of *Withania* consists of a stout main root 20-30cm long with a few (2-3) lateral roots of slightly smaller size. The roots are somewhat tuberous they are occasionally branched and distal ends are tapering and slightly woody.

Distribution and Cultivation

Twenty three known species of *Withania* are known to be distributed in drier parts of tropical and subtropical zones ranging from Canary Islands, Mediterranean region and northern Africa to South-West Asia. *Withania somnifera* and *Withania coagulans* are two species specifically known for their economic and medicinal value (Mirjalili *et al.*, 2009). Wild species has been reported from Pakistan, Afghanistan, Palestine, Egypt, Jordan, Morocco, Spain, Canary Island, Eastern Africa Congo, Madagascar and South Africa (Table 25.1). It also occurs in Australia. As a result of wide growing range there are considerable morphological and chemotypical variations in terms of local species. *Withania somnifera* (Linn.) Dunal is a cosmopolitan plant which grows throughout the drier parts and in subtropical India. It is widely distributed in North-West India, Gujrat, Rajasthan, Madhya Pradesh, Uttar Pradesh and Punjab plains extending to the mountains of Himachal Pradesh and Jammu. *Withania somnifera* has been reported to found wild in the grazing grounds of Mandasur and forest lands at the Bastar district of Madhya Pradesh including all the foot-hills of Punjab, Himachal Pradesh and Western Uttar Pradesh.

effects of allopathic medicines have made them popular all over the globe. *Withania somnifera* is one such important and crucial herb of India. *Withania* exhibits many medicinal properties such as anti-inflammatory, anti-stress, antitumor, antioxidant, antineoplastic effects, rejuvenating, immune-modulatory, cardio-protective and hypothyroid activities (Figure 25.2). Ashwagandha has been utilized in the treatment of several diseases such as chronic fatigue, impotency, constipation, spermatorrhoea, rheumatism and memory loss. Roots of *Withania* are used in the treatment of asthma, bronchitis, anemia, aging, asthenia, infertility, memory loss, arthritis, multiple sclerosis, immune dysfunction and rheumatism. Leaves are used for fever, healing of external wounds, haemorrhoids, tumors, anthrax pustules, syphilitic sores and ophthalmintis (Sharma *et al.*, 2011).

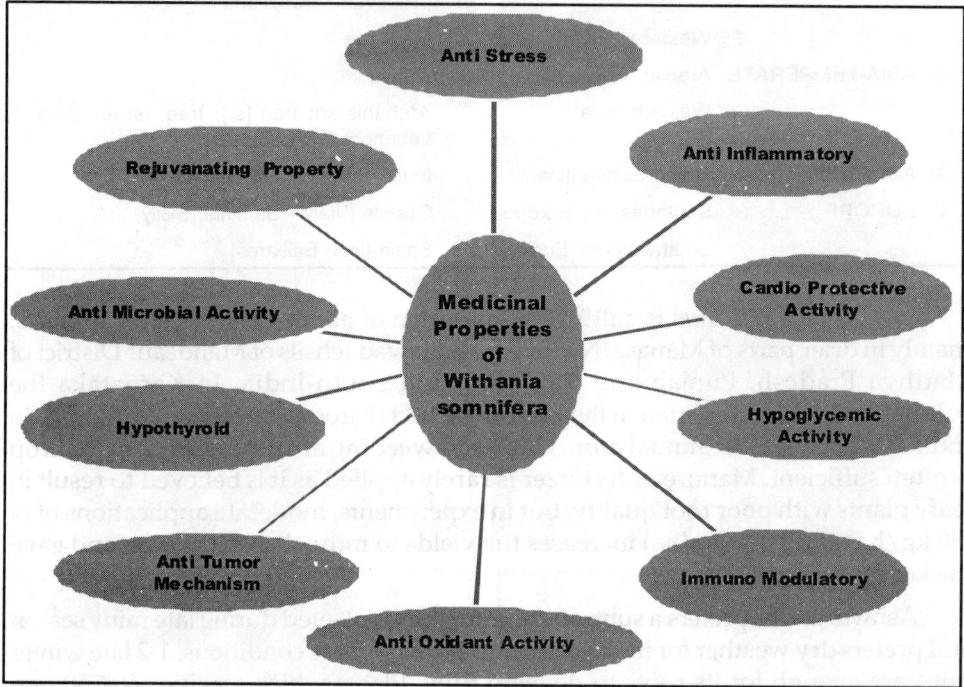

Figure 25.2: Medicinal Properties of *Withania somnifera*.

Ashwgandha has been a crucial component of Ayurvadic Rasayan. Rasayana is a specialty of Ayurveda which mainly deals with the preservation and promotion of health. It promotes longevity and present or delay the aging process. Rasayan promotes resistance against infection and maintains the equilibrium of Vata, Pitta and Kapha. Most of the active principles of Ayurvedic and Unani medicines are found in *Withania somnifera* and around hundred different preparations are already commercially available (Singh *et al.*, 2010, Tripathi *et al.*, 1996).

According to Ayurveda system of medicine, the plant increases sexual desire, rejuvenates the body, increases strength, improves quality and quantity of semen. It is useful in the management of white discoloration of the skin, management of edematous conditions and in emaciation, and under nutritive conditions it is Vajikara, Rasayani, Balya, Atishukrala, Shwitrapaha, Shothahara and Kshayapaha. The tribal Bheel and Garasodia tribe give root powder orally to male patient of asthma and bronchitis (Singh and Pandey 1998). It has been classified in Charak Samhitaa as balya (promoter of strength), and is considered useful in the cases of consumption. In Sushruta Samhitaa, therapeutic preparations (external use) based on this drug have been described valuable for promoting the growth of normal tissues after surgery, for quick healing of wounds and for the treatment of swelling (Sukh Dev 2006).

Use of *Withania Somnifera* as Ayuvedic Rasayana along with the findings from research enlight the antioxidant activity of *Withania somnifera*. The anti-oxidant effect of the components present in the roots of *Withania* also explain the herb as anti-stress, anti-inflammatory and anti-aging effect (Singh *et al.*, 2010, Bone 2009). The anti-oxidant activity of *Withania* is crucial with specific reference to brain and central nervous system, since the tissue of CNS are more susceptible to free radical damage than other tissue because they have high lipid and iron contents, both known to generate reactive oxygen species free radical damage of nervous tissue leads to epilepsy, schizophrenia, Parkinson's disease and Alzheimer disease. The active component of *Withania somnifera* namely sitoindosides VII-X and witheferin – A have been analyzed for their antioxidant activity, using the free radical scavenging enzymes superoxide dismutase, catalase and glutathione peroxidase levels in rat brain. An increase in these enzymes represents increased antioxidant activity and a protective effect on neuronal tissue (Mishra *et al.*, 2000, Bhattacharya *et al.*, 1997).

Studies performed on animals have shown that Ashwagandha has an effect on thyroid activity too, which suggest that the herb can prove to be useful in the treatment of hypothyroidism. This plant is known to have a stimulatory effect on the production of T_4(Verma & Kumar 2011). Numerous research studies have shown direct potential of *Withania somnifera* in preventing cancer and anti-tumor activities. Moreover, it has also been observed that the herb increases the sensitivity of tumor to radiation and chemotherapy along with reducing the side effects of these conventional therapies (Marine Winter 2006). *In vitro* as well as *in vivo* studies have been successfully conducted to reveal cytotoxic and anti-tumor potential of *Withania somnifera*. In an *in vitro* study conducted by Kaur *et al.* (2004) powdered leaf extract was used to treat cell lines of osteogenic, sarcoma and breast cancer. Cells showed reduced proliferation as compared to control and hence the study revealed anti proliferative effect of *Withania somnifera* on human tumor cells.

In an another study conducted by Jai Prashan *et al.* (2003), 13 constituents from *Withania somnifera* have been analyzed for their anti-proliferative activities on lungs, colon, CNS and breast tumor lines. As an outcome of this study anti-proliferation effect depends on the dose administered. The *in vivo* work indicates that *Withania somnifera* is useful as antitumor agent in slowing down the tumor growth and also in increasing the survival time. Christina *et al.* (2004) in their work inoculated Swiss albino mice with Daltons ascetic leukemia. Then a dose of *Withania somnifera* (20 mg/kg) was administered intraperitoneally to the test animals. Animals given with the drug demonstrated $0.9\pm 0.12 \times 10^6$cells as compared to $1.35\pm 0.8 \times 10^6$ cells in the control group. The extract also represented the reduced packed cell volume and tumor weight.

In an another study carried by Leyon *et al.* (2004) the effect of *Withania somnifera* root powder extract and its constituent withanolide was examined on B16-10 melanoma. C57BL mice were injected with melanoma cells and either pretreated or treated simultaneously with 20mg dose/animal/24 hrs of powdered *Withania somnifera* root extract IP. Treatment with *Withania somnifera* or withanolide-D resulted insignificant decrease in the tumor growth as compared to the control. Numerous studies have revealed the ant-inflammatory effects possessed by Ashwagandha. Withanolides have proved to be a potent inhibitor of pro-inflammatory transcription factor NF-KB and AP-1 (Ojha & Arya 2009). Studies have also shown the potential of Ashwagandha to ease the symptoms of arthritis. Natural steroid content of the drug is much more than hydrocortisone, a prescribed anti-inflammatory.

Anti-inflammatory effect of Ashwagandha also contributes to its utilization in rheumatologic conditions. Anbalagan and Sadique (1981) administered powered roots of *Withania somnifera* to rats before an hour of induction of inflammation by injecting Freud's Complete adjuvant, he observed that the drug was effective enough to reduce inflammation. The studies of Anbalagan and Sadique (1984) have demonstrated that *Withania somnifera* suppressed alpha-2 macroglobulin which is an indicator of anti-inflammatory drugs in the serum of rats injected with carrageenan suspension.

Stress has been associated with deterioration in mental and physical health and many studies have proved that stress is one of the factors associated with cardiological disorders. Ashwagandha has been utilized since ages as an anti-stress drug. It is one of the most widely utilized tranquillizer in India and is considered to have importance similar to Ginseng in China. Studies have revealed that Ashwagandha and Panax Ginseng decrease the severity of stress induced ulcers, these herbs reverse stress induced inhibition of male sexual behavior and also inhibits effects of chronic stress. The activity of *Withania* and Ginseng extract has nearby same action. But *Withania somnifera* has an advantage that it does not results in Ginseng – Abuse Syndrome (a condition with high blood pressure, water retention, muscle tension and insomnia). The anti-stress effect of *Withania somnifera* is supposed to increase stress induced changes and provides cardio-protection. Effect of Ashwagandha has also been linked to increase in heart weight and glycogen in myocardium. (Singh *et al.*, 2010, Bhattacharya *et al.*, 2000, Ojha & Arya 2009, Dhuley *et al.*, 1998 2000).

Extract and tonic of *Withania somnifera* have sound effect on immune system. Studies reveal that Ashwagandha acts as an immunoregulator as well as a chemoprotective agent. In a study carried out on mouse the root extract of Ashwagandha was found to increase total WBC count, inhibited delayed-type hypersensitivity reaction and increase phagocytotic activity of macrophages. Stimulatory effect of Ashwagandha on the generation of cytotoxic T- lymphocytes has been confirmed both *in vivo* and *in vitro*. The extract of *Withania* have been found to be effective in decreasing the occurrence and average number of skin lesions when compared to control. Level of reduced glutathione, superoxide dismutase, catalase and glutathione peroxidase in the exposed tissue returned to almost normal values after administration of extract (Verma & Kumar 2011). Beside the eminent medicinal properties, the herb is also known to exhibit anti-bacterial and anti-fungal properties (Bhatt *et al.*, 2011).

As a result of multi drug resistance developed by microbial pathogens due to extensive utilization of antibiotics along with increasing failure of chemotherapeutics and lack of new drugs the screening of medicinal plant for their anti-microbial activity have gained momentum. Medicinal plants have been known to produce compounds with antimicrobial activities. Antibacterial activity of many plant species have also been reported (Pandey & Mishra 2010, Sundaram *et al.*, 2011).

Santhi & Swaminathan (2011) have analyzed anti-bacterial activities of *Withania somnifera* using different extracts namely aqueous, alcoholic and acetonic. Leaves of *Withania* are utilized as raw material for the preparation of the extract. The inhibitory effect of extract was examined against *E. coli, Klebsiella, peumoniae, Pseudomanas* sp., *Proteus mirabilis* and *Salmonella parathyphi B*. Their findings include maximum anti-bacterial activities in acetone extract followed by alcoholic extract. The aqueous extract has been found to possess minimum inhibitory action.

Sunderam *et al.* (2011) have studied and confirmed *in vitro* anti-bacterial activities of crude extract prepared from *Withania somnifera* against bacterial pathogens. They utilized different solvents namely ethanol, ethyl acetate, dichloromethane, hexane (from high to low polarity) to analyze the anti-bacterial activities of the extracts prepared against clinically isolated bacterial pathogens (*Staphylococcus aureus, E. coli, Pseudomonas sp. , Bacillus subtilus*). Utilizing agar well diffusion method minimum inhibitory concentration was determined by analyzing zone of inhibition. Results obtained in the study revealed that polar solvents have high antibacterial property as compared to non-polar solvent against both gram positive as well as gram negative bacteria. The findings also indicated that the alcoholic as well as ethyl acetate extracts possessed strong anti-bacterial activities while hexane and dichloromethane fractions were less effective.

In another work of Arora *et al.* (2004) the methanolic extract of *Withania* both leaves and roots were found to exhibit anti-bacterial activity, whereas only root extract in hexane show antibacterial activity. In a similar kind of another study aqueous extract of *Withania* showed antibacterial activity against Gram negative bacteria *N. gonorrhoea*. (Kambizi and Afolayan 2008).

Studies conducted by different workers by utilizing the extracts of *Withania somnifera* have well established the antifungal properties of the herb. Singh *et al.* (2010) have evaluated the antifungal activities of *Withania somnifera* against common pathogenic fungi. Different extracts such as hexane, ethyl acetate, methanol, had been utilized for analyzing the antifungal potential of the herb. Fungi namely *Aspergillus niger, A. flavus, Fusarium oxysporium* and *F. moniliformis* were utilized as test organism. Using disc diffusion method it was established that all tested extracts inhibited the fungal species with varying degree of sensitivity. Their findings showed that hexane and ethyl acetate extracts have maximum inhibition zone against *A. niger*. Methanol extract displayed maximum activities against *F. oxysporium* and *A.*

flavus while aqueous extract showed maximum activity against *F. moniliformis*. The *in vitro* studies carried out by Prakash *et al*. (2005) evaluated the antifungal potential of two phyto-extracts of *Withania somnifera* (Ashwagandha). Pathogenic fungi *Alternaria alternate, Curvularia lunata, Helminthosporium* sp., *Fusarium* sp. and *Phytophthora parasitica* had been utilized as test organism. Both the extracts were found to inhibit the growth of test fungi. Phytoextract-II of *W. somnifera* was most inhibitory to *Helminthosporium* sp. followed by *Fusarium* sp., *Phytophthora* sp., *A. alternata* and *C. lunata*. The extracts prepared even from the fruit coat (calyx) had also been reported to exhibit antibacterial and antifungal activities *against* various human and plant pathogens such as *Proteus merabilis, Klebsiella pnemoniae, Agrobacterium tumefaciens* and *Aspergillus niger*. (Singariya *et al*., 2012).

Anti-Tumor Mechanism (Marine Winters 2005)

Withania somnifera exhibits many mechanisms through which the herb tends to suppress the development of tumor (Figure 25.3). Tumor bearing animals treated

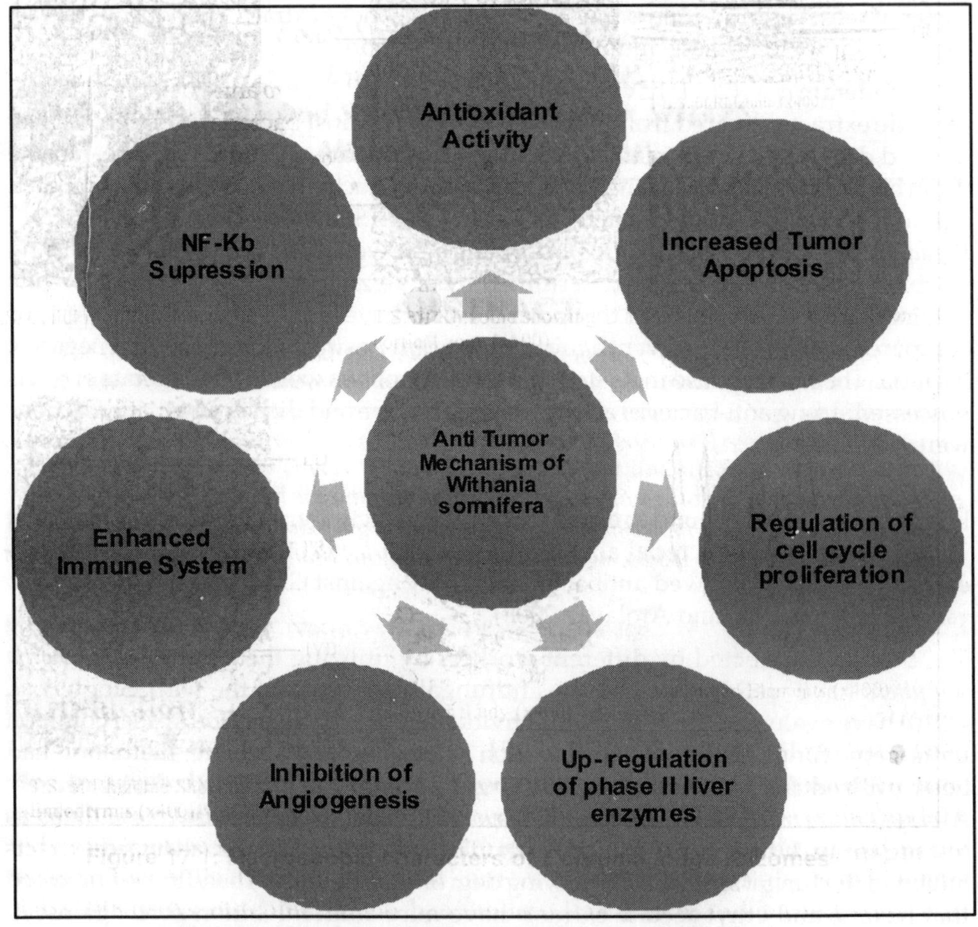

Figure 25.3: Anti-tumor Mechanisms of *Withania somnifera*.

with WS showed increased GSH, SOD, GPX and Catalase in liver and skin. These enzymes repair oxidative damage caused by tumor growth and inflammation.

Antioxidant activity of *Withania somnifera* up regulates phase II liver enzyme. The work carried out by Padamavathi *et al.* (2005) demonstrates that Swiss albino mice fed with 2.5-5 per cent . *Withania somnifera* root extract showed 1.6 - 1.26 fold up regulation of DT-diaphorase (DTD) and GST. Both enzymes have been found to involve in the conjugation of metabolites of cytochrome p450: which helps in liver detoxification of toxic phase I byproducts. WS also mitigate to up regulate the cell growth via the potent tumor suppression gene p53, which regulates cell cycle proliferation . Mathur *et al.* (2004), reported that the cells from Wistar rats exposed to UV-B radiation demonstrated cluster of mutated p53 proteins, a precursor to carcinogenesis. Pretreatment of an extracted constituent of *Withania somnifera* 1-oxo-5beta, 6beta-epoxy with a 2-enolide at 20 mg/kg body weight IP for 5 days prior to irradiation and 12 weeks after the irradiation resulted in no mutant p53. The test animals also exhibited normal dermis and skin tissue, without the evidence of necrosis or carcinogenesis, suggesting a possible role of *Withania somnifera* in conjugation with radiation.

Biochemistry of *Withania somnifera*

Medicinal and therapeutic properties of *Withania somnifera* have urged a need to investigate different metabolites produced by the plant and their pharmacological importance. The need to isolate and identify various secondary metabolites has also gained importance because it has been observed that the metabolites produced vary with the variety of WS and also with the type of tissue from which metabolites have been isolated. Even the growth conditions have contributed to the variation in the production of secondary metabolites.

The biologically active chemical constituents are alkaloids (ashwagandhine, cuscohygrine, anahygrine, tropine etc.) (Figure 25.4), steroidal compounds including ergostene type steroidal lactones, withaferin A, withanolides A-Y, withasomniferin-A, withasomidienone, withasomniferols A-C, withanone etc. Other constituents include saponins containing an additional acyl group (sitoindoside VII and VIII) and withanolides with a glucose at carbon 27 (sitoindoside IX and X). Apart from these contents, the plant also contain chemical constituents like withaniol, acyl steryl glucosides, starch, reducing sugars, hantreacotane, ducitol, a variety of amino acids including aspartic acid, proline, tyrosine, alanine, glycine, glutamic acid, cystine, tryptophan and high amount of iron. Withaferin A, chemically characterized as 4b, 27- dihydroxy-5b-6b-epoxy-1-oxowitha-2, 24-dienolide is one of the main withanolidal active principles isolated from the plant. *W. somnifera* showed chemogenetic variation and so far three chemo type I, II and III have been reported. These are chemically similar but differ in their chemical constituents especially in withanolide content. In the Indian variety, thirteen Dragendroff positive alkaloids have been obtained. Seven new withanolide glycosides called withanosides I-VII have been isolated and identified. Much of *Withania* pharmacological activity has been attributed to two main withanolides *i.e.* withaferin A and withanolide A.

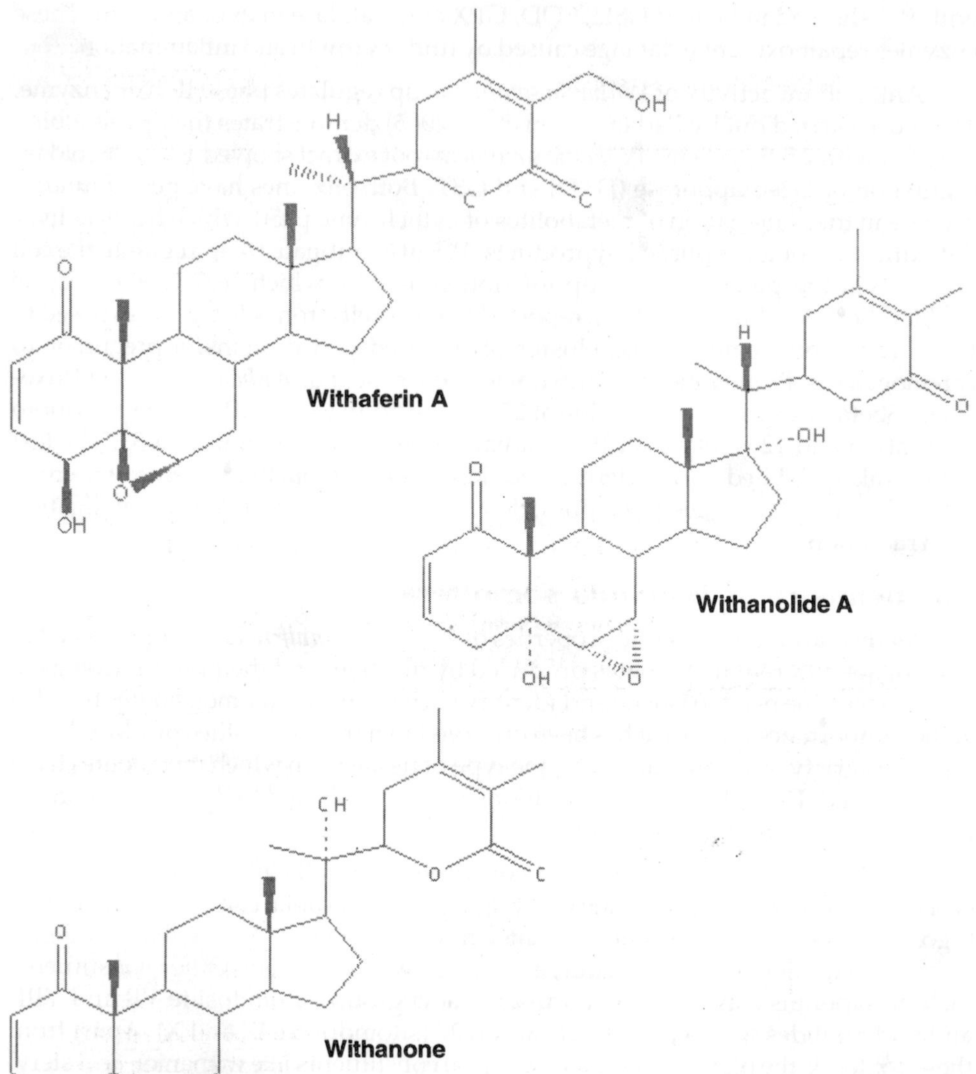

Withaferin A

Withanolide A

Withanone

Figure 25.4: Chemical Structures of Three Main Withanolides Isolated from *Withania*.

The biochemical composition of *Withania somnifera* has been widely studied and researched (Table 25.2).

Chatertjee *et al*. (2010) worked on comprehensive metabolic fingerprinting of *Withania somnifera*. Extract of leaf and root were utilized for the study. As an outcome of the work undertaken, a total of 62 major and minor primary and secondary metabolites from leaves and 48 from roots were identified. Around 29 identified

Table 25.2: Biochemical Composition of *Withania somnifera*.

Sl.No.	Category of the Compound	Specific Compounds
1.	Alkaloids	Isopelletierine, Anaferine, Cuseohygrine, Anahygrine, Tropine
2.	Somniferin	
3.	Steroidal Lactones	Withanolides, Withanoferins
4.	Saponins	Sitoindoside VII and VIII
5.	Withanolides(Carbon At 27th Position)	Sitoindoside IX and X
6.	Others	Resin, fat, coloring matters, reducing sugar, phytosterol, Ipuranol and a mixture of saturated and unsaturated acids

metabolites were common to both tissues. These included fatty acids, organic acids, amino acids, sugar and sterol based compounds. Chatterjee *et al.* (2010) subjected hexane extract to GC-MS after esterification to reveal the composition of fatty acids and sterols. Palmitic acid, Oleic acid, Linoleic acid, Linolenic acid were major fatty acids present in root and leaf extracts. Presence of Campesterol and Stigmasterol were particularly indicated in root extracts.

Withanolides

Withanolides are the major chemical compounds mainly localized in leaves. Their concentration normally range from 0.001 per cent to 0.5 per cent per g dry weight. The withanolides are a group of naturally occurring C28 – steroidal lactones built on ergostane framework, in which C-22 and C- 26 are oxidized to form a six membered ring (Mirjalili *et al.* 2009, Singh *et al.* 2010).

Basic structure of Withanolide

The withanolide skeleton may be defined as a 22- hydroxyergostan- 26 oic acid – 22, 26 – lactone (Figure 25.5). As a result of modification of either carboxylic skeleton or side chain different variants of withanolides are well known. These compounds are often described as modified withanolides or ergostan – type steroids related to withanolide. These compounds are generally poly oxygenated and it is believed that plants elaborating them possess an enzyme system capable of oxidizing all carbon atoms in a steroid nucleus. The characteristic feature of withanolides and ergosane – type steroid is one C8 or C9 – side chain with lactone ring which may either be six membered or five membered, and may be fused with the carbocyclic part of the molecule through a carbon – carbon or through oxygen bridge (Mirjalili *et al.* 2009).

Withanolides are derivatives of cholesterol. Biosynthesis of Withanolides starts with conversion of acetate to acetyl coenzyme- A. Two units of acetyl coenzyme- A combine to form mevalonic acid. The (R) mevalonic acid is converted to iso pentenyl pyrophosphate (IPP). IPP undergoes condensation with its isomer 3, 3, dimethyl allyl pyrophosphate (DMAPP) to form geranyl pyrophosphate (GPP). GPP condenses with another molecule of IPP to form farnesyl pyrophosphate (FPP). Two molecules

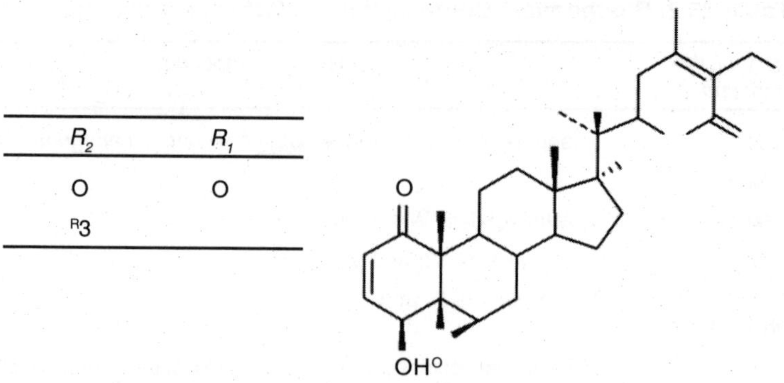

R_2	R_1
O	O
R3	

OH°

		R1	R2	R3
1	Witheferin A	OH	H	H
2	Withanolide D	H	OH	H
3	27-Deoxywitheferin A	H	H	H
4	Dihydrodeoxywitheferin A2,3-diH	H	H	H
5	Dihydrowitheferin A2,3-diH	H	OH	H
6	17 Hydroxywitheferin A	OH	H	OH
7	Withanoside IV	Glc(1-6)-Glc	H	OH
8	Withanoside VI	Glc(1-6)-Glc	OH	H

Figure 25.5: Basic Structure of *Withania* Structure and its Derivatives. Adapted Mirjalili *et al.* 2009.

of FPP condenses to form squalene, which is oxidized to squalene 2,3 epoxide, which converts to lanosterol. Lanosterol is the source of different steroidal triterpenoidal skeletons. (Mirjalili *et al.* 2009).

Withaferin-A (4β,27-dihydroxy-1-oxo-5β,6β-epoxywitha-2-24-dienolide), was the first member of this group of compounds to be isolated from *W. somnifera*. Numerous biological properties and structural features led to further study and research of the compound. Many other similar compounds were also isolated and identified later (Leet *et al.* 1982). Lavie's group (Lavie *et al.* 1965) elucidated the structure of withaferin-A in the leaves of this plant, which is mainly valued for its anti-cancerous properties. The yields of withaferin-A from intact plants of *Withania* spp. (Israel Chemotype) are 0.2-0.3 per cent of DW of leaves (Abraham 1968). Study conducted by Gupta *et al.* (1996) confirmed withaferin-A to be present in leaves but found it is absent in roots, stems and seeds. Ashwagandhanolide, a dimeric thiowithanolide has been isolated from root extract of *Withania* (Subaraju *et al.* 2006).

In their quantitative analysis of Indian chemotypes of *W. somnifera* by TLC densitometry, Gupta *et al.*, (1996) detected alkaloids in all the above mentioned plant parts, with the highest content found in the leaves. This is in contrast to the general belief that tropane alkaloids are restricted to the roots of *Withania* spp. Extraction with 45 per cent alcohol yields the highest percentage of alkaloids. The isolation of nicotine, somniferine, somniferinine, withanine, withananine, pseudowithanine, tropine, pseudotropine, 3á-tigloyloxytropane, choline, cuscohygrine, *dl*-isopelletierine and new alkaloids anaferine and anhygrine has been described (Gupta & Rana 2007). The reported total alkaloid contents in the roots of Indian *W. somnifera* varies between 0.13 and 0.31 per cent , though much higher yields (up to 4.3 per cent) have been recorded in plants of other regions/countries. In addition to the alkaloids, the roots are reported to contain starch, reducing sugars, hentriacontane, glycosides, dulcitol, withanicil, an acid and a neutral compound. The leaves are reported to contain five unidentified alkaloids (yield 0.09 per cent), chlorogenic acid, calystegines (nitrogen-containing polyhydroxylated heterocyclic compounds), withanone, condensed tannin and flavonoids. The berries have amino acids. Four types of peroxidases have been purified and characterized from *W. somnifera* roots (Johri 2005).

Epilogue

With ever increasing developments in the field of medicine the contribution of medicinal plants is most significant. *Withania somnifera* is one of the most widely utilized medicinal plant which has been successfully used in Ayurvedic medicine for centuries as well as in modern medicine system. Immense medicinal potential of *Withania somnifera* as depicted by various studies conducted indicates that modern research should be encouraged for further exploitation of the drug to evaluate the efficacy and determine safe limits of the drug along with the development of more approved drugs. Ashwagandha is either propagated vegetatively by cuttings or seeds. But vegetative propagation is time consuming. The plant has become endangered due to the over exploitation of the plant for medicinal purposes. Therefore novel techniques like plant tissue culture can be used in micropropagation and conservation of *Withania somnifera* (Sharma *et al.* 2012) since it has immense scope in rapid propagation of the plant.

References

Abraham A, Kirson I, Glotter E and Lavie D 1968. A chemotaxonomic study of *Withania somnifera* (L.) Dun. *Phytochemistry*. **7**: 957-962.

Anabalagan K and Sidique J 1981. Influence of an Indian medicine (Ashwagandha) on acute phase reactant in inflammations. *Ind J Exp Bio* **19**: 245-249.

Anabalagan K and Sidique J 1984. Role of prostaglandins in acute phase proteins in inflammation. *Biochem Med*. **31**: 236-245.

Arora S, Dhillon S, Rani G and Nagpal A 2000. The *in vitro* antibacterial/synergistic activities of *Withania somnifera* extracts. *Fitoterapia*.**75**: 385-388.

Bhattacharya S K, Bhattacharya A, Sairam K and Ghosal S 2000. Anxiolytic-antidepressant activity of *Withania somnifera* glycowithanolides: an experimental study. *Phytomedicine* . **7:** 463-469.

Bhattacharya S K, Satyam K S and Ghosal S 1997. Antioxidant activity of glycowithanolides from *Withania somnifera*. *Ind J Exp Biol* **35:** 236-239.

Bhatt S, Solanki B, Pandey K, Maniar K and Gurav N 2011. Antimicrobial activity of Ashwagandh, Shunthi and Sariva against various human pathogens: an *in vitro* study. *Int J Pharma and Biosci* **1**: 772-779.

Bone K 1996. *Clinical Applications of Ayurvedic and Chinese Herbs*. Phytotherapy Press. Queensland, Australia.

Chatterjee S, Srivastava S, Khalid A, Singh N, Sangwan R S, Sidhu P O, Roy R, Khetrapal C L and Tuli R 2010. Comperhensive metabolic fingerprinting of *Withania Somnifera* leaf and root extract. *Phytochemistry* **71**: 1085-1094.

Chaurasiya N D, Uniyal G C, Lal P, Mishra L, Sangwan N S, Tuli R and Sangwan R S 2008. Analysis of withanolides in root and leaf of *Withania somnifera* by HPLC with photodiode array and evaporative light scattering detection. *Phytochem Annal* **19**: 148-154.

Chopra R N, Chopra I C, Honda K L and Kapur L D 1958. *Indigenous Drugs of India*. UMDhu and Sons Pvt. Ltd. Calcutta.

Christina A J, Joseph D G and Packialakshmi M 2000. Anti-carcinogenic activity of *Withania somnifera* Dunal against Dalton's ascetic lymphoma. *J Ethnopharmacol* **93**: 359-361.

Dalavayi S, Kulkarni S M, Itikala R L and Itikala S 2006. Determination of withaferin A in two *Withania* species by RP-HPLC method. *Ind J Pharma Sci* **68**: 253-256.

Dhuley J N 2000. Adaptogenic and cardioprotective action of Ashwagandha in rats and frogs. *J Ethnopharmacol* **70**: 57-63.

Farooqui A A and Sreeramu B S 2004. *Cultivation of Aromatic and Medicinal Plants*. Univ Press, Hyderabad.

Ganzera M, Chaudhary M I and Khan I A 2003. Quantitative HPLC analysis of withanolide in *Withania somnifera*. *Fitoterapia* **74**: 68-76.

Gupta A P, Verma R K, Misra H O and Gupta M M 1996. Quantitative determination of withaferin-A in different plant parts of *Withania somnifera* by TLC densitometry. *J Med Arom Plant Sci* **18**: 788-790.

Gupta G L and Rana A C 2007. *Withania somnifera* (Ashwagandha): A review. *Pharmacog Rev* **1**: 129-136.

Johri S, Jamwal U, Rasool S, Kumar A, Verma V and Qazi G N 2005. Purification and characterization of peroxidases from *Withania somnifera* (AGB 002) and their ability to oxidize IAA. *Pl Sci* **169**: 1014-1021.

Jayaprakashan B, Zhang Y , Seeram N P and Nair M G 2003. Growth inhibition of human tumor cell lines by withanolides from *Withania somnifera* leaves. *Life Sci* **74:** 125-132.

Kambizi L and Afolayan A J 2008. Extracts from *Aloe ferox* and *Withania somnifera* inhibit *Candida albicans* and *Neisseria gonorrhoea*. *African J Biotechnol* **7**: 12-15.

Kaur K, Rani G and Widodo N 2004. Evaluation of the anti-proliferative and anti-oxidative activities of leaf extract from *in vivo* and *in vitro* raised Ashwagandha. *Food Chem Toxicol* **42**: 2015-2020.

Lavie D, Glotter E and Shvo Y 1965. Constituents of *Withania somnifera* Dun. part IV the structure of withaferin-A. *J Am Chem Soc* **30:** 7517-7531.

Leyon P V and Kuttan G 2004. Effect of *Withania somnifera* on B16F-10 melanoma induced metastasis in mice. *Phytother Res* **18:** 118-122.

Mahesh B and Satish S 2008. Antibacterial activity of some important medicinal plant against plant and human pathogens. *World Agri Sci* **4:** 839-843.

Mathur S, Kaur P and Sharma M 2004. The treatment of skin carcinoma induced by UB B radiation, using 1-oxo-5beta-6beta-epoxy-with a-2-enolide,isolated from the roots of *Withania somnifera*, in arat model. *Phytomedicine* **11**: 452-460.

Marie Winters, N D 2005. Ancient medicine, modern use : *Withania somnifera* and its potential role in integrative oncology. *Alt Med Rev* **11**: 269-277.

Mirjalili M H, Moyano E, Bonfill M, Cusido R M and Palazon J 2009. Steroidal lactones from *Withania somnifera*, An Ancient Plant For Novel Medicine. *Molecules* **14**: 2373-2393.

Mishra L C, Singh B B and Dagenais S 2000. Scientific basis of therapeutic use of *Withania somnifera* (ashwagandha):a review. *Alt Med Rev* **5:** 334-346.

Nasreen S and Radha R 2011. Assessment of quality of *Withania somnifera* Dunal (Solanaceae) pharmacological and phyto physicochemical profile. *Int J Pharma Sci* **3**: 152-155.

Ojha S K and Arya D S 2009. *Withania somnifera* Dunal (Ashwagandha): A promising remedy for cardiovascular diseases. *World J Med Sci* **4**:156-158.

Padmawati B, Rath P C, Rao A R and Singh R P 2005. Roots of *Withania somnifera* inhibit forestomach and skin carcinogenesis in mice. *Evid Based Comlement Alt Med* **2**: 99-105.

Prakash V, Agarwal A, Sharma S, Gupta S and Gupta A. Antifungal activity of *Withania somnifera* (L.) Dunal on some phytopathogenic fungi. *Bull Nat Inst Ecol.* **15**: 215-218.

Praveen N, Malik P S, Manohar S H and Murthy H N 2010. Distribution of Withanolide A content in various organs of *Withania somnifera* (L.) Dunal. *Int J Pharma Biosci* **1**: 1-5.

Sangwan R S, Chaurasiya N D, Mishra L N, Lal P, Uniyal G C, Sharma R, Sangwan N S, Suri K A, Qazi G N and Tuli R 2004. Phytochemical variability in commercial herbal products and preparations of *Withania somnifera* (Ashwagandha). *Curr Sci* **86**: 461-465.

Santhi M and Swaminathan C 2011. Evaluation of antibacterial activity and phytochemical analysis of leaves of *Withania somnifera* (L.) Dunal. *Int J Curr Res* **3:** 010-012.

Senthil K M, Vinodh K D, Saravana K A, Aslam A and Shajahan A 2011. The phytochemical constituents of *Withania sominifera* and *Withania obtusifolia* by GCMS analysis. *Int J Pharma Phyto Res* **3**: 31-34.

Sharma N, Koshy E P and Dhiman M D 2012. *In vitro* induction of flowering in *Withania somnifera.* Abstr in USSTC held in Dehradun, pp 107.

Sharma V, Sharma S, Pracheta and Paliwal R 2011. *Withania somnifera*: A rejuvenating ayurvedic medicine herb for the treatment of various human ailments. *Int J Pharm Tech Res* **1**: 187-192.

Singariya P, Kumar P and Mourya K K 2012. Antimicrobial activity of fruit coat (calyx) of *Withania somnifera* against some multi drug resistant microbes. *Int J Biol Pharm Res* **3**: 252-258.

Singh A, Duggal S, Singh H, Singh J and Katekhaye S 2010. Withanolides: Phytoconstituents with significant pharmacological activities. *Int J Green Pharm* **4**: 229-237.

Singh B, Gahoi R and Sonkar A 2010. A quality assessment and phytochemical screening of selected region of *Withania somnifera* Dunal. *Int J Pharma Sci Res* **1:** 73-77.

Singh G, Sharma P K, Dudhe R and Singh S 2010. Biological activities of *Withania somnifera . Ann Bio Res* **1**: 56-63.

Singh V 2009. Phenology and reproductive biology of *Withania somnifera* (L.) Dunal (Solanaceceae). *J Pl Repro Bio* **1**: 81-86.

Singh V P and Pandey R P 1998. *Ethnobotany of Rajasthan, India*. Scientific Publishers, Jodhpur, India.

Subaraju, G V, Vanisree M, Rao C V, Sivaramakrishna C, Sridhar P, Jayaprakasam B and Nair M G 2006. Ashwagandhanolide, a bioactive dimeric thiowithanolide isolated from the roots of *Withania somnifera*. *J Nat Prod* **69**: 1790-1792.

Sukhdev 2006. *A Selection of Prime Ayurvedic Plant Drugs: Ancient Modern Concordance.* Anamaya Publishers, New Delhi.

Sundaram S, Diwedi P and Purwar S 2011. *In vitro* evaluation of antibacterial activities of crude extract of *Withania Somnifera* (Ashwagandha) to bacterial pathogens. *Asian J Biotech* **3:** 194-199.

Tripathi A K, Shukla Y N. and Kumar S 1996. Ashwagandha, *Withania somnifera* Dunal (Solanaceae): A status report. *J Med Arom Pl Sci* **18:** 46-62.

Verma S K and Kumar A 2011. Therapeutic uses of *Withania somnifera* (Ashwagandha) with a note on withanolides and its pharmacological actions. *Asian J Pharma Clinic Res* **4:** 1-4.

Index

D